T. M. (Theodore Minot) Clark

Building Superintendence

A Manual for Young Architects, Students, and Others Interested in Building

Operations

T. M. (Theodore Minot) Clark

Building Superintendence
A Manual for Young Architects, Students, and Others Interested in Building Operations

ISBN/EAN: 9783743686939

Printed in Europe, USA, Canada, Australia, Japan

Cover: Foto ©berggeist007 / pixelio.de

More available books at **www.hansebooks.com**

BUILDING SUPERINTENDENCE

A Manual

FOR YOUNG ARCHITECTS, STUDENTS, AND OTHERS
INTERESTED IN BUILDING OPERATIONS AS
CARRIED ON AT THE PRESENT DAY

BY

T. M. CLARK
FELLOW OF THE AMERICAN INSTITUTE OF ARCHITECTS

THIRTEENTH EDITION

New York
MACMILLAN AND CO.
AND LONDON
1895

All rights reserved

COPYRIGHT, 1889,
BY JAMES R. OSGOOD & CO.

Transferred to Macmillan & Co., March, 1894. Printed May, 1894.
Reprinted February, 1895.

Norwood Press:
J. S. Cushing & Co. — Berwick & Smith.
Norwood, Mass., U.S.A.

PREFACE.

This is not a treatise on the architectural art, or the science of construction, but a simple exposition of the ordinary practice of building in this country, with suggestions for supervising such work efficiently. Architects of experience probably know already nearly everything that the book contains, but their younger brethren, as well as those persons not of the profession who are occasionally called upon to direct building operations, will perhaps be glad of its help.

CONTENTS.

	PAGE
INTRODUCTION,	3

CHAPTER I.
THE CONSTRUCTION OF A STONE CHURCH, ... 10

CHAPTER II.
WOODEN DWELLING-HOUSES, ... 107

CHAPTER III.
A MODEL SPECIFICATION, ... 219

CHAPTER IV.
CONTRACTS, ... 261

CHAPTER V.
THE CONSTRUCTION OF A TOWN HALL, ... 269

INDEX, ... 333

NOTE.

THE observing reader will notice a curious discrepancy between the factors of safety used in calculating the strength of wooden beams in the first and last parts of the book, three being given as the proper factor in determining the size of the girders in the stone church, while six is used in calculating the roof timbers of the town-hall, which forms the subject of Chapter IV, the constant being the same in both cases. As this inconsistency exactly represents the change which took place in the received standard of strength for beams of timber between the sterotyping of the pages of the first and last chapter, it is perhaps best to let it stand, with a proper warning. The factor of safety, three, used in the earlier part of the book, is that still required by the New York Building Law, and the constant is taken from Trautwine's Handbook, a very conservative authority. The factor subsequently used, six, corresponds with that adopted by what were then the latest and most cautious writers. Within a few months, Professor Lanza's experiments at the Massachusetts Institute of Technology, upon large timbers, have shown conclusively that with the constants hitherto given in the text-books even a factor of safety of six is too small. In his experiments, upon beams ranging from four to eight inches in width, by twelve and fourteen inches in depth, the average breaking weight of selected, straight-grained timbers of spruce and Georgia pine was about *one-half* that given in Trautwine's tables as the proper constant for each, and not much more than one-third the constants adopted from Laslett and others by so high an authority as the South Kensington "Notes on Building Construction." In the light of these results, the New York law is evidently dangerously misleading, and the floors calculated in accordance with it, so far from being three times as strong as necessary, are on the verge of collapse; while, with the constants at present in use, the nominal factor of six, as employed by the most prudent constructors, is really reduced to something like two or three. Between the varying and perfectly arbitrary factors of safety used by different authorities, and their almost equally diverse constants, the whole subject of strength of timber has hitherto been in a state of most unscientific confusion, to which Professor Lanza's investigations have added the finishing touch. It is much to be hoped that the result of these will soon be to bring order out of the chaos, by establishing our rules for the strength of heavy beams upon definite knowledge instead of hasty inferences. Meanwhile, we will not bewilder our readers by attempting to improvise any new standard of our own, but will simply say that until a better rule is offered, by using one-half of Trautwine's constants as the true one, and a factor of safety of four, they can obtain, in calculating their future beams, a result in reasonable accordance with the present state of knowledge on the subject.

BUILDING SUPERINTENDENCE.

THE DIRECTION OF BUILDING OPERATIONS.

ALL who have had any experience in the supervision of building operations know the importance of having a systematic plan in pursuing their examination of any given work, and the difficulty, without such aid, of giving adequate attention to all the innumerable points of construction which require notice at their proper time, and before they are covered up or built over, so as to make changes inconvenient or impossible, and there are few who cannot recall instances of vexatious mistakes, costly alterations, or buildings left insecure through want of attention at the right moment to defects which an hour's labor would then have remedied. To the young architect, especially, judging from the writer's own experience, a manual which may help to direct his attention to all the various details which should be noticed, and put him in mind of the defects to be looked for at each stage of a given construction, cannot fail to be of use, and such a manual, it is hoped, the following pages will supply. No doubt there will be imperfections and omissions; but the writer trusts they will not be found very numerous, and that such as may exist will be duly criticised and corrected by those competent to do so. Reference will be made to the different modes of building which prevail in various localities, so far as the writer's knowledge permits, as much for the purpose of comparison and criticism as for the sake of extending the usefulness of the work.

The general subject of superintendence will be considered under the three heads of Stone Buildings, Wooden Buildings and Brick Buildings.

In the division of Brick Buildings will be special reference to the distribution of weights, and those details of boiler and steam work, and ventilation, which are more commonly to be considered in connection with city structures than others; and under Wooden Buildings the questions of grading, cellar-work and drainage will be dwelt upon, which continually call for settlement in country construction.

In each class the progress of a typical building will be described, from the first breaking of ground to the completion of the work, showing the successive stages of construction and the order of delivery of material on the ground, such as they would be found by the superintendent in his periodical visits. At each imaginary visit occasion will be taken to call attention to points which need to be considered at that particular stage, although the execution of them may belong to a subsequent period, and some general directions for judging of the quality of materials will be given at the time they make their first appearance on the premises.

From a sense of its great importance, the writer, although not considering himself peculiarly qualified for the task, will endeavor, in treating of the different kinds of work, to present a standard by which each sort can be judged. Nothing is more embarrassing to the young architect than to be called upon to decide a dispute between contractors and owner on the question of what constitutes a "good, substantial, and workmanlike manner" of executing any particular piece of construction. He is not likely to commit the error of being too lenient in his requirements, but he may, if unaccustomed to practical work, be led into unreasonable exactions or untenable positions which he will subsequently have to abandon, to the detriment of his reputation and authority. It often happens that the owner has formed *a priori* an ideal of what the work should be, which does not agree with the actual execution, and he consequently refuses to accept it, or claims an allowance in the price on account of what he asserts to be defective workmanship; and the architect is called upon to decide between the parties. In such cases it is of the utmost importance to him to be able to show such thorough familiarity with common practice as will command the respect of both. Without this he is continually liable to do injustice either to the contractor or his client, as well as to incur their ill-will. Take, for instance, a question which sometimes actually occurs in the case of buildings faced with freestone ashlar. The owner, perhaps, has been reading some textbook on construction and has come across the familiar direction that "all stones used in masonry should be laid on their natural bed." On his next visit to his building he looks for the application of this rule and is surprised to find every stone in the facing set up edgeways. He sends for the contractor, who professes never to have heard of any rules about beds, and declines the owner's request either to take down the masonry and do it over again, or to make a discount from the contract price of the work. The disputants then betake themselves to the architect, who finds all his resources of tact and experience required to convince the owner that although his rule might be theoretically correct, universal custom justified the contractor in violating it. He is sure to be sharply questioned, and if his practical knowledge proves defective, all his rulings in favor of the builder expose him to suspicion and discredit.

It will be of advantage, however, in connection with the discussion of the usual practice in the different kinds of work, to point out such improved methods as are sometimes used, but with a caution to the reader, that under the common contract the builder cannot be compelled to use them at his own expense, unless they are recognized in the best ordinary practice of the locality to which he belongs, or are particularly mentioned in the specification, although most good mechanics are glad of any suggestions for the improvement of their work, even at some extra cost to themselves, and will be equally grateful for any help toward a uniform standard of practice, to which both they and the architects can refer.

For the discomfiture of bad workmen, the young architect will be warned against some of the ways in which defective materials or construction are covered up, and will be reminded to look for bad work before the building arrives at so advanced a stage that it can no longer be detected or remedied.

To save space, it is supposed that the reader is familiar with the principles of construction as given in the text-books, and with the common forms of specifications and building contracts. Any one in need of such elementary information will find Dobson's "Art of Building," in Lockwood's series, the best cheap work on the general subject, and Brooks's "Dwelling Houses," in the same series, will afford some additional details in its special branch. As a complete and authoritative text-book, however, no work within our knowledge can compare with the three volumes of "Notes on Building Construction," prepared for the use of the English Science and Art Department at South Kensington, and published by Rivingtons, Waterloo Place, London.

For those who read French, Ramée's "Architecture Pratique" will be an extremely practical and useful little book, while those who have sufficient leisure will find the time necessary for the study of Rondelet's standard work, the "Traité Théorique et Pratique de l'Art de Bâtir" well repaid.

There are many other useful works, both in French and English, which treat of various portions of the subject. Among the best are Wightwick's "Hints to Young Architects," published in Lockwood's series, with Hatfield's "American House Carpenter," published by Wiley & Son, New York.

The specifications in the English books are useless in this country, our materials, modes of construction, and technical terms differing completely from those in use across the water, but models which will serve very well as guides for the non-professional reader in the construction of brick or stone buildings will be found in the printed specifications for public structures of various kinds, copies of some of which can generally be had by inquiring of the proper officers. For wooden buildings, the best work in print is perhaps Mr. Wm. T. Hallett's "Specifications for Frame Houses," published by Bicknell & Comstock, New York. If a form of contract is desired, a very good one can be found in Mr. Hallett's book, or blank agreements for building, as well as a special form for "mechanics' work," can be obtained of any stationer. These forms are not all that could be desired, but do very well for ordinary cases.

One or two preliminary remarks may be made before describing the actual processes of inspection.

In the first place it is necessary to be as familiar as possible with the plans and other drawings of the building to be constructed. Nothing is of so much service in rendering the labors of the superintendent valuable to his employers and himself as the thorough understanding of the projected building which will enable him to foresee the consequences of every step, to judge of the position and workmanship of each part of the edifice, while in process of construction, with reference to its final use and finish. His duty is only half fulfilled if he trusts blindly to the accuracy of the plans. He should examine, compare and correct them minutely, thoroughly, and frequently. It is impossible in the architect's office to avoid all mistakes in drawing or figuring, but such as escape the eye of the busy professional man can easily be detected by a little care in comparing them with the work on the ground, and this duty clearly belongs to the superintendent. Not only should he make sure of the accuracy of the plans, but he must also look out beforehand for other points which may affect or hinder the construction when once begun. If, for instance, the drawings show stone and brick work bonded together in elevation, it should be his duty at once to procure bricks of the kind to be used in the facing, and lay them up with mortar joints of the usual or specified width, in order to ascertain with certainty the height which a given number of courses of brick will lay. It is common in such work to assume that five courses of brick will lay one foot in height, and the detail drawings for the stonework are often made and figured accordingly. If, then, as often happens, the particular brick used is a little thicker or thinner than the standard, the stone once cut from an incorrect assumption will fail to bond properly, and, if it cannot be recut, must either be thrown away or inserted as best it may, the wide joints and irregular lines bearing witness to the incompetency of the one who directed the work.

Even supposing the plans to be correct, the superintendent will find many opportunities for saving both contractor and owner from the annoyance and expense caused by the carelessness of workmen. It is impossible to get the ordinary mechanic to concern himself about the future matters which will depend upon his work, and a little foresight in supervision will prevent many careless deviations from the drawings or specifications, which, although the contractor would be bound to correct them, he will be glad to have detected in season to save him that expense, and will show his gratitude by special endeavor to please. Another important point in efficient supervision is, after inspecting the materials delivered, to make sure that those rejected are removed from the premises. If they are marked for rejection, as they should always be, let the mark be on the *face* of the cracked or thin stone, or on the *upper side* of a "waney" floor-board, so that it can be recognized if it should be afterwards put into the work. If this precaution is not taken, and if the marks are not made in such a manner that they cannot be rubbed or planed off, materials once discarded will be smuggled into the building in spite of the injunctions of the superintendent or contractor.

It is impossible to be too thorough in each periodical inspection of the building. It will not do to examine one portion one day and another the next; the proper way is to go all over the structure at each visit. Wherever a man is working, or has been working since the last inspection, go and see what he has done. In this way it will be possible to gain that definite knowledge of every portion of the structure which is the only security against concealed vices of workmanship.

One other precaution must be observed: let not the young architect put too much faith in what workmen say to him. The best of them dislike to pull down or change what is already done, and if inadvertence or temporary convenience has led them into palpable violations of the specifications, they will often stretch the truth considerably in their explanations and excuse. Some are much worse than this, and will deliberately avail themselves of the credulity and inexperience of the young architect under whose authority they come, to obtain from him, both before and after the execution of the work, such concessions from and interpretations of the strict letter of the specification as will be most to their profit. It is difficult for one who is not quite certain that he knows how to distinguish between fresh and damaged cement, for instance, to persist in rejecting a lot of which he has suspicions, but which the builder declares to be not only of the best quality, but the only lot of that quality which can be procured without seriously delaying the work, and if the superintendent is found to be accessible to such representations, his credulity is sure to be tested on many other points. The only way in which young architects can escape being occasionally made victims of such practices, is for them to make up their minds what is right by the best light that they can, and insist on their directions being followed. With a little thought, and assiduous study of other buildings in process of construction, it need rarely happen that their orders will be unreasonable, and a firm stand on such occasions as arise in the early stages of the work will not only make their subsequent duty much lighter and pleasanter, but will often save them ultimate discredit and regret.

These preliminary remarks are applicable to every kind of building. Others might be added, but as they may perhaps be better understood and remembered in connection with some practical illustration, it will be as well to defer their consideration until they present themselves in treating of a particular class of work.

THE CONSTRUCTION OF A STONE CHURCH.

CHAPTER I.

THERE are many advantages in beginning the study of construction with stone buildings. Such structures require a minuteness of attention, and a precision and accuracy of workmanship, superior to that which is generally bestowed upon buildings of any other material, and the careful and continual measuring, levelling and verifying which they demand, together with the foresight which must **Construction of Stone Buildings.** be exercised in order that all parts of the masonry and wood-work may come together without error, tend to form habits which are of great value in the direction of any kind of building operation. Moreover, the construction of most stone buildings involves calculations of thrust; of pressures, vertical and oblique; the resistance of materials, and many other technical problems well fitted to exercise and develop the capacity of the young professional man, while the comparative slowness with which they are erected affords him sufficient time for a careful consideration and solution of such questions, which is not always possible in the case of lighter structures.

If it is considered also that stone-masonry of some kind enters into nearly all building operations, and that foundation-works, even of light erections, sometimes require the overcoming of serious difficulties, the propriety of introducing the general subject by its most intricate and technical branch will be sufficiently plain.

Fig. 1.

The church, of which we propose to follow the construction, is intended to be situated on an elevated ground, descending irregularly toward the east. The soil appears to be gravelly, but variations in the appearance of the grass over the lot indicate that the sub-soil is not uniform. The total difference in the level of the ground in the length of the proposed building is about six feet.

The church is to be cruciform in plan, with clerestory and nave aisles; the chancel to be without aisles, as wide as the nave, and with a semicircular apsidal termination. The principal entrance is to be through the tower, which stands at the south-west angle, forming the termination of the south aisle of the nave, and there is a second entrance,

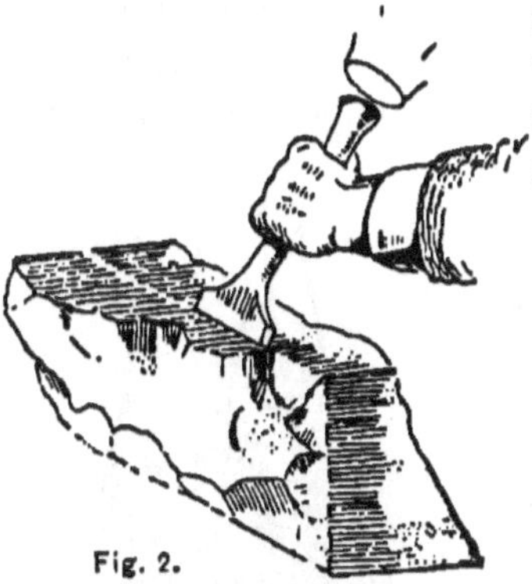

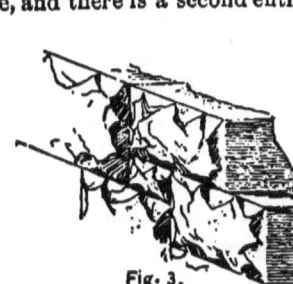

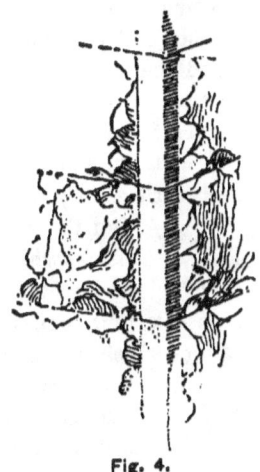

Fig. 2.

Fig. 3.

opening into the transept through a small enclosed porch which occupies the angle between the transept, which has considerable projection, and the south aisle of the nave. East of the north transept is the organ-chamber, with wide openings into transept and chancel. The corresponding angle between the south transept and the chancel is occupied by the robing-room, with private entrance from the outside. Under the chancel is a society-room, entered directly by steps from the outside. The rest of the space under the building is used for a cellar, reached by a circular stone staircase in a staircase-turret attached to the main tower. The same stair ascends also

Fig. 4.

to the bell-chamber. The whole building, including the spire, is faced with brown freestone from the floor-line upward. The basement wall, from the floor-level to the ground, is faced with granite. The foundation-walls, and the backing of the walls above ground, are of a slaty local stone. The spire and staircase-turret are backed with brick.

In the interior, the clerestory arches, and the caps and bases of the columns which carry them, are of Ohio freestone. The plinths and shafts of the columns are granite, the shafts being polished. The chancel steps are of marble; the vestibules are tiled with marble; and the chancel floor is laid with encaustic tiles. The outside steps are granite, and the winding stair in the turret is of hard brown sandstone.

Fig. 5.

Ashlar. The exterior stone facing is specified to be "neat random ashlar, quarry-face, with pitched joints." This means that all the exposed surfaces are to be freshly split, without weather-worn faces, such as would be admissible if *rock-face* were specified, and that the joints, instead of irregularly projecting, as is permissible in some engineering work (Fig. 1), are to be "pitched off" to a line previously drawn around the stone, all parts of which lie in a true plane at right angles with the surfaces formed by the joints. This is done with a "pitching tool" or wide chisel with a very thick edge (Fig. 2), **Modes of Dressing Stone.** and the result is to furnish blocks which, however rough and projecting in the centre, all possess four well-defined edges, by means of which they can be placed upon **Pitched Joints.** and beside each other with as much accuracy as the smoothest-faced stones, presenting a surface like that shown in Fig. 3.

Other parts of the exterior stone-work are differently treated. the spire is to be a "broach" (Fig. 5), and the arrises, or edges, of the octagonal spire, as well as of the broaches, are to have chiselled draft lines (Fig. 4) two inches wide. There will be also draft lines one and one-half inches wide around all openings, and the reveals of openings, together with all cornices, copings, weatherings of buttresses, washes of sills, roof of staircase-turret, and bands in spire, will be crandled; that is, brought to a plane surface by means of the crandle (Fig. 6), which is a toothed hatchet composed of eight or ten pointed chisels wedged tight into a frame with a handle. With this tool the surface of the stone is obliquely struck

Crandle.

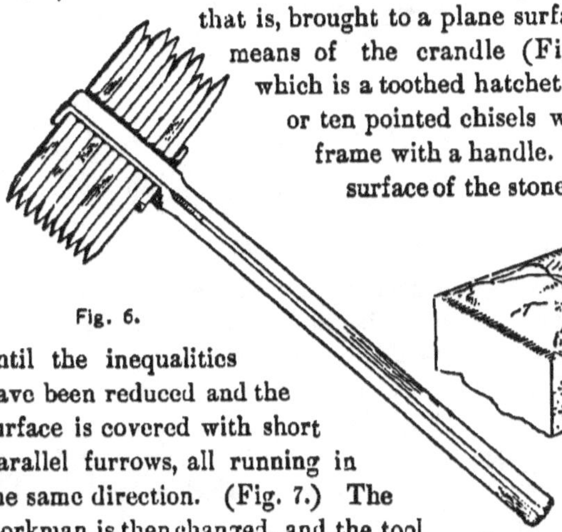

Fig. 6.

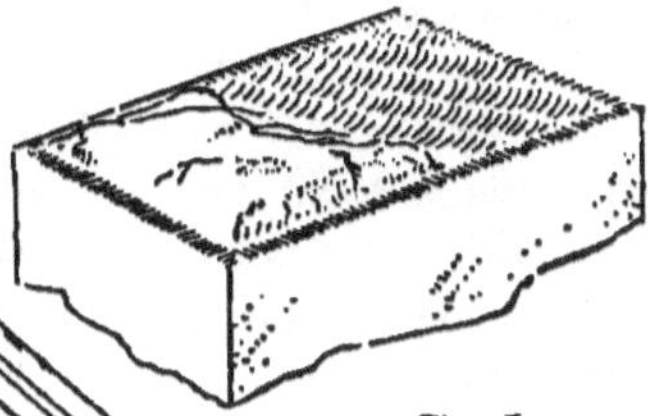

until the inequalities have been reduced and the surface is covered with short parallel furrows, all running in the same direction. (Fig. 7.) The workman is then changed, and the tool applied in the other direction, so as to make furrows crossing the first. By this means a comparatively smooth surface is formed, covered with a network of short lines. (Fig. 9.) Before beginning the work a line is drawn around the stone, and the joints either pitched off to this line, as before described, or a chisel draft sunk all around the face until the line is reached. (Fig. 8.) Whichever mode is adopted, by removing the rough projections of the central portion first with a toothed chisel

Fig. 7.

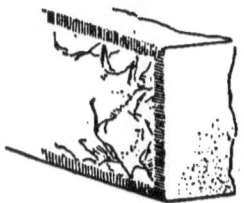

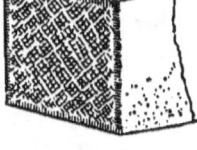

and then with the crandle until it is reduced to the plane of the draft line, a true surface is obtained. It need hardly be said that the first operation in all stone-cutting is to form the joints, and from these

Fig. 8.

Fig. 9.

four surfaces, at right angles to each other (Fig. 10), all the other bounding planes of the finished stone are derived.

Droving.

The tracery of the windows is droved; that is, finished by working over the entire surface with a wide chisel, so as to cover it with parallel lines, all running in the same direction, except around arches, where they radiate from the same centre as the curve of the arch. (Fig. 11.)

Of the granite, the ashlar is quarry-faced, with pitched joints like the sandstone. The "wash" or upper inclined surface of the water-table is pene hammered; that is, pounded

Pene Hammering.

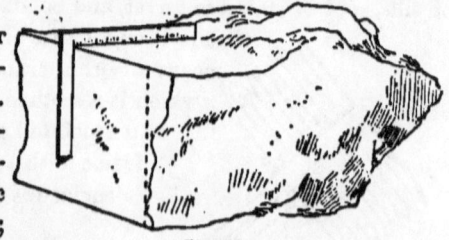

Fig. 10.

with a heavy, double-edged hatchet, or "pene hammer," until the inequalities are chipped away and the surface is reduced to a plane, determined by draft lines previously drawn, and covered with coarse parallel lines. The outside steps, and the reveals and sills of basement openings, receive a further finish, by being pounded all over, after the pene hammer has done its work, with a "bush hammer," or as some say, "patent hammer," in which several parallel blades are bolted into an iron frame with a handle.

Bush Hammering.

This instrument (Fig. 12) also leaves parallel marks on the stone, but much finer than those of the pene hammer. For the outside work, a tool having six blades to the inch will give a sufficiently smooth surface, which is specified as "six-cut." The plinths of the interior columns are first pene hammered, then six-cut, and finally "ten-cut," with a bush hammer having ten blades to the inch, which leaves the stone very smooth. Eight blades would give an intermediate finish. The joints and beds of the granite are "pointed," or brought to a roughly plane surface by means of the "point," a short, thick chisel, with a very short edge, often less than a quarter-inch in length.

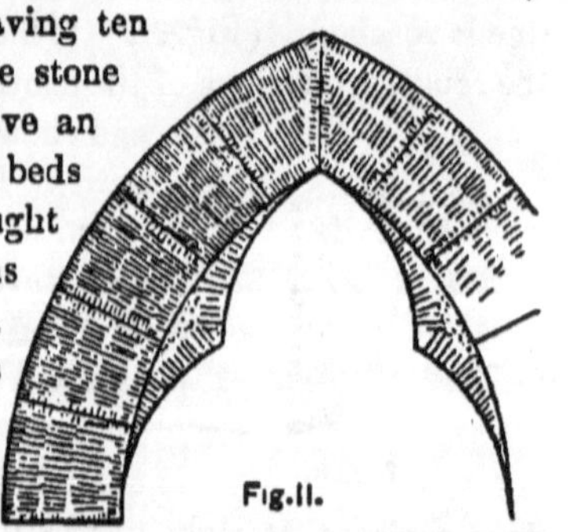

The interior capitals and a portion of the arch mouldings, as well as the finials and some other outside work, are carved. The marble steps to the chancel are polished. In the carved work it is important to leave all the marks of the chisel untouched. Inferior workmen

Fig. 11.

sometimes smooth and rasp the carving, to the destruction of its crispness and beauty. The rest of the interior freestone, on the contrary, will be rubbed. This is usually done by laying stones sawed to a suitable thickness, either with steel blades supplied with sand and water, or with the much more rapid diamond saw, face downward on a large mill-stone revolving horizontally. This smooths off the lines left by the saw, and the surface thus given is left to form the face of the work. The granite shafts are polished by machinery, emery of different grades being first used, after the stone is brought to a smooth surface with tools; followed by "tin putty," or precipitated oxide of tin, which produces a glass-like lustre. For red granite, the so-called jewellers' rouge, or crocus, which is really an oxide of iron, is used for polishing, and the particles which find their way into the crevices of the granite are allowed to remain, for the sake of heightening the color. Occasionally the final gloss is given with wax, but this fraud can be detected by the apposure to the weather, which quickly removes the lustre so obtained.

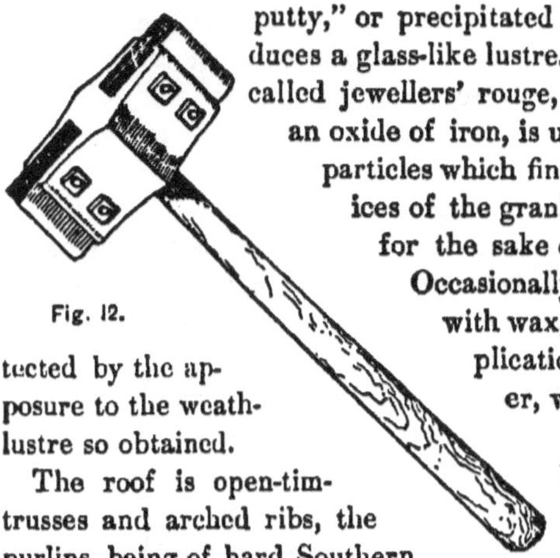

Fig. 12.

The roof is open-timbered, with hammer-beam trusses and arched ribs, the trusses, as well as the purlins, being of hard Southern pine. The trusses are covered with matched pine boards, tarred felt, and slate, and the panels formed by the purlins and trusses are lathed and plastered. The interior is furred with studding, lathed and plastered. A panelled wainscot, four feet high, runs around the room.

The interior of the tower, which forms a vestibule, is lined with a four-inch wall of face-brick, separated by a two-inch air-space from the outside wall, but tied to it with iron anchors. This lining extends the whole height to a ceiling of moulded girders, forming panels filled with matched-and-beaded sheathing, and covered with tin on top. The space above this, though accessible from the staircase-turret, is left rough, up to the "bell-deck," also covered with tin, on which stands the framing for the peal of bells, which should be as far as possible independent of the tower walls.

The interior of the society-room beneath the chancel is lined with a four-inch brick wall, separated by a two-inch air-space from the outside masonry, and tied to it with iron, and is plastered on the inner brickwork. The walls of the robing-room and organ-chamber are furred, lathed and plastered. The tiled floors of the vestibules and chancel are set on bricks laid upon boarding cut in between the beams. All the other floors are double, the under boards of spruce throughout, and the upper flooring of Georgia pine in the society-room; elsewhere of white pine. Interior finish is of hard wood.

The foregoing description contains a brief abstract of the specifications. In accordance with these the contract has been signed, and the builder, or the contractor for the excavation, as the case may be,

Setting out the Building. meets the architect upon the ground to lay out the work. If a special plan has been prepared, with the exact dimensions of the walls and the lengths of the diagonals in plain figures upon it, the setting-out will be much simplified, and there will be less danger of mistake, but the cellar plan is often the only drawing provided for the purpose, and the additions or subtractions necessary to give the dimensions of the trenches must be made on the spot.

In important works it is best to call upon an engineer to set out the lines, but circumstances may make it necessary to do without such professional aid. In any case, the architect should always be present.

Some builders, and a few architects, possess an engineer's compass, by means of which they can "turn off" the various angles around the building with precision; but in most cases the parties will arrive at the spot furnished only with a tape-measure, a few stakes, and a "mason's square," consisting of three pieces of wood nailed together in such a way as once to have formed, perhaps, a right-angled triangle. Even these rude instruments will, however, suffice, if the tape is accurate, and especially if the architect has had the forethought to bring a second tape, which will much facilitate matters.

The first thing to be done is to stake out the outside lines of the main walls of the building, the "ashlar lines," as they are called. Many persons neglect these in the first laying out, and run only the exterior lines of the cellar-wall, which usually projects three or four inches beyond the ashlar lines, but the danger of mistakes is much lessened, and the work in the end simplified, by setting out at the beginning the perimeter of the main walls, and then drawing parallel lines outside of these to indicate the face of basement-walls, or the line of excavation, as the case may be. By this means the exact relation of foundation and superstructure is shown at a glance, and there is no danger of a line drawn to indicate the face of the foundation being mistaken for the ashlar line, and the wall being built several inches outside of it, as sometimes happens where only one line is given at first, much to the detriment of subsequent work.

The staking out of the plan must begin by the establishment of some main rectangle or triangle, to which the various projections can be added one after the other. In this instance, the preliminary figure may be the rectangle formed by the nave, including the aisles, from the west front as far as the transepts. **Preliminary Rectangle.** Having fixed upon the location of the building, the line of the west front is determined, and upon this line is measured the width of the nave with the aisles, to the outside of the aisle walls. A short stake should be driven into the ground at each end, and a copper tack in the head of the stakes will indicate the exact points. The next step is to run lines from these points at right angles with the first line, which will represent the face of the aisle walls. If a surveyor's compass is at hand, the right angle is easily turned off; if not, the best way of proceeding is that shown in Fig. 13, where A B is the line of the west front, X Y marking the width of nave and aisles. Find from the plans the distance Y T, from the west front to the angle of the transept, and calculate the diagonal distance from X to T, by adding the squares of X Y and Y T, and extracting the

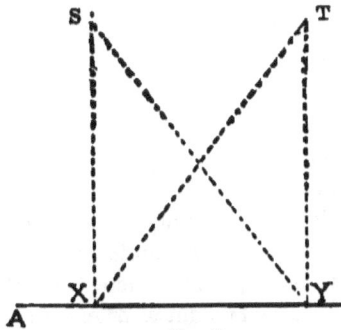

Fig. 13.

square root of their sum. (*The square of the hypothenuse of a right-angled triangle is equal to the sum of the squares on the legs.*) Then take two tapes and placing the ring of one at X, and of the other at Y, measure from X on the first the length of the diagonal just found, and on the second the distance Y T, and bring the two points together. Mark the place where the measurements on the two tapes coincide with another stake and copper tack.

Measuring Diagonals. Change the tapes over, and measure X S equal to Y T, and the diagonal Y S equal to X T, and mark the spot where the points again coincide with a fourth stake and tack. Verify the work by measuring the distance between the two eastern stakes, S and T, which should equal that between X and Y. If the measurements have been accurately made, the four stakes will form the corners of a perfect rectangle. Care must be taken to keep the measuring-tape level. If the ground

Tapes to be Level. falls in any direction, the tape should at the lowest point be held up so that the ends will be at the same level, and a plumb-line dropped from the correct point on the tape to the ground will give the position of the stake. It will save time and trouble to have the lengths of the diagonals calculated beforehand and marked on the plans; not only for the main rectangle but for minor ones.

Perhaps two tapes will not be at hand, or the mathematical attainments of the party may not extend to the extraction of square roots. In that case, after setting out the line of the western front as before, measure off one of the two long sides of the preliminary rectangle, as nearly at right angles with the base line as possible. By stretching a string for the base line, with another for the long side of

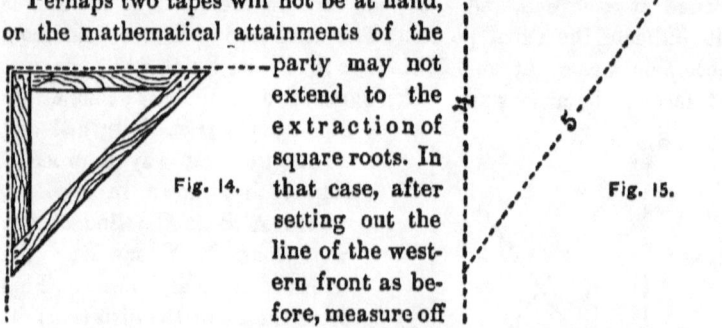

Fig. 14.   Fig. 15.

"Three, Four, Five" Rule. the rectangle, the latter can be brought approximately to the proper direction by the help of the mason's square (Fig. 14) or by the application of the "three, four and five rule"; which consists in marking a point on one string three feet from the angle, and on the other four feet from the same angle, then adjusting the relative position of the two until a straight line drawn diagonally across the angle, between the two points, measures exactly five feet. (Fig. 15.) Then the two strings will form a right angle, since 5 is the hypotenuse of a right-angled triangle whose other sides measure 4 and 3 feet.

Having found the angle approximately by these means, a stake should be temporarily driven at the extremity of the long side of the rectangle so found, and the other side set out parallel with this, by measuring the distance C D (Fig. 16) equal to A B, from the point C, distant on a second line, A C, as far from A as D is distant from

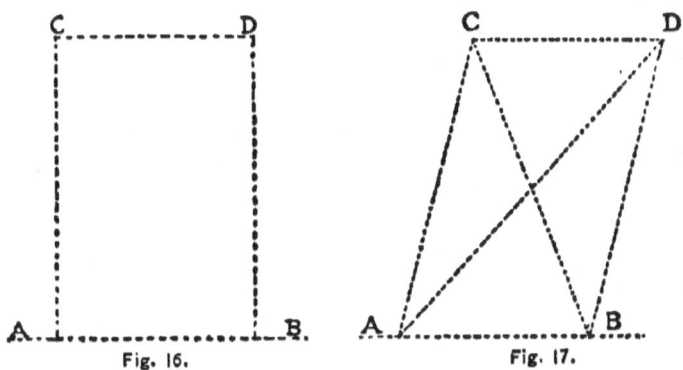

Fig. 16.   Fig. 17.

B. Then the figure A C D B will be a *parallelogram*, since its opposite sides are equal; but it may not be a *rectangle*. As a test, the diagonals should be measured. If they are unequal, the figure, instead of being rectangular, has the form of which Fig. 17 is an exaggerated representation. To remedy this, move the stake D, at the farthest extremity of the longest diagonal, toward the stake C, a distance about equal to two-thirds of the difference in length between the two diagonals. Then move the stake C an equal distance in the same direction, so as to keep the length of that side of the rectangle correct. Measure the diagonals again, and if now found to be equal, verify all the measurements, then drive the stakes firmly and set copper tacks as before. This preliminary rectangle having been accurately marked out, the rest of the work follows easily.

Lay out the transepts by stretching a line through the eastern side of the rectangle just found, prolonging it indefinitely at each end. Prolong also the north and south sides of the rectangle indefinitely eastward. Lay off the projection of the transept, as figured on the plan, on the extension of the east side of the rectangle; from this point, parallel to the main axis of the building, measure the proper width of the transept. Measure the same distance on the prolongation of the north and south sides of the rectangle, and from the points thus found lay off again, approximately at a right angle, the projection of the transept. Join, if they do not already coincide, this point to the corresponding one found by measuring parallel to the axis; the place of meeting will be the true north-east or south-east angle of the transept. Drive stakes at all the angles thus found, and set copper tacks to mark the exact points.

Secondary Rectangles.

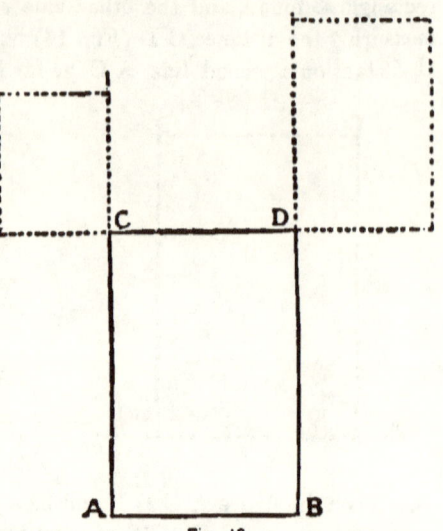

Fig. 18.

These secondary rectangles, representing the transepts, will have their angles correct without further trouble, since A C D B having been made truly rectangular, the angle, F D E, included between the prolongations of the lines, D B and C D, which include a right angle, will itself be a right angle, and one angle of the parallelogram, D F G E, of which the opposite sides are measured equal, being a right angle, the others are also right angles.

To determine the apsidal curve, both for the walls and the excavation, it is necessary to fix the centre, from which an arc of the requisite radius can be struck whenever needed. (A, Fig. 19.)

The usual way of striking such an arc is to drive a strong stake at the centre, with a nail or spike inserted at the proper place in the head of the stake, and using a long rod with a notch fitting against the central spike and cut to the requisite radius to describe the curve. At the first setting-out, such a stake may be driven for temporary use, the apsidal semi-

Centre of Curve.

circle described and marked with a number of small stakes or a line cut in the turf with a spade. Then the excavation can begin at this line, but as the central stake would soon be dug away, it is necessary to provide means for recovering it at pleasure. This can best be done by fixing points entirely outside the excavation from which lines can be stretched which will by their intersection indicate the point required. For our purposes four such points will be necessary. Two of these, J K, may with advantage be situated on the prolongations of the lines which show the northern and southern ashlar lines of the transepts, and the distance from the angles, G and H, of the transepts should be twice the distance from the centre of an imaginary line connecting G and H to the point, A. Then two cords, one stretched from H to J, and the other from G to K, will cross each other over A. One more step remains to be taken. As the stakes which at first indicate the actual points, G and H, will

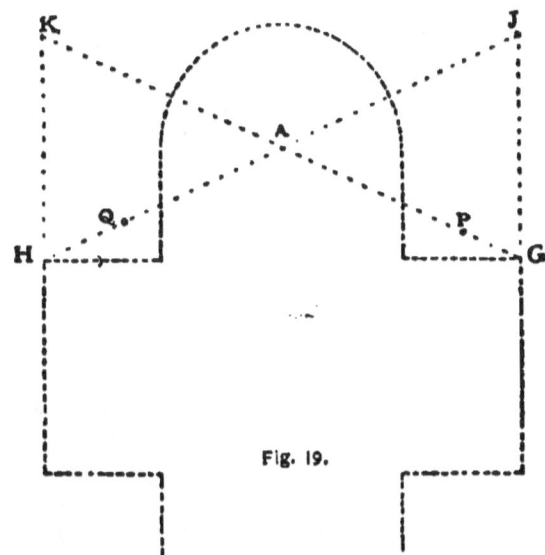

Fig. 19.

be removed by the excavation, it is necessary to drive supplementary stakes, P and Q, with copper tacks in their heads at any point on the lines just found. Then cords, stretched from these intermediate points, will serve the same purpose as if drawn through to G and H, and the stakes which mark them will not be disturbed. The secondary rectangles, which are formed by the aisles, tower, porch, robing-room, or organ-room, can be easily laid out in the same manner as that described for the transepts.

As the stakes which now mark the corners of the building will be dug away as soon as the excavation is begun, it is necessary to provide some further means by which their positions can still be shown when the stakes themselves are removed. This is done in much the same way that the centre of the apse circle was fixed, — by making the point to be marked fall at the intersection of two straight lines drawn from some more distant stations, which in the case of the corners of a rectangle may best be situated on the prolongations of its sides, as shown in Fig. 20, where the point, A, is marked with as much precision by the intersection of the lines stretched from the indefinite points, X and Y, as it would be by a stake.

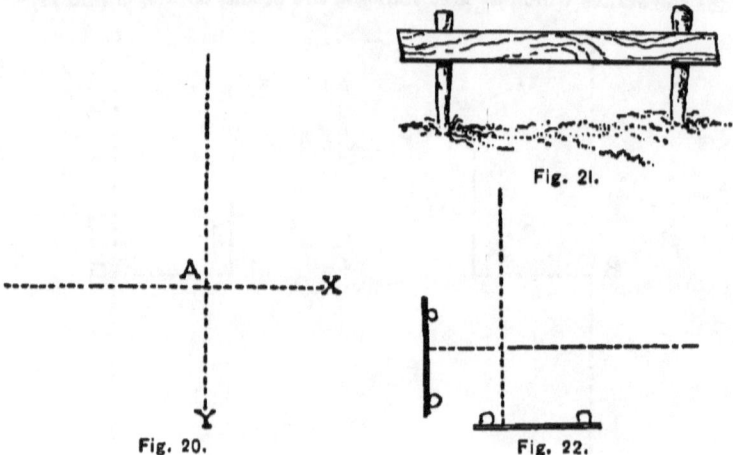

Fig. 21.

Fig. 20. Fig. 22.

In practice, the points, X and Y, are usually given, not by stakes, but **Batter-Boards.** by notches, or horizontal "batter-boards," which are nailed to stout stakes, five or six feet apart. (Fig. 21.)

Two of these batter-boards are necessary to determine each angle (Fig. 22), and as it is of great importance that they should be firm and permanent, they must be set up four or five feet back from the edge of the excavation, or more, if this is to be very deep, and the stakes set firmly in the ground. They must, moreover, be as nearly

BUILDING SUPERINTENDENCE.   23

on a level as possible, those which stand at the lower portions of the site being raised on high but strong stakes. Having set up these batter-boards opposite the corners already found, lines should be stretched between them (Fig. 23) so as to coincide with the lines already marked by the stakes. Plumb-lines suspended from the cords over the stakes will show when they pass exactly over the copper tacks which mark the precise points to be transferred, and as each line is accurately fixed, a small notch should be cut in the board where the string passes over it. When all the lines are fixed upon the batter-boards, the first set of notches will, if the ashlar face of the walls was taken for the measurements, serve at any time to fix the line of that face for any wall. The projection of the water-table or base course beyond the ashlar surface, the projection of the outer face of the foundation, the thickness both of the foundation and the superstructure, can be indicated by notches on the batter-boards, measured from the original notch, and marked so that they can be readily distinguished, as shown in Fig. 24.

Fig. 23.

Fig. 24.

It is well to fix one batter-board, at least, at the height of some given level of the building, as for instance, at the top of the water-table. If an engineer is at hand, or if the architect or contractor can use a level, there is a great advantage **Bench Mark.** in setting the top of all the batter-boards at exactly this height: then the masons will be able, when the time comes for setting this course, to level across from the batter-boards at various points, and thus obtain a line more perfectly horizontal, and with less trouble than would be possible by starting from a single point.

The fixing of the batter-boards completes the setting-out of the building, and a line may then be cut with spades, some fifteen or sixteen inches outside of the original stakes, which can then be removed and the work of excavation begun.

All the business of staking out the building belongs properly to the contractor, unless an engineer is employed, but the architect should always be pres-

Importance of Correct Plantation.
ent to see it done and verify the measurements, and especially to observe the accuracy of the angles. The crucial test of the rectangularity of the lines is the measurement of the diagonals, which should be insisted upon, and the stakes shifted patiently until they are correct.

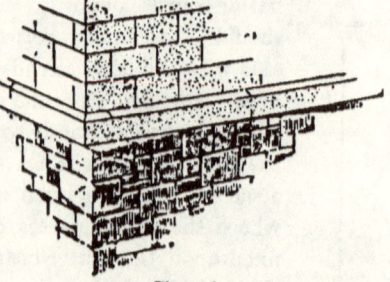

Fig. 25.

If this is not attended to at the outset, the superintendent is very likely to see subsequently the upper walls here and there overhanging the foundation, in the attempt to bring back the superstructure to its proper shape (Fig. 25); or, in the finishing, to find doors and windows unexpectedly cramped or thrown out of centre, the pattern of the tesselated pavement tapering off to nothing at one end, the frescoed ceilings misshapen, carpets fitting badly, or some other of the innumerable vexations and irremediable consequences of incorrect setting-out.

While the workmen are cutting the sod in long strips and rolling it up, which may be desirable if it can be used for improving portions of the lot, it is necessary to choose places where the loam from the surface and the gravel from beneath may be piled

Disposal of Excavated Material.
up separately in "spoil-banks," out of the way of building operations, but near at hand for use in the subsequent grading. Nothing is more common than to see the earth from a cellar thrown out at one edge of the excavation, to be soon after shovelled over again in order to dig a place for some pier or post; then, perhaps, the middle of the heap turned over a third time, to cut a trench for drain or other pipes, and finally the greater part of the mass shovelled again into wheelbarrows, and transported halfway around the building, to be used for grading up in some place on the other side, where it might just as well have been thrown in the first place.

Let the young superintendent, therefore, think where the gravel is likely to be used for grading or road-building, and how the heaps

can be most conveniently arranged, in reference both to its future use, and to the least laborious mode of bringing it from the cellar. The amount of material which the cellar will furnish can be approximately estimated with very little trouble, as well as the quantity which will be needed for the grading about the building, so that if there proves to be a surplus which must be provided for, places may be arranged for disposing of it, either in filling up hollows about the lot, grading the approaches, forming terraces, or other improvements; and the earth may then be excavated from the cellar, hauled to the place designated, and dumped in immediately, without having to be twice handled. Wherever grass is intended to cover the new surface, loam must be piled near, or brought from other parts of the excavation.

The location of the avenues and paths, with the drainage works necessary, should be determined at the outset, and marked upon the plan which the superintendent must have always at hand to refer to.

A large space on the most level part of the ground, and not far from the principal entrance, should be reserved for unloading and piling up timber, and for framing. Room must also be left for delivering stone at various points around the building, and for other materials. Especially should the area about the main entrance be kept free of obstruction. These cares properly belong to the contractor, since the extra expense and delay caused by re-handling and improper disposal of material must be paid by him, but the efficient and intelligent conduct of such works is indirectly advantageous to all parties, and the superintendent, by his greater familiarity with the plans, as well, perhaps, as his superior skill in interpreting their indications, is able to foresee future contingencies more clearly even than those who will suffer most by want of due precaution. *Space to be Reserved for Materials.*

Applying these observations to our present building, we notice the conditions to be as follows : The street runs along the south side of the lot; and the ground sloping gently toward the east, the best place for the entrance roadway will be near the western end of the building, arranging it, if no other considerations oppose, so that the grade will ascend slightly from the street. This will be of advantage in securing an outlet for water into the street-gutters during heavy rains. The building is not on the crest of the *Grading.*

hill, the ground rising continuously westward. It will therefore be necessary, in order to prevent the water running down the hill from reaching the walls, which it would soon saturate, to grade up at the west front of the church sufficiently to turn it back by a slight slope in the reverse direction; and it will improve the appearance of the structure to have this graded surface somewhat extensive, so as to form a plateau in front of the building, nearly level, and spacious enough to turn a carriage easily. This level will be continued along the south wall of the church as far as the porch opening into the south transept, forming a terrace, wide enough for a foot-path, and regaining the natural surface southward and eastward by easy slopes.

The avenue crosses a small natural basin before reaching the plateau, and beyond it continues along the north side of the building to the sheds, which are situated at some distance to the north-east. A separate foot-path from the street leads to the entrances of the robing-room and the society-room in the basement.

The gravel from the excavation will therefore be principally needed for the plateau and terrace on the west and south sides, and for the paths and avenues; and the loam will nearly all be used on the south side, where a deep soil is desirable to insure a good growth of grass. It should therefore, as it is stripped from the surface of the excavation, be piled in a heap south of the south transept.

It will be very advantageous for the avenue to have it hardened by the traffic of heavy teams bringing materials, and equally an advantage for the teams to have a practicable road for wet weather, instead of being obliged to go over the grass-land, which is soon cut into a mass of mud. Hence while the turf and loam is being stripped off the site of the church, the driveway should be staked out, together with the plateau in front of the church, and the terrace, and in order that the new material may unite with the subsoil beneath, the surface should be ploughed, the loam taken off, and added to the main "spoil-bank."

Avenue Building.

As fast as this is done, a gang of laborers should dig a trench eighteen inches wide and two feet deep on each side of the roadway, throwing the gravel into the middle; and then fill these trenches half full of stones, put six inches of hay or straw over the stones, and throw back gravel enough to fill the trenches.

The French drain, or trench filled with stones, should be continued around the side of the plateau next the hill, and made considerably deeper in that place — three to six feet, according to the springiness of the soil. The object of it is to prevent water from working in at the sides and softening the gravel of the road just below the surface, breaking this up, however hard it may have become. If stones cannot be had, agricultural drain-tile may be laid in the trench, and the joints covered over with a piece of paper before filling up with gravel. The road thus defended against the undermining influence of water from the sides will soon be dry and hard, though below the general surface, and ready for the gravel-carts which will by this time be ready to bring their loads from the cellar excavation, coming to the surface by a runway at the eastern end, where the height is least, and passing along the rudimentary avenue at the north side of the building to deposit them, first on the plateau, until that is brought up to the height required, and then upon the avenue, a sufficient quantity being dumped in the hollow near the end to bring up the grade to a uniform slope with the rest.

Besides the deep intercepting drain around the upper edge of the plateau, it is best in all but the most porous soils to make another French drain under the plateau itself, in the shape of a V, the vertex of which points up the hill, while the extremities of the legs end in the road drains, one to the north and the other to the south of the western end of the church. This will prevent the water which falls upon or gets in beneath the fresh surface of the plateau from running along the comparatively hard, sloping stratum of natural soil beneath toward the foundation walls of the building, which will soon be penetrated by its persistent flow. It will be best to leave the avenues at a grade about six inches below the final level. If well drained they will soon become hard under the heavy traffic, and a final coating of six inches of screened gravel at the completion of the building, brought to a neat surface, will give a good and durable finish to the work.

The terracing along the south side may be provided for most economically, if the ground is firm, by directing the excavated material on that side to be thrown out on the bank without loading into carts; but if the soil is soft or sandy, the edge of the excavation must not be weighted with material until the cellar walls are built, or it is liable to cave in.

The contractor for the building is not obliged to do all these works, unless they are mentioned in the specification or contract. In general, if there is no agreement otherwise, the builder is expected to take off and reserve the sods, transport and pile the loam and gravel separately wherever directed within the bounds of a reasonably large lot, but he would not be expected to spread and level the material, except about the building, unless such levelling were mentioned specifically. The other works mentioned, although they are best carried on at the same time, would be included in a separate contract in connection with the subsequent terracing, planting, sodding, or gardener's work which might be determined upon for the general adornment of the lot.

Recapitulation. To recapitulate the most important things to be remembered, and precautions to be observed in first laying out the building and starting the work: —

Examine the figures on the plans, to see if they are correct.

See that the steel or other tapes used are divided into feet and *inches*, not into feet and tenths of a foot.

Stake out provisionally the actual ashlar lines of the building.

Measure the diagonals of the principal rectangles.

Transfer the lines given by the stakes to batter-boards, permanently fixed. If the ground is not level, or nearly so, the horizontal dimensions must be measured level, and transferred to the stakes by a plumb-line. After determining the ashlar lines, the foundation-walls should be marked on the batter-boards, and the lines of the excavation given about eight inches outside the face of the foundations.

Set some permanent mark representing either the top of the floor-beams, the water-table, or any other convenient level, providing carefully for the change in the surface of the ground which will be made by the subsequent grading. Write distinctly on this stake or "bench mark" the level which it is intended to represent, and also the depth of the cellar bottom below it, allowing three inches for concrete.

Consider and decide about the laying-out of the lot, and if any of the avenues can be used in the building operations have them immediately staked out, cleared of loam, and drained.

Confer with the contractor in regard to the most convenient place for delivering the materials and dumping gravel. Explain to him the future plan of grading, and interest him in your provisions for avoiding unnecessary handling of the earth, so that your directions may not seem arbitrary. Remember to leave spaces as follows: —

>Not less than 2,500 square feet for piles of lumber.
>Two or three plots of 500 square feet each for brick and rubble.
>About 500 square feet for other materials.

These to be near the building, but leaving space for heavy teams to drive up, unload, turn, and go out. Space may also be needed near the street as follows: —

>1,000 square feet for stone-cutting sheds.
>1,000 "   "   rough blocks of stone.

Determine the position of the main provisional entrance, and keep the approach to it clear. This entrance need not necessarily be one of the regular church doors. In this instance, the main access to the church, through the tower, is somewhat tortuous for the introduction of long timbers or other bulky materials, so that it will be best to provide a temporary one, which can be done by leaving the arch of the large west window open down to the ground. When the necessity for so large an entrance is over, the wall can easily be built up as high as the sill.

SECOND VISIT.

We will suppose that the matters treated of in the foregoing notes have been satisfactorily disposed of, and the laborers have been some days at work. By this time it will be necessary to appear again upon the ground. The second visit finds the loam removed, and the excavations completed down to the cellar bottom at the western end. Teams are hauling rubble-stone for cellar walls, and a car-load of staging lumber has arrived on the ground.

Let the superintendent begin by going all around the outside of the building, comparing the excavated lines with the marks on the batter-boards. Any mistakes should be pointed out at once, before they are driven from the mind by other matters. *Tour of Inspection.* Next, let him make the tour of the inside of the excavation, examining the bank carefully. He finds the ground at the lower, or eastern end to be a fine gravel mixed with stones. The bank at the upper end, for five or six feet below the surface, is composed of a similar gravel, but below this appears a stratum of greenish clay, hard *Appearance of the Excavation.* at the upper part, but softening into mud below. The clay continues along the bank to the site of the tower, where it ends in a large mass of loose slaty rock, from which water trickles rapidly. The south side of the tower excavation shows only gravel.

These appearances demand careful consideration, for they indicate a state of affairs involving both serious dangers to be overcome, and costly extra works to be planned, and the payment for them satisfactorily arranged.

The operations which will be necessary in the present case will be best understood by going through a process of reasoning similar to that which should occupy the superintendent's mind on viewing the circumstances.

The appearance of the hard clay stratum a short distance below the gravel at the upper side of the excavation warns us that it is first **Clay Stratum.** of all necessary to cut off the water, which in rainy weather will soak through the gravel, and collecting on top of the clay will follow it down hill in a wide, shallow sheet, so that the foundation wall, at the line where the clay stratum comes against it, would soon be soaked. We have had the trenches cut eight inches wider than the wall, expressly for the purpose of allowing this to be built up smooth and independent on the outside, and protecting it by filling the vacant space with gravel, which will intercept a part of the descending sheet, but this is not enough; a trench must be dug a few feet in front of the west wall of the church, deep enough to cut into the clay stratum the whole length. If the clay bed slopes northward or southward, as well as eastward, the trench may follow it downward, discharging into one of the road drains. (Fig. 26.) If its section in that direction is horizontal, the trench may take the form of a shallow V, with an outlet at each end. These trenches are to be half filled with loose stones, covered with straw or hay, and filled up with gravel.

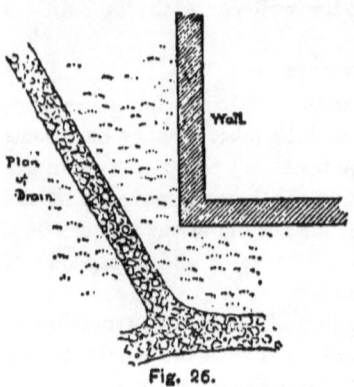

Fig. 26.

The water descending the hill on the surface of the clay stratum is thus completely intercepted, and if it were not for the ledge of rocks, which, as we have ascertained, extends beneath the clay bed, there would be no need of any further precautions, unless to give a little extra depth to the trenches in the clay, to insure against the effects of frost, which will penetrate a foot deeper in a clay soil than in dry gravel.

But by closer attention we shall find that the clay bed is uniformly hardest at the top, and grows softer downward, being softest just above the ledge. This means that the rain falling on the hill above, filtering down through the gravel and clay strata till it reaches the rock, is there arrested and compelled to descend along its surface, working its way between the stone and the overlying clay, which it reduces to a mud so thin, after protracted wet weather, that the superincumbent material will slide along the top of the ledge thus lubricated, if any way is open for it to escape. The indications are that the ledge extends along the whole western line of the building, and by excavating the church cellar, a space will be open into which the clay, pressed upon by the weight of the walls which come over that portion of the foundation, will be able to force its way, the cellar floor rising and the wall settling, as the soft mud beneath is squeezed out under the load. (Fig. 27.) This is no imaginary danger, but is the certain consequence of the operations proposed. How can the programme be changed to meet the difficulty?

Clay over Sloping Rocks.

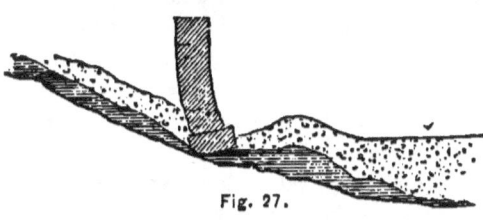
Fig. 27.

It might be possible to cut an intercepting trench in the ledge, similar to that by which we propose to cut off the water descending on the surface of the clay; but this would be expensive, and there might be seams of loose rock, like that found under the tower, through which the streams would run in the interior of the ledge, and coming again to its surface below the trench, would render this useless.

Might not the foundation be carried through the clay to the rock itself?

This would be effectual in preventing settlement, but from the inclination of the surface of the ledge exposed under the tower at the north-west corner, we can roughly estimate that the rock under the north-west angle would be at least ten feet below the cellar bottom, and to carry the foundation to this depth would add materially to the cost of the building. Besides, it would be hardly wise to dam up the descending water by a continuous wall, which would be in danger of being gradually undermined by the flow. Rather let us bring up piers from the rock, with arches below the cellar floor level, and the wall can then rest on these arches without danger of settlement, while the water will find its way between them and continue its course, far enough below the cellar floor to do no harm. For these piers we will use concrete, which will be much cheaper than brick or stone masonry, and will resist better the undermining action of the subterranean water.

A short deliberation is needed to convince us that this is the best mode of overcoming the difficulty, and the arches are marked out on the foundation plan, through the whole extent of the west front, leaving the largest opening in the middle of the wall, under the great west window. Whether similar arches shall be continued under the north and south walls is next to be considered; but these are so much lighter than the west gable wall that their effect on the clay bed would be far less; moreover, this bed is here at a much greater distance from the surface, and, most important of all, the effect of a vertical pressure would be to press it outward, instead of inward toward the excavation, so that the weight and inertia of the whole depth of gravel above it will operate to keep it in place. There is some danger of unequal settlement at the junction of these walls, which stand on slightly compressible gravel, with the west wall, which by its piers extends to the incompressible rock; so we will enlarge the footings of these walls near the junction, thereby spreading the weight over a large surface, and reducing the land on each square foot of gravel so far that it will be borne without any yielding. An additional six inches on each side, obtained by adding one course of footings, will suffice, if the stones are reasonably flat, and making a memorandum of this we proceed to consider the foundation of the tower.

Here there is no doubt as to the support, the rock being everywhere above or near the level of the footings. The point requiring most attention is the spring of water flowing from the seam of loose rock. There is only one thing to be thought of; that is, to collect this water in a covered receptacle which cannot overflow, and convey it by a tight conduit to a safe outfall beyond the walls of the building. This will be somewhat costly, and the builder, and perhaps the church committee, desirous of avoiding needless expense, will quote the example of other structures in the neighborhood which have uncovered springs in the cellar, and where they are allowed to flow away by an open channel, but the architect or superintendent should not allow himself to be persuaded by these arguments. He will find, if he cares to inquire, that in every one of the buildings mentioned, the cellar walls and ceilings are dripping with moisture, the first story beams are blackening with incipient decay, the structures themselves are chilly and difficult to warm in winter, and a penetrating smell hangs about them in summer, especially after rains.

Water Vein in Rock.

Wet Cellars.

These are the houses where the young people die of consumption, one after the other; or the churches that one enters with a sudden depression of spirits, and leaves with a headache or a cold. Let the architect claim the authority due to superior knowledge, and refuse to sanction anything short of absolute security against water within the cellar walls.

Fortunately, there is but one spring, although that is a copious one, flowing some fifty gallons per hour. We will therefore excavate, by pick and by blasting, a rough well on the line of the seam, just outside the cellar wall, and carried to a depth of at least 2½ feet below the cellar floor. If there had been several water-bearing seams, we should have been compelled, instead of excavating the well outside the wall, to make it beneath the wall itself, by cutting the trenches two feet or more below the cellar floor, and putting in the first foundation stones dry, without mortar, so that the water could collect in the vacant spaces. This would keep the moisture from invading the cellar, but the wall might be damp from the water standing beneath it, so that a well entirely outside of the wall is preferable. We will, however, to provide for the possibility of water coming in wet weather through seams now dry, deepen the trenches about a foot, and lay the first course of stones dry, bringing this trench into communication with the well, so that water entering under any part of the wall will find its way to the well. A channel is then to be made, and a tight pipe laid with cemented joints below the cellar floor from the well, across the tower and under the opposite wall to the outside of the building, until the gravel is reached, where the pipe may end in a pit filled with loose stones.

Having plainly indicated the arches, piers, drains and well, the question is to be settled, — Who shall pay for all this extra work?

The principle to be kept in view in the decision is that, unless some special agreement has been made, the builder cannot be obliged to pay the cost of extra foundation, concrete, or other works rendered necessary by peculiarities of the ground which could not have been reasonably expected or foreseen when the contract was signed : he is, however, assumed to have examined the ground where the proposed building was to stand, and to have included in his contract price the risk of common defects, such as clay beds in gravel, rock in a spot where the ledge appears on the surface near by, or of springs in any soil.

Of course the best way would have been to bore at different points around the building, to find out the depth and nature of the soil, and by the light of these tests to draw the foundation plans and specify the various works, but this is rarely done in ordinary buildings, and the object of this treatise is as much to come to the rescue in the common cases of forgetfulness, omission, or unforeseen difficulty, as to point out the course which would be absolutely the best for all buildings.

In the present case, it is decided that, the clay seam being a common occurrence in gravelly soils, the contractor shall bear the expense of the trench and drain for cutting off the water which would flow over its surface toward the building; and shall also **Apportionment of Cost.** pay for the drain to carry off the spring-water from under the tower; but that the cost of carrying down the foundations of the west wall beyond the point shown on the drawings, the extra width of footings under the north wall, and the necessary blasting under the tower, shall fall on the church, for the reason that the ledge did not appear above ground anywhere near the site of the building, and therefore its existence so near the surface as to interfere with the foundations would not reasonably be inferred with sufficient certainty to form an element in the contract price; the same rule applying also to the clay bed resting on the rock.

The new work should be clearly described in a supplementary specification, giving the proportion of cement, sand, and pebbles in the concrete, mentioning the large flat stones which should form the upper part of the piers, and from which the arches will spring; requiring that the arches shall be built on centres, of good, hard brick, in mortar made with one part sand to one of cement, in four rowlocks for the small arches, and five in the large one; with any other particulars which may make the meaning more clear. The drawings should be rectified by notes and diagrams in the margin, a record of all the facts, with copies of all instructions and orders, kept by the superintendent, and a price agreed upon, setting off against the cost of blasting under the tower the amount of earth excavation saved, and the value of the rough stone obtained. An additional agreement, embodying the supplementary specification and the extra price agreed upon, is drawn up, signed by both parties, and attached to the original contract, and the work is ready to proceed.

All this sounds long, but it is time well spent, for a few hours more in arranging the preliminaries will save days as well as dollars to both parties in the final settlement. A glance at the rough stone delivered, with an admonition to the builder to get the footing-stones as flat as possible, and the summary sending off the ground of a lot of staging lumber which has just arrived, containing a number of knotty and shaky poles, calling the builder's attention at the same time to their rejection, may terminate our duties for the day.

THE THIRD AND FOURTH VISITS.

Very soon after these affairs have been agreed upon it will be necessary to make another visit to the building, to see that the execution is rightly begun.

We find the excavation finished, the blasting done under the tower, the collecting well and pipe completed, and the foundation-wall under that part already some five feet high. There is no other stone-work started, and thinking it a little strange that this, the most difficult point of the work, should have been begun first, we examine the wall minutely. The inside face looks all right, the stones being perhaps a little small; but that may be the way the stone runs. Outside, the gravel has been filled in nearly to the top of the stone-work. We borrow a crowbar and force it into the gravel outside the wall in several places. Except a softness of the material, which shows that it has not been properly rammed, or "puddled" by wetting it thoroughly, so as to pack it closely into its place, we observe nothing out of the way until we approach the corner, where for some feet the bar, instead of sinking its full length, close to the outside of the wall, strikes against the solid rock not far down. We call the foreman and ask him if the foundation-wall stands entirely clear of the ledge. He hesitates, and finally replies that the ledge, after being cut away through part of the thickness of the wall, showed such a "nice flat top" that it seemed a pity to excavate it any farther, and he had therefore built up a thin wall against it as high as the top, and then built out over the rock to make the full thickness of the wall as shown on the plans. He adds, with great apparent confidence, that "nothing can be better for part of the wall than the solid rock."

This explanation is specious, but in practice is dangerously misleading. In a wall so built, the water will find its way either through the imperceptible seams of the ledge or over its top into the body of the masonry, keeping it constantly damp. Moreover, there is a serious risk that under the heavy weight of the tower, the thin lining wall built up against the ledge, but in no way bonded to it, would separate from it and fall away, leaving the superincumbent masonry most insecurely supported.

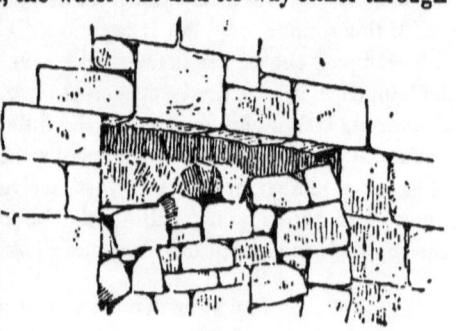

Fig. 28.

There is, besides, the certainty that the foundation-wall, built partly on unyielding rock, and partly of small stones laid in compressible mortar, will settle unequally, and crack, perhaps dislocating the masonry above, and at least opening an inlet for moisture. The work must therefore be immediately taken down to the very foundation, and the ledge cut away so as to leave ample space for the whole thickness of the cellar wall down to the footings, with seven or eight inches additional room outside the masonry to enable it to be properly pointed, and for packing in behind it a screen of gravel, which will intercept and carry safely down to the drains whatever water may ooze through the veins of the rock. The workmen will probably profess never to have heard of a foundation in which the bank was not intentionally cut just to coincide with the outside line of the wall, so that this could be built up directly against it, thereby saving them all the trouble of selecting stones for this side, so as to have it smooth; plumbing it, to get it vertical; and pointing it, so as to have it impervious. It is true that this is the common method of cheap builders, but it is not, and should not be, countenanced in work of any importance, even in dwelling-houses, except of the lightest and cheapest kind. **Outside Face of Foundation-Walls.**

A wall built in this way (Fig. 29) is neither safe nor satisfactory. The joints at the back, being concealed, are usually devoid of mortar, or if any is put in, it falls out again, so that a gradual compression of the outer portion is liable to take place, as the weight of the superstructure increases, bulging the inner face of the wall toward the cellar, and the unfilled cavities next the bank collect the water which trickles down by them, and conduct it into the heart of the masonry, while the projecting points of the larger stones imbed themselves in the earth, so that when this freezes and expands, the wall is often lifted as if by a number of short levers, dislocating the joints and making channels for moisture through them. 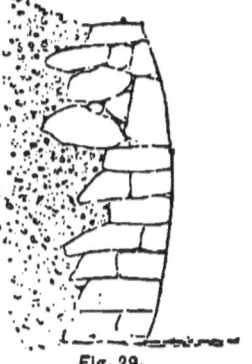 Fig. 29. It is actually much more important to have the outside of a cellar wall smooth than the inside. If the stones are selected so as to show a good face on the outer surface, the joints well filled with cement mortar, and pointed with due care as the work proceeds, holding the trowel used for pointing *obliquely*, so as to "weather" the joint, as the workmen say (Fig. 30), any moisture which runs out from the bank, or descends from above, and flows down over the outer face of the wall, will, when it meets a joint, drip off, falling on the inclined surface of cement, by which it will be conducted safely over the edge of the next stone to run down and drip off again, until it reaches the bottom, where it passes off in the drain, without having been able anywhere to penetrate into the masonry. This essential point in construction is one of the hardest to enforce. It is so habitual with ordinary workmen to neglect the portions which will be concealed, and expend their skill

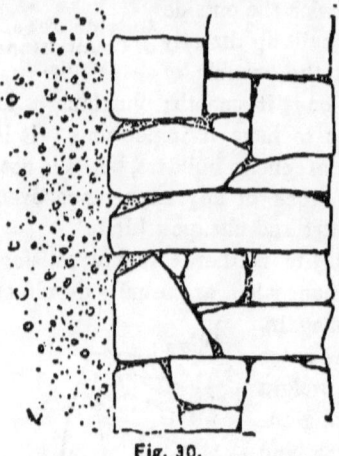

Fig. 30.

on the visible inner surface, that some explanations given to individual men will be necessary, especially of the proper method of pointing, besides a good deal of watching, to see that the directions are followed.

Having given the requisite orders for taking down the objectionable masonry, excavating the ledge properly, and rebuilding in the manner described, we will pursue our tour around the building. Close by we come to the pits prepared for the concrete piers which are to extend the foundation of the west wall down to the rock. The sides of the deepest holes are sustained by a shoring of planks and beams, and the contractor is awaiting orders to put in the concrete. We examine the pits to make sure that the ledge is exposed at the bottom, and clear of clay, which would prevent the concrete from attaching itself to the rock. The deepest excavation, we find, has struck a spring, which runs copiously over the surface of the ledge at the bottom, and the contractor says, with reason, that the cement will be washed out of his concrete as fast as he puts it in. There is a remedy for that; but before beginning the concreting we must test the quality of the materials. Meanwhile, we send a boy to fetch a dozen yards of oiled cotton cloth.

In accordance with our previous directions, the contractor has screened the gravel which he proposes to use, and the finer part is heaped up on one side of a large plank mortar-bed, while the coarser pebbles are piled on the other. The fine gravel, or rather sand, when rubbed in the hand gives a dry, crackling sound, and is prickly to the skin. We wet some of it, and grasp a quantity. On opening the hand it will not retain its shape, but falls down loosely, and does not soil the

Concrete Making.

skin. It is therefore sharp and clean, suitable to be used for mortar or concrete without washing. If it should happen that the sand is very fine, it may still be used, if sharp and clean, but the proportion of fine sand by measure to a given quantity of cement should be less than that of coarse, and the coarser kinds are much to be preferred. The rounded pebbles screened out of the gravel are free from earth or clay, but some dust clings to them, which would prevent the perfect adherence of the cement to their surface, and we direct them to be thoroughly washed by throwing buckets of water over them.

Sand.

Some officious individual has added to the heap a quantity of the angular fragments of disintegrated rock from the tower foundation, but these, although excellent in shape, we find to be somewhat coated with the clay which has been washed into the rock seams, and therefore unfit for concrete unless washed clean. This would be a long process, since any admixture of clay clings very persistently to sand or stone, and is very injurious unless entirely removed, so, as we have an ample supply of cleaner material at hand, we order all the clayey fragments to be taken away.

The cement is next to be passed upon. Of this we find ready for us a large number of barrels, bearing a great variety of brands, and gathered from the stocks of all the local dealers within reach. Among them are casks from the Newark Lime and Cement Company, F. O. Norton & Co., Connolly & Shafer, the James Cement Company, besides a few barrels with other marks. The mason is in a hurry to begin, so we tell him to use at once any of the F. O. Norton or the Newark cement that has not been damaged. The first cask of these which is opened contains a crust of hardened cement three or four inches thick, but the enclosed portion remains in its normal state of fine powder. We order the crust to be rejected, but allow the inner portion to be used. The other barrels seem uninjured. Meanwhile, casting an eye now and then on the mortar mixers to see that they put conscientiously one shovelful of cement to two of sand, and thoroughly mix the dry sand and cement before adding water, we proceed to test roughly all the brands of cement before us with which we are not familiar.

Masons have various ways by which they profess to form an opinion of the goodness of cement. Some dip their hands or arms

into the barrel, and if the powder feels warm they pronounce it good; others taste it, and if it bites the tongue they call it suitable for use, the strength of the cement being supposed to be proportional to the intensity of the bite; and there is another common belief that the dark colored Rosendale cement is stronger than the light. It is needless to say that all these tests are simply worthless; in fact, they are principally employed to impose upon modest young architects, who can sometimes be deceived by such mysterious performances, the result of which is sure to be in accordance with the interests of the party applying the tests.

Testing Cement.

Let us cast aside these divinations, and taking a handful of cement from an average barrel of each of the brands, mix it with water into a cake, put it in the sun, or in any dry place, for half an hour or more, till it acquires such a consistency as to be barely indented by the pressure of the end of a match or a stick of equal size, cut square, and weighted by resting a brick upon it. Place the cakes in some regular order, so that the different varieties of cement of which they are made can be distinguished, and as fast as they reach the requisite hardness, put them into a tub of water, till all are immersed. Note the time required for each one to reach its first "set" in the air. Finally, make a second series of cakes, and leave them exposed to the air, without immersion.

By this time a batch of concrete is mixed; the sand and cement have been thoroughly mingled until no lumps of cement or sandy streaks can be discovered in the heap, water is then added, not in too great quantity, but enough to give a pudding-like consistency to the mass, and the whole is well stirred and shovelled over again; then the stones, which have been well wet before putting them into the mortar, are added, and all mixed quickly, but thoroughly. If well mixed, the bulk of pebbles may be double that of the mortar. The object of wetting the stones before adding them is to wash off the light dust which very rapidly settles on them, and prevents the adherence of the cement. The moisture of the mortar would wash the stones clean by long stirring, but time is of importance, and it is best and easiest to dash on a few buckets of water at first.

We begin with the deepest pier of concrete, — the one which has to be laid in a stream of running water. Taking the oiled cotton, which has by this time been brought, we fashion it into a large, rude bag, nearly water-tight, which is taken down into the hole and filled with concrete. The water rises around the edge, but not enough to overflow the mass, and after packing the con- **Laying Concrete in Water.** crete solidly down into its place by means of a wooden rammer, we leave this, and proceed to put a layer into each of the other excavations, throwing it in from the top, so as to compact the mass by the momentum of the fall, as it is not easy to reach it with wooden rammers. After dividing the material already mixed among all the piers, the concreting should be stopped for the day, and the men put on other work, as we are more likely to obtain a compact mass by putting it in the pits in twelve-inch layers on successive days than by filling in the whole body at once. The holes should be covered with boards to prevent rain from washing in sand on top of the layers already deposited.

The next day, after our regular preliminary tour, first outside and then inside the building, we examine the samples of cement which we made up the day before and laid aside. The specimen of the Newark Lime and Cement Company's cement left twenty-four hours in the air is found to be quite hard, and breaking the cake with a slight pulling strain, much as if it were a stick of candy, we find it to possess a very sensible tensile strength, and the two halves separate with a clean fracture instead of crumbling. The cake of the same brand left in water retains its shape, and has increased considerably in firmness. The cement may therefore be pronounced good, as is usual with that brand.

The F. O. Norton samples, both in water and in air, show similar qualities in at least an equal degree; and all the barrels of this brand which on being opened show no signs of caking are accepted.

The sample in air from the barrels marked "James Cement Co." is quite hard, harder than either of the two preceding, and a slight bluish efflorescence, like mould, has already begun to appear on its surface, but the specimen left under water has crumbled into a soft heap.

Of the Connolly & Shafer manufacture, the portion left in the air retains its shape, but has not acquired much consistency: it crushes in the fingers like clay. The sample in water is nothing but mud.

It does not necessarily follow, because this last variety sets slowly, that it is essentially bad, but it will be unsafe to use in our concrete, and inconvenient in the masonry, so that unless a second sample should show much better qualities, we will discard all the barrels of that brand.

The James cement, which sets quickly and hard in air, but under water breaks up and crumbles, should be rejected for the concrete, but may be used for the masonry. If the rapidity of its setting should interfere with its convenient use, as will very likely be the case, especially in hot weather, it should be mixed with a small portion of lime.

If none but James cement should be obtainable, or other brands having similar characteristics, let the superintendent try whether a sample of it mixed with half its bulk of slaked lime and made into a ball will set hard under water; if so, it may be safely used in that way, even for concrete.

Some of the quickest-setting Rosendale cements, when immersed in water without having previously acquired a certain degree of hardness in the air, will set rapidly, and immediately crumble again, and never acquire any subsequent consistency. With such it is often found that the addition of a small quantity of lime will confer upon it the qualities of the better cements, causing it to set perfectly under water, and improving it for use in air by retarding the setting slightly. With the very slow-setting cements little can be done unless there is time to wait for them. They may do for adding to lime mortar in stonework above ground, where it is desirable to harden the mortar, but for foundations, on which the weight is to be rapidly added, or in work under water, it is best to avoid the use of any cement whose setting is found to be uncertain or long delayed.

It is unnecessary to say that these tests are by no means such as would be used for engineering work of importance, but they will do well enough for rough determinations, and an ample margin of strength is, or should be, always left in the smaller operations of construction. More accurate methods of judging will be described in treating of city buildings.

In the course of our preliminary tour around the works we noticed with surprise that one of the concrete piers was already finished, and the top nicely smoothed over, and having completed the tests of the cement, we return to inquire into the matter, taking the foreman with us. We examine the ground closely, and notice some stray pebbles dropped around the edge of the hole, and some such dialogue takes place as the following: —

Superintendent: — " Mr. Foreman, how did you get this pier done so soon ? "

Foreman: — " Well, sir, we hurried a little on this pier, because we wanted it to git set before it rained, and " —

Superintendent: — I left word to put in only twelve inches of concrete at a time in each pier."

Foreman: — " O law, sir, that aint no way to build a pier. There aint no one can tell me nothin' about concrete. That's as nice a job of concrete as ever " —

Superintendent (remembering the scattered pebbles) : — " You didn't put the stones in dry and then grout them, did you ? "

Foreman (slightly taken aback) : — " Well, sir, perhaps, — yes, we did: you see, that is the best way to do where you have such coarse sand, and then " —

Superintendent: — " Get some one here and take that all out. It is impossible to tell now how much cement there is in it, but it has not begun to set, so if you will take it back to the pen and add a shovelful of cement to every two shovelfuls of this, and mix it well, I will let it pass to put into all the piers twelve inches thick."

Foreman (deferentially) : — " Yes, sir, anything you say, sir."

Pursuing our way after this little episode we come to a squad of men laying footing-stones for the clerestory wall on the gravelly bottom. These must be carefully looked after, for the weight of the clerestory wall being concentrated on the piers of the arcade will try the strength of the foundation very seriously. The drawings show a continuous foundation-wall, but no inverted arches, it being impossible to get the necessary abutment for these without considerable additional expense, and it is therefore necessary that the masonry of the foundation should be well bonded together longitudinally, so as to receive the pressure as a solid mass; otherwise the settlement will be greater under the piers, and the work will be dangerously dislocated.

Setting Footing-Stones.

Most of the footing-stones on the ground are good flat pieces, but here and there are some misshapen lumps, and one of these, just as we come up, is suspended to the derrick boom, ready to lower into its place. The men have tried faithfully to hollow out a basin in the gravelly bottom of the trench to fit the irregularities of the stone, but when this is lowered into its place, it rocks unsteadily. It is raised again, and the bed remodelled. This time the stone fits better, but is still unsteady. The men are discussing whether to let it go as it is, or try again, when the superintendent comes up, and stepping upon the stone rocks it until he is satisfied that there are no large cavities beneath. Sending for buckets of water, he directs **Puddling with Water.** fine gravel to be heaped around the stone, picking out all pebbles and lumps, and the water to be then thrown on, pailful after pailful, or, still better, a stream from a hose to be directed upon the mass. The water settles away through the sand, searching out all cavities into which it can flow, and carrying particles with it wherever it goes, which gradually compact themselves in the hollows under the stone until it can no longer be moved. This puddling process is continued a little longer, to make sure that a full and perfect bed is formed under the stone, and directions are given to do the same with all the levellers which have uneven beds.

Where the bottom of the trench is clay or rock, a thick layer of cement mortar should be spread to bed the footing-stones in, for the purpose of filling up all cavities between the substratum and the stone, but in gravelly soil the puddling with water is often much better than the bed of cement, especially with stones of very irregular shape. This expedient for filling in cavities under and around masonry is capable of still more extended use. The writer once knew a case where a church tower had been nearly completed upon a foundation badly built and with joints only half-filled with mortar. The tower began to settle, and the contractor for the superstructure, a man distinguished for his boldness and ingenuity in emergencies. sent for the town fire-engine and a quantity of fine sand, and putting the sand into the tower cellar, kept the engine playing upon it for half a day. The floods of water found their way out through every crevice, and wherever the water went the sand followed, until all the cavities were packed full. It was heroic treatment, certainly, but effectual; the settlement ceased, and the tower stands perfect to this day.

Let us look at the stone delivered on the ground for foundation-walls. It is of various kinds, some pieces being slaty, some tough, with rounded surfaces, like fragments of boulders, as they probably are. Many blocks of the greenstone from the tower excavation are to be seen, and these should be examined with suspicion, for fear of almost invisible cracks, which will let the water soak slowly through, besides unfitting the stone to resist a strain. To test them, they should be struck with a hammer. If they ring clearly, they are good; a seam, even if invisible, will betray itself by the dull sound which follows the blow. The boulder stones are usually good, if not too much rounded. One side, at least, should be quite flat, to form the bed. Slate stones vary in different localities. In some places they are of immense strength if placed flat in the wall, and form admirable material; in others, especially in eastern Massachusetts, the tendency to cleavage in the secondary planes, across the laminæ, is so decided that the stones, although apparently sound and strong, will break across after being placed in the wall, as soon as the weight of the superstructure comes upon them. This is a most annoying defect, as the stones cannot then be taken out, and the parts often separate a quarter of an inch or more, making a seam which it is difficult to close by pointing. The only way, when a tendency is noticed in the stone to break up into fragments of regular crystalline form, is to avoid using it in long flat pieces, for lintels, bond-stones or templates, or in any other position where it will be subjected to a cross strain. The softer lime and sand stones, when used for foundations, are much less liable to such defects, but being somewhat absorbent, special pains must be taken to isolate them from the banks of the excavation by a backing of sand or gravel, and to provide for thoroughly intercepting and draining off the moisture which might come in contact with them.

Foundation Stone.

An inspection should now be made of the lime, and opportunity should be taken to inquire into and criticise the methods that the foreman proposes to employ in mixing the mortar for the upper portion of the masonry. The barrels have, we find, been piled on a slightly elevated spot, the ground descending in all directions, so that water may not during heavy rains run down against them. Boards have been placed underneath, to keep them from the dampness of the ground, and a covering of boards has been laid on top, to shelter them from storms. This would not be sufficient protection in ordinary cases, but the contractor tells us

Lime.

that he intends to build up the foundations of the chancel at once, and lay the floor over them, which will give him a dry place for storing materials, and we acquiesce in this arrangement. Two or three of the casks in the pile have burst open, and looking in we see some of the lumps in them crumbled down into soft powder, while others are hard, but remain inert when dipped in water. These are damaged casks, and must be rejected as worthless, however good the original quality of the lime may have been. A large part of the barrels are marked "ground lime," and contain a dingy-colored lime in powder. This, if not damaged, will make good mortar, although it slakes quietly, and if not pulverized will not slake at all. The mortar of common ground lime is slightly hydraulic, and will harden under water. In general, the hydraulic limes, which will harden under water, or in damp situations, without admixture of cement, slake quietly, and need to be ground after burning, while the fat limes, such as are used for plastering, slake energetically, and are better kept in lumps, so that they may not be slaked by the moisture of the air, while the mortar made from them hardens slowly in the air, and under water, or in damp soils, never, unless cement is added to the mixture, which is usually done where they are used for masonry.

There is one singular exception to this rule among the American limes, of which we shall perhaps find some examples in our miscellaneous heap of barrels. Let us open that old flour-barrel, without mark of any kind. We discover it to be filled with a black substance, in lumps resembling cinders. This black lime is made from a beautiful pink marble, and slakes fiercely in water, making a dark-colored mortar, which sets like a strong cement. Two or three other casks, containing white lime, are destitute of brands, and their contents should be tested by putting a few lumps from each into water. The lime from one slakes quickly, but only superficially, leaving a hard core. It is therefore underburnt, and must be rejected and sent off the ground. The lumps of core, if allowed to get into the mortar, would be liable to swell afterwards and crack the joints or throw off the pointing. Another barrel contains overburnt lime, which remains inactive for a long time in the water, even when powdered, at last slaking slowly. This must also be discarded; it is less valuable than so much sand.

The other barrels are stencilled "Rockland," "Rockport," "Canaan," "Glens Falls," "Thomaston," or other well-known brands, and if not damaged by water or by gradual air-slaking, are probably all good enough for making stone mortar with an admixture of cement.

In mixing the mortar the foreman should be persuaded, if possible, to put on all the water for a batch of lime at once, instead of by successive buckets, with intervals of stirring between, thereby chilling the lime as fast as it begins to heat. Even filling the pen with a hose is too long a process for securing the best results; with very active lime the most successful mode is to pour it in a mass from a large cask. The proper quantity of water is one-and-a-half barrel to each barrel of average lump lime, and this should be measured as accurately as possible. If too much is added, the mortar will be thin; if too little, it will be thick and become difficult to work as the slaking proceeds, so that the mixer will add more water to the mass, thereby chilling it and putting a stop to the slaking process, and a granular, lumpy mortar will be the result.

Mortar.

Much of the labor of stirring would be saved, and the quality of the mortar improved, by covering the pen, as soon as the lime lumps have been evenly spread over the bottom and the requisite proportion of water added, with a canvas or tarpaulin, and leaving it to itself for half an hour or so, during which time the confined steam and heat, aiding the action of the water, will reduce the whole to a smooth, uniform paste.

On no account should the lime be slaked on the bare ground or in a hollow made by an embankment of sand. Such practices, though they still linger in country districts, have long been obsolete in all places where good workmanship is held in honor. A water-tight pen of planks, about four feet by seven, must be made, with plank bottom, and sides about ten inches high. This will give room for treating one cask of lime at a time.

After the lime is slaked and all the lumps reduced to smooth paste, it should stand as long as possible before mixing with the sand, which may, if the lime is good, be added in the proportion of two parts of sand to one of the lime paste, or five to one of dry lime.

There is a common error that cement will take more sand than lime. This arises from the fact that in mixing cement it is generally allowable, unless great strength is required, to add sand to the dry cement powder in the proportion of three to one, or, as the mixture

is usually effected, one shovelful of cement to three of sand, whereas for lime mortar the rule of two parts by measure of sand to one of lime *paste* cannot be exceeded without injury; but as the crude lime swells in slaking to about two and a half times its original bulk, a quantity of sand equal to double the amount of hydrated paste would represent five times the bulk of the original lime lumps. If the cement were mixed with water before adding the sand, which would be impossible, on account of its rapid setting, the result would be the same, since the cement expands very slightly, if at all, in slaking.

Our specifications require that the mortar for foundations should be made with "one-half cement." Let the contractor and his men understand that this means one cask of cement to each cask of lime; not one-half a barrel of cement to one of lime, as some masons pretend to interpret it.

The lime mortar alone will stand for weeks unchanged, but the addition of cement causes it to set in a few hours; it should therefore be mixed only as required for immediate use. Some judgment and observation will be needed to make sure that the cement is added in the proper proportion, unless it is mixed with the sand previous to adding this to the lime, which is not practicable unless it can be used immediately, since the cement would slake by absorption of moisture from the air on standing. The mixture should be thoroughly made, which will be shown by the uniform color of the mortar and the absence of streaks or spots.

Specifications are sometimes so loosely drawn as to omit all mention of the mortar. In such cases the character of the mixture and the proportion of materials will depend greatly upon local custom, but the practice of mixing the lime for mortar of foundation-walls with at least one-half its bulk of hydraulic cement is so universal and so necessary, where ground lime or some other variety having hydraulic properties cannot be procured, that it should be required in all cases where mortar is used below the ground surface. In wet or springy soils, or for heavy buildings, the dose of cement should be equal to that of lime.

As for the quality of materials, neither law nor custom presume any but the best to be intended where nothing is said to the contrary in the specifications. Under no pretext can damaged or inferior lime or cement, or loamy sand, be imposed by a builder upon his employer.

THE FIFTH VISIT.

BEFORE his next visit, let the superintendent provide himself with a light steel rod: steel wire three-sixteenths or one-fourth of an inch in diameter can be obtained at the hardware stores in pieces about five feet long, which answer very well. Let him divide his inspection as before: first a tour around the walls outside, then inside; next a survey of the materials inside the excavation, and lastly of those outside. The concrete piers are found to be nearly done. They should be completed and left for a week or two to harden. The footings are nearly all in, the masonry under the north and south exterior, or aisle walls, is three or four feet above the cellar-bottom, and the clerestory foundation several courses high. The tower foundation is also started, and the drainage-well and trench in that place ready. We notice dust on the surface of the concrete, which otherwise appears well mixed, and is about as hard, and of much the same consistency, as ordinary sweet chocolate. By breaking off a piece and rubbing it on the hand, a rough judgment can be formed of its composition. If it contains too much sand, it will lack coherence, and crumble away. We direct that the dust shall be washed clean off the surface of the piers before the next layer of concrete is added; otherwise the two layers will not adhere.

Outside the aisle walls the gravel has been filled in as fast as they were built up. This is customary and proper, but we will have a hole dug to satisfy ourselves that the outside of the wall has been pointed as we directed. If not, we order the whole to be dug out, the wall thoroughly wet, and the pointing done in a proper manner. If all is satisfactory, we will remind the workmen to puddle the gravel filling as fast as it is put in, by throwing on water. If water is difficult to get, the gravel may be packed with wooden rammers. In a clayey soil, the filling next the outside of the wall should not be of the excavated material, but gravel or cinders should be obtained, at least for the lower part.

As we pass around the building, we take care to look at the lines from which the walls are being built. They will probably be shown by cords stretched between the batter-boards, from which plumb-lines hanging at intervals serve to transmit the required points to the cellar-bottom where the men are at work; and the batter-boards should be examined to see that the cords are attached at the proper notches; if these are correct, we must observe whether the masonry is being laid exactly to the lines so given. Nothing is easier than to make mistakes in these respects at the outset, which will be very difficult to remedy afterwards. It may often be observed that a few courses in a wall have been built incorrectly, and the line having been soon afterward rectified, the masonry is built out, overhanging these courses either on one side or the other, so as to recover their proper position. Any such work should be immediately taken down, and rebuilt correctly from the bottom. A want of firmness and decision in this respect on the part of the superintendent will be the source of much greater troubles afterwards.

So far as we observe, the workmanship of the wall is tolerably good; the horizontal joints are well broken, giving a good longitudinal bond, and the cross bond is maintained by a sufficient proportion of stones extending through the whole thickness of the wall. (Fig. 31.) If any portions had come to our notice where vertical joints came one above the other (Fig. 32) through three or four successive courses, we should at once have ordered them

Bonding.

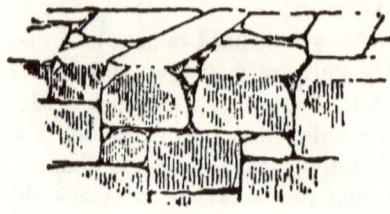

Fig. 31

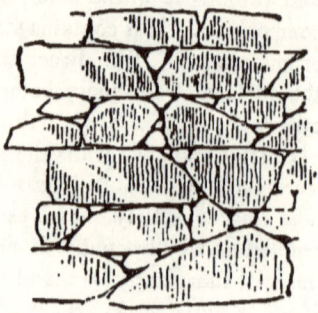

Fig. 32.

torn down and rebuilt, but none appear. It will, however, be well to watch the men from time to time, and observe their manner of working. The acquaintance with their individual characters thus formed will save much time subsequently, by showing us in what quarters to look most sharply for careless indifference to orders, intentional shirking, and well-meaning ignorance, as well as where to expect intelligence, faithfulness and skill.

As we pass along by the aisle wall we notice a mason at a little distance haranguing his companions on some subject about which he seems to have a flow of words, if not of ideas. Seeing us coming, he hastily shovels up a trowelful of gravel and stone chips from beside him, and throws them dry into a cavity in the stone-work before him, then dashes a quantity of mortar on the top, and smooths it over. To all appearance, his part of the wall is done just like his neighbors' work, but our suspicions have been aroused, and we approach and thrust the steel rod down into the fresh masonry. The supple wire insinuates itself among the stones far down into the wall, meeting now and then with the slight resistance due to the soft mortar, but penetrating many void spaces of considerable size, which are instantly detected by the feeling. One or two other trials give the same result, and as masonry so laid is liable to settle under the weight of the walls above, besides being permeable to water, we order the man to take down his work and rebuild it with joints properly filled. He grumbles, but begins with a very poor grace to remove the stones, while we remain near to see that our direction is strictly complied with, testing meanwhile the walling laid by the other men, which proves reasonably satisfactory. It is too much to expect that all the voids will be completely filled, but the steel rod will quickly show the difference between good and bad work. Every man whose workmanship is once found to be carelessly or intentionally defective should be noted, and the portion of wall on which each is engaged should be continually tested.

There are other qualities in rough masonry besides a large proportion of mortar which are essential to its good quality, and about many of these also the steel wand will inform us.

The usual practice of masons in rough walling is, after setting the larger stones, to fill the interstices with "chips," or even pebbles, more or less carefully fitted, put in dry; then to dash in mortar, trusting that it will work its way into the crevices. It does so to a great extent, especially if the wall is grouted occasionally with thin mortar, but the dishonest or indifferent men shirk the trouble of fitting in the smaller stones one by one, and content themselves with throwing in a lump or two of any shape, and then a quantity of small chips, which catch in the crevices and hang long enough to allow a fair bed of mortar to be spread over them, hiding the empty cavities below. This sort of work is immediately detected by the steel rod, which can be felt to shake and dislodge the loose pieces. The very best workmen avoid either of these methods, and place no stone, even the smallest chip, except in a bed of mortar prepared to receive it, rubbing it well in, and settling it with blows of the trowel or hammer, again driving smaller fragments into the mortar which is squeezed up around it, so that nowhere does stone meet stone without a cementing layer. The men who do work of this kind should be remembered, and the others incited to imitate them as far as possible.

The following points should be constantly and carefully observed. In laying the larger stones, the workmen will of themselves set the smoothest face toward the visible side of the wall; the superintendent must see that the outer side, which will be buried in the ground, has also a good, smooth face; that the bottom bed of each stone is level, or nearly so; or if not, that the masonry on which it is to rest is brought up with mortar and stone chips to fit its concavities before it is laid; that there are plenty of headers, or bond stones, extending across the wall from side to side to prevent its splitting; that long stretchers running lengthwise of the wall are sufficiently numerous; and that all angles are tied by long stones laid alternately in either wall. (Fig. 33.)

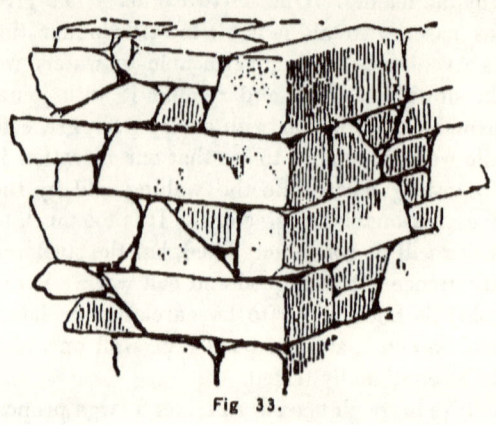

Fig. 33.

Fig. 34.

Care must be taken in building up the wall to keep in mind the position of the window and door openings which are to come above. The tendency always is for masonry below a pier to settle under the excess of weight, down to the very footings, tearing itself away from the less compressed portion under the opening (Fig. 34), so that long stones should be built in, extending from the part under the pier to that under the opening, to carry the weight out and distribute it uniformly over the whole foundation. For the same reason, if the sills of openings are built into the wall, instead of being "slipped" in afterwards, they must be pinned up only at the ends, a clear space of half an inch being left between their under sides and the masonry below them, which should not be pointed up until the completion of the building: otherwise the settlement of the piers will carry down the ends of the sill more than the middle part, and it will be broken.

The proper proportion of headers varies according to circumstances, but in an ordinary foundation one stone at least in every space of five feet square should extend through the whole thickness of the wall.

An opening should be left for drain-pipes to pass out, and for water and gas pipes to enter the building, covered with strong stone lintels. Neither contractors nor their men ever think of this, unless reminded by the superintendent, and in consequence, when the time comes for laying the pipes, ragged holes have to be broken through the wall, at the imminent risk of causing settlements.

A very important element in determining the character of the foundation walling is the height of the masonry above it. Not only will a high and heavy wall compress a loamy or other yielding ground beneath it more than an adjoining light wall, but the mortar joints in the high wall, if laid in the same way as those of the lighter wall, will be squeezed into a smaller compass by the greater weight; a very considerable inequality of settlement resulting from the combined effect of the two causes, with consequent dislocation of the masonry.

To illustrate this by the example before us: The tower at the south-west corner of the church, disregarding for the present the circumstance that it will stand on a rock foundation, while the adjacent walls rest upon gravel and clay, is to be so high, and the masonry near the bottom so heavily weighted, in comparison with the light and low aisle wall which adjoins it on the south-east, that if the lower portion of the two were to be built up together, in the same manner, with stones of the same size, in ordinary mortar, the compression of the mortar in the tower under the increasing weight would be so much greater than in the aisle wall that by the time the spire was finished, of two stones, one in the aisle wall and the other in that of the tower, originally set at the same level, the latter might be forced down two inches or more lower than the other; a movement which would cause dislocation the whole height of the aisle wall. Many stone church-towers show this effect, which can, however, be avoided by proper care. There are three ways in which the difficulty may be met. One is to make the tower masonry of the largest stones, as high as the top of the aisle wall, making the aisle wall of small stones. In this way the number of joints in the high wall will be so much less than in the low one that although the compression of each will be greater, the aggregate settlement will be about the same. Another expedient is to make the mortar joints in the high wall thin, and those in the low adjoining wall thick. The third is to lay the high wall in cement, and the low one in mortar made mostly of lime; then the contraction of the cement joints being relatively much less than with lime mortar, the total settlement can be kept nearly equal in the two walls.

The *rationale* of the last method depends upon the distinction, which should never be lost sight of, between the " setting " action in lime and cement. Strictly speaking, pure lime mortar does not " set." The soft paste resulting from the slaking process if exposed to the air, or placed on a piece of blotting paper, or between dry and absorbent bricks, will lose a little of the water used in mixing, leaving a firm, damp mass of hydrate of lime, which consists of pure lime holding in a loose chemical union about twice its bulk of water. This water still continues to evaporate slowly, and the paste to diminish in bulk, during a period of months, years, or even centuries, if the wall is very thick; and if the hydrated lime forms the cementing medium between the courses of a wall, the wall will settle as long as the evaporation and shrinkage continue. The superposition of a heavy weight increases the settlement, partly by the forcing of the semi-plastic material out of the joints, and partly by the pressing out of the water of hydration more rapidly than it would pass off by natural evaporation. This indefinite shrinkage of the lime is the principal, perhaps the sole reason for the addition of sand to mortar. The particles of sand being incompressible, and divided from each other by thin layers of lime, the contraction of these layers exerts a comparatively small influence on the total mass, so that a joint of half an inch in height, in mortar of lime and sand, will usually settle less than a sixteenth of an inch; while if made of lime only, it might shrink half its width.

With cement the action is quite different. When mixed into paste with water, a few minutes only elapse before the soft paste suddenly assumes a firm consistency, so as to resist the impression of a pointed instrument. This is the "set," and forms a true chemical reaction, by which a portion of the water enters into close combination with the cement, from which it cannot afterwards be separated except by heating to redness. With the help of this combined water, the constituents of the cement enter upon a series of reactions by which they gradually form a hard stone, little less in bulk than the original cement paste, and with some cements even equal to, or greater than the volume of paste. This characteristic quality of cement gives it great value in controlling the settlement which forms an important element in the consideration of stone structures; and by mixing cement and lime in different proportions a whole range of mortars can be obtained having any desired quality as to diminution of bulk in hardening.

Returning to our tower, for which we have to choose among the three methods of keeping the masonry at the same level in its walls and in the comparatively low aisle wall adjoining; we reflect that to lay the tower walls in cement and the adjoining wall in lime mortar would be sufficient, but the contrast in color between the brown cement joints and the white of the lime would be objectionable in the walls above ground. The same would be the case, in a less degree, if we were to lay the tower wall with thin joints, using thick joints in the aisle wall. If we lessen the number of joints in the tower, instead of diminishing their width, by building it of large stones, the same end will be attained, and the contrast of the massive masonry in the one with the small stones in the other will be rather piquant and attractive than otherwise.

But we must not forget the difficulties presented by the ground beneath the tower and the adjacent walls. The trenches show that under the tower we can reach the rock everywhere, at least by going down two or three-feet below the general bottom of the trenches in one corner. All this foundation will then stand on the solid ledge. To get a rock foundation for the aisle-wall would, however, require very deep digging, the ledge sloping rapidly eastward; and yet if one wall is built on rock and the other on compressible ground, the latter will settle and tear itself away. The soil overlying the rock under the aisle wall, as shown by the trench, is gravel, which has the advantage of being practically as incompressible as the rock itself if not loaded beyond a certain point. In general, it will not yield perceptibly under a less load than five tons to the super-

Compressibility of Gravel. ficial foot, but to make sure, we will take three tons as the limit, as a soil of so little depth over a ledge is less reliable than if it were deeper. We will reckon up roughly the weights with which the soil is to be loaded. The foundation, allowing an extra foot for the excess in width of the footings over the rest of the wall, is 12 feet high, 2 feet thick. The wall above is 20 feet high, 20 inches thick; total, $12 \times 2 = 24$

$$20 \times 1\tfrac{2}{3} = 33$$

— 57 cubic feet of masonry to each linear foot of the aisle wall, which at 150 pounds per cubic foot, an average weight for such masonry, will amount to 8,550 pounds.

Of the roof, which slopes at an angle of 45°, about two-thirds the weight will come on the aisle wall, the rest being borne by the clerestory wall. The roof, measured on the slope, is 15 feet wide, and the weight of rafters, boarding and slate, and plaster-

Calculation of Weight on Foundation. ing on the under side, may be taken at 30 pounds per superficial foot. Adding the *possible* weight of wet snow and ice, 40 pounds per superficial foot, makes 70 pounds per square foot, which multiplied by the width, 15 feet, gives 1,050 pounds to each linear foot of roof. Two-thirds of this, or 700 pounds, is to be added to the previous weight of the masonry, 8,550 pounds.

One more burden must be calculated and added to the rest to find the whole load which will need to be sustained by the subsoil beneath the wall: that is, the floor, which being level rests half on the aisle and half on the clerestory wall. The span is 10 feet, so that 5 square feet, weighing 30 pounds per foot, with a *possible* additional load of

120 pounds, will give 5 × 150 = 750 pounds more, to be added to the total pressure on the footings, the whole amounting to 8,550 + 700 + 750 = 10,000 pounds, or exactly five tons, on each linear foot of wall. As the wall is two feet thick, the weight on each *linear* foot is divided over two *superficial* feet, making two and one-half tons on each. We have increased the spread of the footings from the usual six inches on each side to twelve; this will divide the burden at the point of contact with the earth over four square feet instead of two, making the pressure but one and one-quarter tons to the superficial foot of soil. This is sufficiently far within the limits of resistance to compression to give assurance that no settlement of the substratum is to be feared.

We have then only to direct that the largest stones shall be selected for the tower and its foundations, that every stone shall be hammered well down into its bed, so as to bring the surfaces as nearly as possible into contact, and that all the crevices shall be thoroughly filled with stone chips and mortar. The aisle wall adjoining is to be built of smaller stones, and tied into the tower wall every few feet in height with long stones as well as with iron anchors.

In bedding the tower footing-stones upon the rock, any little ridges or projections on the surface of the ledge must be hammered off, so as to give a moderately even bed, and small stones and mortar must be built up to fit the irregularities of the under side of the footing-stone, and finally, a thick bed of mortar spread over all, so that there will be no cavities under the stone. All the heavy blocks should be laid with a derrick, so that they can be held suspended over their place while the bed is being prepared, and **Derrick-laid Stones.** if they are very irregular, lowered into place and then raised again, so that the impression made in the soft mortar will show whether the bed is exactly fitted. The practice of rolling the stones into their place with crowbars must never be permitted in a heavy wall. The bars tear up and dislocate the bed of small stones and mortar to such an extent that it is impossible to be sure that the stone when so laid does not rest on the edges of two or three little chips, which will crush and cause serious settlements when an increased load comes to be placed upon them.

It will not be so easy to make a neat outside and inside face to the foundation-wall of large blocks, but it should be done, especially on the outside, even if some of the stones have to be dressed off It is dangerous in a heavy building to leave, as is often done, the larger pieces in a foundation-wall projecting outside, to save the trouble of cutting away the excess of size. Not only will water get into the wall by running along the top of such a stone, but hard earth, or a pebble, may be wedged under the projecting end, so as to keep it up while the wall settles under the increasing load, causing a bad crack beneath it, and throwing the whole weight on the inner end of the stone, which is likely either to give way altogether, or to break up the masonry about it.

Our tour outside the walls being now completed, that inside may be short. We must see that the drain under the tower does not get obstructed, and that a good opening, spanned by a strong stone, is left for it to pass beneath the walls. Workmen have not so high a respect for drain-pipes as architects, and will often cover up a choked or broken pipe, saying nothing about it, thinking that they will be out of the way before the trouble is discovered, and careless of the very great expense which may be necessary to replace it. The clerestory foundation must be sharply watched: long stones are the first requisite for this wall; everything depending on the efficiency with which the concentrated load on the piers is spread out laterally over the foundation, till the pressure on the footings is uniform under piers and openings alike. The

Ventilating Apertures. workmen must be cautioned to leave an opening two or three feet square at the top of this wall, under the arches (Fig. 35), and the same precaution must be observed wherever any portion of a cellar is cut off by walls from the main part, in order to secure circulation of air. If deprived of this, the beams of the floor above are sure to rot before many years, and will sometimes fall in all at once after a few months.

Fig. 35.

The derricks, of which two or three are probably now set up, must be examined. Let the superintendent see that the ropes are not

BUILDING SUPERINTENDENCE.

fraying out, and that neither the mast nor the boom is cracked or sprung out of perfect straightness. He must also observe where each of the guy-ropes is fastened. Every one should be secured to a growing tree, or a post set five feet or more into a hole in the ground, and the earth refilled and packed around it. If any guy-rope has been carelessly fastened to a fence-post, which is very likely to be half rotted off at the ground, or to a curbstone, or a boulder, or any other anchorage not perfectly secure, orders must at once be given to have it changed, and all guys must be strictly required to be drawn up taut. A loose derrick rope, or an insufficient anchorage, is terribly tried when a heavy stone on the end of the boom is swung around so as to bring the strain suddenly upon it, and although that is properly the contractor's affair, a little attention on the part of the architect will do no harm, and may save loss of property, and even life. The foundation-walls of the chancel, it should not be forgotten, are to be lined with brick, and anchors must be built in to hold the lining as described below. We will remind the contractor of this in good season.

Derricks.

Anchorage of Guy-ropes.

The last thing to be done is to inspect the materials delivered since our previous visit, which will end our duties for the day.

A quantity of granite for the face of the basement wall above ground has been sent, already cut; it being very common to cut the harder stones at the quarry, while the softer freestones are cut at the building. The blocks are of random sizes, and vary much in thickness, some being one or more points less than two inches thick. These should be at once rejected, no matter how thick they may be at the other edges (Fig. 36), since their corners are liable to break off under the weight, and disfigure the work. Still worse are the stones which, though of sufficient thickness around the edges, are hollow in the middle. (Fig. 37.) Not even the most skilful backing can make these secure. Usually, the minimum thickness admissible in the facing blocks is mentioned in the specification, but if not, nothing under six inches should be allowed in the basement wall, and not that unless all such stones are anchored to the backing.

Fig. 36.   Fig. 37

It is rather advantageous to have the stones large on one bed and small on the other, provided they do not come to too thin an edge. (Fig. 38.) Such stones bond well together and to the backing.

It is common in specifications to require that granite shall be "free from knots, sap, shakes and rot." Rotten, or crumbling granite is easily detected, as are also the brown stains known as "sap," and the black or white lumps called "knots." Shakes, if very bad, are shown by their discolored edges, but we are likely, with some kinds of granite, to find stones with seams through them, which are tight enough to hold together while the stone is cut, but will, after they are placed in the wall, open by the effect of the weight upon them, and allow rain-water to penetrate. Where the stone is thick enough to extend nearly through the wall, a great deal of water will often in heavy rains blow into the building through a seam which may have been quite imperceptible when the stone was set in place. The most certain way of detecting blocks so affected is to strike them with a hammer, rejecting the stones which do not ring clearly.

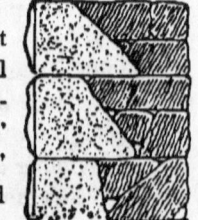

Fig. 38.

The only other new material which we find delivered is brick, of which two lots are on the ground; one near the west end of the building, which we find to be mainly composed of small bricks, of a dark color, the ends of many being black or bluish, and generally crooked or irregular in shape. In the middle of the cellar others are piled, larger, and more regular in shape, but light-colored, at least one in four being very pale. We will test the hardness of these by actual trial, since the color by itself is an unreliable indication, many clays yielding a light-colored brick of very good quality. Selecting two of the paler ones, we strike them together; they meet with a dull sound, and the edges crumble at the point of contact. After a few blows one breaks in two, showing an earthy fracture, destitute of the compact, hard look of good brick.

Brick.

The reddest-looking of this lot ring quite clearly when struck together, and their good shape is in their favor, but the men could not be depended on to pick out the best ones, and it is safest to order the whole lot sent back. If the contractor is honest, he will have ordered good brick, and if the superintendent rejects them, the loss will not be his, but the brick-maker's. If the superintendent passes them, through negligence or complaisance, not only is the contractor deprived of his support in attempting to compel the dealer to furnish better materials, but he is likely to think that he need not himself be too scrupulous in other respects. Let the superintendent make a note of having rejected the bricks, and give the contractor a memorandum of it; following up his action by a close watch to see that his directions are carried out. Let him never allow any brick of which he can crumble the edge with his fingers to remain on the ground. The hard, but crooked bricks, if not too much distorted, may be utilized in the backing of the stone-work, or in forming the jambs of basement openings, but must not be used in any pier or arch.

The staging lumber will need occasional attention. Although with us the architect takes no responsibility about the scaffold, his directions in regard to it will be listened to with respect, and he has an undoubted right to control its construction where that may influence the execution of the building.

SIXTH VISIT.

At our next visit we find the first staging up, the drain-pipes on the ground, centres ready for the arches under west wall, and the arch bricks delivered. We go around the outside of the walls, then inside, then examine the materials inside, and lastly, those outside. This should be the regular routine of each visit, as the surest way of observing whatever may be new in the work.

The concrete piers are firm enough to build the arches upon them, the foundations are going on well and nearly finished, with good bond, and neatly pointed outside; the long stones under clerestory wall have not been forgotten, and trials with the steel rod reveal only a few places to be taken out and filled up. In the tower foundation the opening for drain-pipe is properly formed, and the large stones well laid and bonded. On our way to the tower, we watch the setting of the capstones to the concrete piers. The stones have been cut square, with "skew-backs" formed for the arches to spring from, and all the faces "pointed" to a uniform surface; the top of the concrete is well wet, a layer of cement mortar made with an equal bulk of sand is spread over it about an inch thick, and the stone lowered into it, and beaten down by blows of a sledge-hammer, not applied directly on the stone, but on a piece of timber interposed.

The cement will soon set, and the centres can be placed in position at once and the arches commenced. These should be built in sepa-
Brick Arches. rate concentric rings or rowlocks, four inches thick, rather than in the fashion called "bonded," where each ring is tied to the others by bricks set the eight-inch way. The latter has some advantage in point of appearance, but the arch of separate rowlocks possesses a certain elasticity, and power of accommodating itself to the weights upon it, which make it much better in heavy constructions. The bricks must be very regular in shape, well *soaked* in water, — not merely sprinkled, — and laid with mortar of equal parts of cement and sand. The arches spanning the basement openings, behind the straight lintels which terminate them outside, should be built in a similar manner.

None of the granite facing of the basement wall has yet been set, although the grade lines which mark its commencement are set out
Ashlar. by strings stretched between stakes outside. We take advantage of the opportunity to question the contractor about the manner in which he proposes to anchor the face to the backing. The specification indefinitely requires it to be "well anchored," without further details. Much depends on the character of the ashlar. If the stones are thick, with many of them extending through the whole thickness of the wall, as is common in Europe, no other ties will be necessary to keep the weight of the superstructure from forcing the facing stones off the wall, but with us such ashlar varies from two to twelve inches in thickness, according to the value of the material, backed by a rubble masonry of rough stone or brick, to which it must be held by iron ties.

A four-inch ashlar, and still more a two-inch, which is used only for facings of marble, must have, for a high wall, at least one anchor
"Clamping" of Ashlar. in every stone. When the ashlar is thicker than this, they may be much less numerous. We find our granite blocks to average eight inches in thickness, and being assured by the contractor that the stones to come will be of a similar character, we agree with him that if the anchors are so distributed that there shall be at least one to every three feet in length, and two feet in height, the work shall be accepted as satisfactory, stipulating also that the last course of ashlar, under the water-table which marks the transition from granite to freestone at the first-floor

level, shall have an anchor in every stone. The brownstone ashlar for the upper part of the wall will be of about the same thickness, and the same proportion of anchors is directed for that also; every stone under the horizontal string-courses and cornices to be anchored,

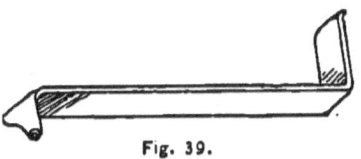

Fig. 39.

in the same way as under the water-table. The anchors are made of wrought-iron strips about one inch wide, and as much as one-twelfth-inch thick. Iron of one-sixteenth-inch thickness is sometimes used, but is too light. One end is turned up about two and a half inches, and the other is turned down about one and a quarter inches. This end is heated by the blacksmith and driven by a blow into a round hole made in the anvil, which rolls it into a tubular shape suited for insertion into a hole drilled in the top of the stone to be anchored. (Fig. 39.)

The drill-hole should be one and a half or two inches from the face of the ashlar block, and the length of the anchor should be so measured as to extend entirely through the wall, the other end turning up close against the inner face.

As these ties would be soon destroyed by rust if used in their natural state, they must be protected by tarring or galvanizing. The latter is most expensive, and perhaps best, but the former is generally employed.

In setting the first course of stone above ground, it is advantageous to have it overhang the foundation-wall about an inch: then the rain-water, which flows in sheets down the exposed surface during storms, when it reaches this point drips off, and is absorbed by the ground, instead of continuing its journey down the face of the foundation-wall. (Fig. 40.) Of course, this must be arranged for in making the detail drawings if it is to be done systematically. At all events, the construction sometimes seen, where the base course is set back from the face of the foundation, leaving a narrow level strip on top of it, should

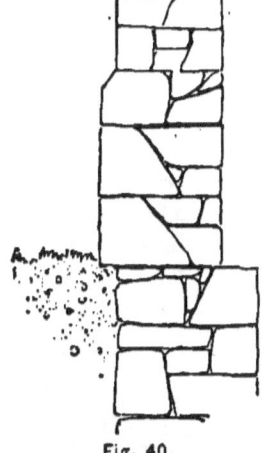

Fig. 40.

not be countenanced. Such a shelf serves only to catch the water streaming down from above and conduct it into the masonry, and if the plans require such a relative position of the foundation and superstructure, the former should terminate by a surface sloping back to the line of the wall above it.

In supervising the facing work, attention should be paid to the appearance. With random ashlar, much of the beauty depends upon the frequency with which the horizontal joints are broken. It is common to specify for such work that no horizontal joint shall extend more than six feet, and this is a good rule to follow in all cases. The difference between a neat and slovenly walling is illustrated by Fig. 41, a and b.

Random Work.

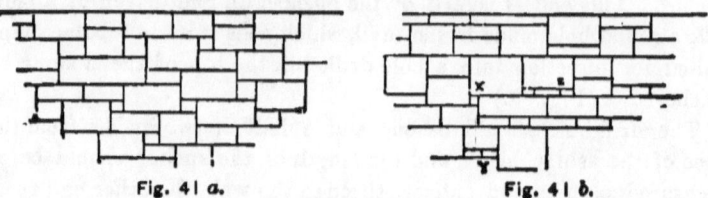

Fig. 41 a.   Fig. 41 b.

In b the effect is injured not only by the long horizontal joints, but by the frequent occurrence of small stones, as at X, Y, Z, inserted to fill awkward vacancies.

The work now goes on without intermission until the granite-faced wall is ready to receive the water-table or bevelled course which terminates it. (Fig. 42.) This, like all horizontal string-courses, particularly if projecting, should be composed of long stones, running back as far into the wall as practicable. They are often specified to have the top bed not less than eight inches in the wall, and this is a

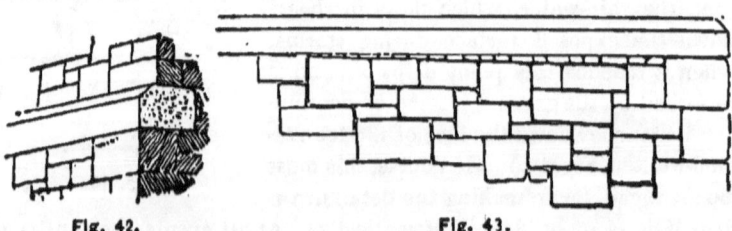

Fig. 42.   Fig. 43.

good standard, though narrow string-courses near the top of the wall may perhaps have an inch or two less. Care is necessary to ascertain the exact level some time before the wall has reached the required point, and it is best to build up all the corners of the building to the line in advance of the rest, and set the corner-stones of the water-table, levelling them carefully with an engineer's instrument, afterwards bringing the intermediate por-

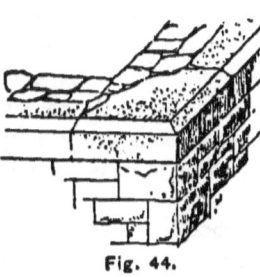

Fig. 44.

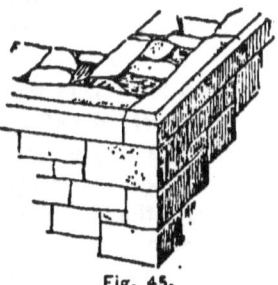

Fig. 45.

tions up to the line. This will prevent an appearance like this (Fig. 43) caused by the attempt to regain a true level after the wall has been carried up nearly to the top.

The quoins, or corner blocks of the water-table, as indeed of all the stone-work, must always, for appearance sake, show a wide head on both sides the angle (Fig. 44), instead of being cut out of a stone of the same thickness as the rest (Fig. 45).

The water-table indicates the level of the main floor, and while preparations are making for laying it, the beams may be placed in position. Stock for these has been delivered at intervals previously, and carefully examined, several loads having been rejected for containing timbers *considerably* less in size (since the timber shrinks after sawing) than the specification calls for, while others have been thrown out on account of

Flooring Timber.

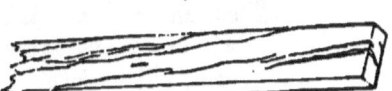

Figs. 46, 47.

pieces badly "shaken" (Fig. 46); or "waney" (Fig. 47), through having been sawed too near the outside of a crooked piece, so that a part of the wood is lacking; or weakened by large knots near the middle of the span. The nave and chancel floor beams are divided into two spans, the inner ends of each span being carried on a line of girders running through the middle of the building, and supported by brick piers under the nave, and an iron column in the society-room under the chancel. The girders of main floor are of Georgia pine, eight inches by ten, those in chancel being eight by twelve, and all are already on the ground. It is necessary to have these properly set before the beams can be put in position; but if the piers were to be built up first, and the heavy timbers laid upon them, there would be danger of overturning or displacing them, so it is best to support the girders by temporary wooden shores until the floor is on, and afterwards build up the piers between the shores. As there is ample head-room in the cellar, the beams are simply notched upon the girders, instead of framing them in, and thereby weakening the girders with mortises.

The carpenters are already cutting the notches, and the foreman hastens after us to ask whether he shall "crown" the beams, and if so, how much. Nothing is said about it in the specifications, and a little reflection is necessary before a reply can be given.

"Crowning."

The crowning, as now usually practised, consists in trimming off with an adze the upper edge of the beams, so as to form a curve, the convexity of which may be one inch or more, as required. (Fig. 48.) Nothing is taken off the middle of the timber, so its strength to resist a distributed weight is not impaired, and as all ordinary beams sag a little under their own weight, and still more when

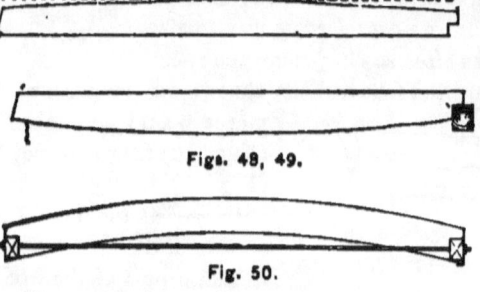

Figs. 48, 49.

Fig. 50.

loaded with flooring and plaster, the crowning enables this sagging to be compensated, and such a beam, when suspended at the ends, will be level or slightly convex on top, the bending due to the weight showing itself on the under side. (Fig. 49.) Formerly the same effect was sought by shoring up the beams strongly in the centre, so as to bend them upward, and then either building them into the walls, or confining them by timbers placed against their ends, and connected by iron tie-rods passing between the beams (Fig. 50), but this method is objectionable, and is now rarely used.

In order to determine whether crowning by the other mode is desirable, we can easily calculate the probable bending of the timbers, or, what will be still better, experiment on the spot, by placing one of the beams, the top of which has been previously ascertained to be straight and true, on supports at the proper distance apart, and loading it at the middle with a weight equal to *half* the load which will come upon it after the building is completed. In this case the beams of the main floor are of spruce, three inches by ten, and of chancel floor three by twelve, all placed sixteen inches from centres, and twenty feet long, the clear span being nineteen feet four inches. They will be plastered one heavy coat underneath, and covered with a double boarding. Each beam carries the whole weight of the flooring above it and the plastering below it, from the central line of the interval between itself and the next beam on one side to the corresponding line on the other side, the distance between these two lines being the same as that between the centres of the beams themselves, or sixteen inches. Each beam will then sustain, independent of its own weight, an area of flooring 16 inches wide and 20 feet long, and an equal area of plastering, amounting to 16 inches ($1\frac{1}{3}$ feet) $\times 20 = 26\frac{2}{3}$ square feet of each. One square foot of dry pine board an inch thick weighs about three pounds, and as the boarding is double, six pounds will represent the weight of flooring per square foot. Plastering of the kind mentioned will weigh about six pounds per square foot, making a total load per foot of 12 pounds, which multiplied by $26\frac{2}{3}$ will give 320 pounds as the distributed load on the beam, exclusive of its own weight. One-half of this, or 160 pounds, applied at the centre will produce the same effect as the whole distributed load, and by loading our experimental beam in this way the amount of bending can be at once ascertained. A bucket hung over the middle of the beam, and loaded with thirty-five to forty bricks according to size, or with a man standing in it, will form a ready means of applying the weight, and a string tightly stretched between the ends of the beam will show the deflection, which we find in this case to be less than a quarter of an inch. So slight a deflection is not worth the trouble of correcting by crowning the beams, especially as the bridging of the floor, which is required by the specification, affords protection against *unequal* deflection; and we therefore inform the foreman that it will not be required. The tops of the beams must, however, be well levelled; and as they are likely to vary somewhat in depth, the gauge for notching them to fit upon the girder must be taken from the *top*, not from the bottom edge. The ends which rest on the walls need not be notched, but are "pinned up" with chips of stone or slate to the required height. Wooden chips must never be used for this purpose, as they soon rot out, and allow the beam to settle.

Concerning the manner in which the beams should be built into the wall, there is much diversity both of opinion and practice. On one point, however, all are agreed; that the ends should be cut on a bevel, thus (Fig. 51), the variation of the inclined line from the vertical being two to four inches, according to the depth of the beam. The object of this is to prevent the destruction of the wall, in case of fire, by the fall of the floor when burnt through. If the beams are bevelled

Bevelling Ends of Beams

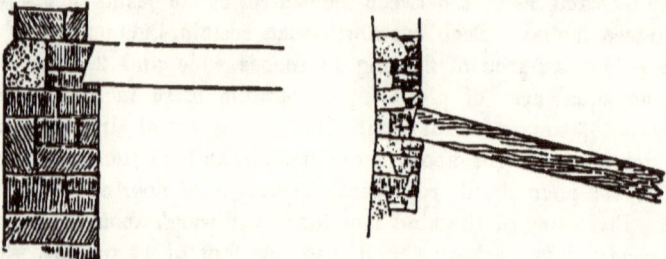

Fig. 51.　　　　　　　Fig. 52.

as shown, they drop out quietly when their outer ends are consumed; but if left square, the portion in the wall acts as the short arm of a powerful lever, whose outer end, being depressed as the floor falls, pries the wall outward with immense force. (Fig. 52.) But besides this, precaution should be taken lest the ends of the beams absorb moisture from the foundation and perish by dry-rot. The most important security against this is the avoidance of dampness in the cellar and its walls by the precautions suggested above, without which the other expedients avail little. If reasonable dryness is maintained about them, sufficient further protection can be obtained by leaving a small open space about the ends of the timbers for circulation of air. Some simply build up vertically behind the ends of the beams, filling in solidly between them, and trusting to the subsequent shrinkage of the wood to open a slight but sufficient communication between the triangular hollow behind the timber and the air of the cellar. Others increase this communication by tacking a piece of zinc or felt paper over the end of the beam, letting it hang down at the sides, so as to keep the masonry at a small distance from the wood. Still more careful constructors build up a recess around each timber, leaving it free on all sides, but with a dry foundation of hard stone, laid in half cement, and the floor two feet or more above the ground, so great precaution is unnecessary except for heavy girders. In the present case, being very confident of the dryness of our wall, we will compromise by filling in closely between the beams with the stone-work, but will leave a space behind them, and will span the end of each beam by a stone or a couple of bricks, so as to open a communication above the beam between the interior space and the air.

Protection against Rot.

When the first tier of beams has been laid and the ends built up with masonry, the work of the superstructure may properly be said to begin, and a variety of new cares will come upon us. Here also commences a considerable divergence between the practice of different localities in respect to many details of construction. In the Eastern States, and to an increasing extent in others, the next step after laying the first-floor beams, and bringing the walls up to a level with them, is to cover the whole floor with cheap inch boards of hemlock or spruce, generally "thicknessed,"— that is, planed on one side so as to reduce them to a uniform thickness, and firmly nailed down in place. This furnishes a convenient starting-point for future operations; materials are stored upon it, the roof is framed upon it, stagings are erected on it, and the men move freely over it.

Superstructure.

Under Floors.

Whenever it becomes necessary to reach the space between the beams, for nailing bridging, running gas-pipes, or other purposes, a board is easily taken up and replaced again, and at the very end of the work the whole is brushed clean, the holes and broken places repaired, and an upper flooring of new, clean, fresh boards put down over it, one or two thicknesses of soft felt being laid between, making a strong, handsome, impervious floor. Among the old-fashioned builders in New York and other places this method is thought unrea sonably costly, and a single flooring only is used, generally of one and a quarter inch boards, the laying of which, of course, in order to preserve them from injury, is delayed as long as possible, all the operations of building being meanwhile carried on over a skeleton of beams, traversed in different directions by lines of planks which have continually to be taken up and changed, while supplementary planks must be brought and laid down to hold all the material used in each story, and if a tool is dropped, or a bolt or anchor-iron rolls from its place, it descends to the cellar, where the workman must go to find it, or let it be altogether lost. To one familiar with both systems, the latter seems to involve a great waste of time and labor, and the testimony of the best builders is that the double flooring, though costing more for material, is so much more economical in these respects as to save more than its extra cost in the completed building.

We find this construction specified for the structure under our charge, and in less than a day after the beams are on and levelled, they are covered with a smooth, firm floor, which can be put down with extreme rapidity, being laid without much attempt at unnecessary neatness. The boards, being of hemlock, are full of shakes; some are cracked in a dozen places, while in others the annual rings have separated, and can be peeled off by layers. Some hemlock lumber is much better than this, but there is no need of being very particular. As soon as convenient after laying the floor, the boards are taken up in a line through the places marked for bridging

Bridging. the beams. There are to be two rows of bridging in each span, but the kind is not specified. Occasionally a builder is found who imagines that a floor can be bridged by fitting in square bits of plank between the beams and nailing them in place; a device as costly as it is perfectly useless, the planks answering no purpose whatever except to burden the floor. The

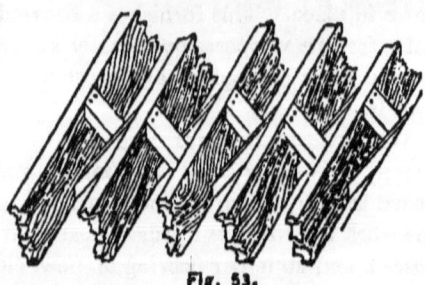

Fig. 53.

proper way is to fit in strips diagonally between the beams, of sufficient length to reach from the upper edge of one to the lower edge of the next, nailing them firmly in place with two large nails at each end (Fig. 53).

A double row is necessary, so arranged as to abut in pairs against the upper and under sides of each beam, and the effect, if arranged as they should be, in lines made perfectly straight, marking the position of each piece by a chalk-line on the edges of the timbers, is to connect the beams together by a kind of truss (Fig. 54), which prevents any one from bending downward without carrying down with it a number of others, greater or less according to the perfection of the workmanship, but ordinarily about ten.

Fig. 54.

Nothing is added to the absolute strength of the whole floor, but any single overloaded beam is enabled to divide its burden with six or eight of its neighbors, which is a great gain. The bridging in important buildings is usually specified to be of two-inch by four-inch pieces, and such may sometimes be necessary; but for ordinary dwelling-house beams, not more than sixteen inches from centres, strips one inch or one and a quarter inches thick and four inches wide will, if well fitted and nailed, prove quite sufficient.

The ends of the beams which rest on the girders should be nailed in place, and in addition "dogs" of round bar-iron three-fourths inch in diameter and about eighteen inches long, turned down at each end and fashioned into a rough, chisel-like point (Fig. 55), should be driven into the abutting ends of two opposite beams, at intervals of about eight feet, the whole length of the building, so as to connect the beams strongly together; and wall anchors of flat bar-iron, one-half inch thick, and at least one and a fourth inches wide, not less than four feet long, and turned up four inches at the end, should be spiked, by means of holes punched for the purpose, to the sides of the same beams that are tied at their other ends to the beams beyond, leaving the turned-up end to be built into the masonry of the wall above the water-table. This end is often split into a fork and each branch turned up, much improving its hold, and

Tying Beams.

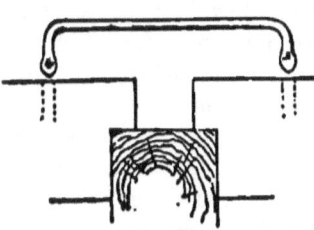

Fig. 55.

Wall Anchors.

the inner end is sometimes turned down and driven into the beam, without being spiked to it. These different anchorages form a continuous tie from wall to wall, and although not far from the ground will assist materially in keeping the masonry upright.

Other anchors must be provided, for tying the angles of the building in the superstructure. These should be of iron similar in thickness and width to those just described, turned up at one end and down at the other, and from four to six feet long. They should be laid alternately in each of the walls forming the angle, and at intervals of from four to eight feet in height, according to the amount of bond obtained by the stones alone. Young architects are often surprised to find that so much iron is needed in masonry, which they imagine by the descriptions given in their text-books to be quite capable, when well bonded, of holding itself together without such aid; but our ordinary structure of thin ashlar facing, backed with incoherent rubble of small stones, is a very different matter from the combination of squared blocks to which the books refer.

Still another set of anchors must be built into certain portions of the walls as they progress, to hold the brick lining which is subsequently to be built up inside. A part of these are to

Ties for Hollow Walls. be used in the society-room under the chancel, and others in the vestibule which forms the first story of the tower. The former of these lining walls is to be plastered, but not the latter, and the anchors needed will differ a little in consequence; those for the plastered wall being arranged to project entirely through the lining, turning up two inches on the inner side, while the others must stop just behind the inner face, so as to lie concealed in the joint. They should be made of iron not less than one-sixteenth inch in thickness and one inch wide, tarred or galvanized, turned up at the end in the outside wall, and extending halfway through it. They are to be built into the outside wall at intervals of two feet both horizontally and vertically, setting them carefully to the proper length by measuring from the *outside*, and left projecting until the time comes for building the lining wall, when they will be found all ready at the proper place. Various forms are given to these ties, all intended to prevent condensed moisture, or water driven through the outer wall, from being conducted by them across the air-space into the lining wall, where it would show itself by a permanent spot in the plastering. The part which crosses the air-space is often bent downward into a V-shape, from which the water drips to the bottom of the cavity, and this, if the depression is deep enough, is moderately effectual, but if too shallow, the mortar falling from above collects on the ties in quantity sufficient to bridge the air-space in some places, and convey water through to the lining. Another form is made by twisting the ties so that the portion in the air-space is vertical, and collects neither mortar nor water. The iron is sometimes used without twisting, but set in the vertical joints, thus attaining the same result at less expense, but with greater trouble; since the irons, being set at random in the wall, never coincide exactly with the joints of the brickwork, and have to be bent more or less to reach one. When inserted in the horizontal joints, little bending is necessary, hardly more than an inch in any case; but when placed in the vertical joints, the lining wall being composed entirely of stretchers, a lateral movement of four inches may often be required. On the whole, the V-bent ties seem to be most suitable in our case, but we direct that they shall be placed vertically one above another, so that those above may shelter those below from falling mortar, and that in building the lining every row shall be cleaned off as soon as the one above it is fixed in place.

While these matters are being settled, the delivery of freestone, both cut into ashlar for the facing and in rough blocks for working into mouldings and arch-stones, has been going on. This should be attentively watched, and if possible a visit should be made to the quarry to inspect the different beds of rock. *Freestone.* There is usually much variation in the quality of stone from the same quarry, one end frequently running into extreme hardness, while the other is too soft for use, or superposed strata may show different properties. Variations in color usually accompany differences of texture, and each different tint of stone on the ground should be tested by chipping off a thin piece, and crumbling the edge in the fingers. The hardest sandstone will resist this treatment like flint, but most building stones crumble slightly at the very edge, while the poorer varieties crush easily. Few of the ordinary freestones show a much greater hardness than common loaf sugar, but those that are softer should be rejected.

Some kinds of stone have an extremely annoying defect in the shape of "sand-holes," which are small formations in the interior of the block, often similar in appearance to the rest of the stone, but

Sand-Holes. destitute of cementing matter, so that on being exposed by cutting, the sand falls out, leaving an unsightly hole, which cannot be successfully concealed by filling with cement. Any variety of stone which proves to be affected in this way should therefore be entirely rejected, as sand-holes may exist in any of the blocks, undiscovered until the stone is cut. Some stones contain similar cavities filled with clay, which are equally pernicious.

Seams injure the quality of many classes of freestone, but are usually more easily detected than those in granite. All sandstone is

Seams. stratified, the beds varying from ten feet to a small fraction of an inch in thickness, and the divisions between the strata show themselves as dark streaks on the edge of the stone. Where the beds, though distinct, are strongly cemented together, the stratification rather improves the quality of the material by the beautifully figured appearance which it gives to a smoothly-rubbed surface, as in the stone from Portland, Connecticut, but the seams are not always tight, or if they are, may not remain so, and the stone scales or "shells" away, so that blocks, especially large ones, of stratified rock should be "sounded" all over with a hammer before setting in the building, to detect any separation between the interior surfaces.

The stone-cutting shed will now become a point of special interest at each visit, since here will be carried into execution the ornamental details of the building, whose accuracy and perfection of workmanship will do much to enhance its beauty.

Among the first things done in the cutting-shed are the jamb-stones for the windows. The sills are to be slip-sills; that is, inserted between the jambs at some subsequent period, and the detailed sections of them are not yet at hand, but a number of jamb-stones

Jamb-Stones. are cut. We have taken care to make ourselves perfectly familiar, not only with the position, height above the floor, and exact dimensions of every window, but also with the depth of the various reveals; and we begin at once to measure the smaller stones to see if they are of the required depth, finding immediately that many fall short an inch or more. These must be marked and laid aside, to be recut for other purposes, making sure that the are not likely to be put into the wall, for the replacing of an unsatisfactory quoin is a disagreeable matter, and it is still worse to leave it in and endeavor to fill out the deficiency by patching.

The other details should be sharply looked to; sketches made to explain obscure drawings; the moulded work frequently compared with the zinc patterns or templets, and these with the sectional drawings; the carving criticised, and the straight-edge and square frequent'y applied to the work of men who seem unskilful or negligent.

The regular routine of a tour around the outside of the building, followed by one inside, and an examination of the new materials, first inside and then outside, should still be kept up, but the inspection should take place at the level of the staging where the men are at work. The steel rod may occasionally be used to examine the workmanship of new men, and a continual watch must be kept to see that the ashlar is properly and sufficiently anchored to the backing, and that all exposed angles or projecting parts of the stone-work, as corners of openings, string-courses, sills and cornices, are well covered up with boards, to protect them from accidental blows, and from falling objects.

Before the walls reach the top, it is necessary, whether the specifications mention it or not, to build in long bolts at every angle, and at intervals of about ten feet along the walls, to hold the wooden wall-plate. The bolts should be of one-inch **Anchoring the Plate.** round iron, three or four feet long, with a short plate at the lower end to give them secure hold, built in vertically, so that the upper end, threaded for a nut, may project an inch at least above the top of the wall-plate when laid. In setting this, holes are bored for the bolts, and nuts with large washers are put on and screwed down firmly.

The setting of the cornice and coping stones must be watched anxiously. It is customary to specify that all cornice or other projecting stones shall balance on the wall, or in other words, that the weight of the portion of the stone on the wall **Cornice-Stones.** shall exceed that of the projecting part, by one-fourth, one-half, or more, as the case may be. Whether specified or not, a considerable excess of weight on the wall is necessary to a good job of stone-work. Heavy cornice-stones are often seen, where the vigilance of an inspector is wanting, which would fall immediately into the street if they were not held back by thin iron straps to the timbers of the roof, and do so fall when the straps rust, or the timbers are burnt away.

Copings must also be strongly secured against sliding off. The "kneelers" or corbels at the foot of the slope, which ostensibly sup-

Copings. port them, are seldom so designed as to be capable of anything more than holding themselves in place, and usually require even for that to be anchored back to the masonry, so that the coping-stones must be held by irons at the foot of each. Finials also, which commonly decorate the peaks of gables, are not

Finials. secure unless dowelled to the adjacent stones, and if composed of several pieces should be drilled completely through, and strung upon a long iron rod previously built into the masonry, with nut and washer at the top to hold the joints firmly together.

The young superintendent will hardly see the finials placed on his building without having undergone several conflicts with the con-

Patched Stones. tractor as to the stones which are to be permitted in the work. He may find his orders about seams and sand-holes cheerfully complied with, but stones accidentally broken after cutting will be used wherever they can escape his eye; the fractured portions, if conspicuous, being neatly cut out, and a new piece inserted and glued in with melted shellac. In dry weather, and while still fresh from the tool, such patches are hardly noticeable unless near the eye, and they should therefore be looked for from the stagings before these are shifted; but when the stone is wet by rain, the inserted piece becomes painfully conspicuous, and as the shellac is slowly destroyed between the stones, the joint becomes more and more evident, and the piece may even drop out if situated, as sometimes occurs, on the under side of a lintel.

Both firmness and consideration are necessary in dealing with these cases. While a patched jamb-stone, sill or lintel, unless of extraordinary size, is inexcusable, it would be harsh to condemn a piece of rich and costly carving for a small mutilation which could be easily concealed; and some fractures may occur after the stone-work is completed which can only be repaired by patching in this way. The walls are by this time ready for the roof, and the tower has been carried up to the same height. The contractor, anxious to save all the time possible, is still at work on this, although the season is getting late, and the nights frosty. One unusually cold morning we arrive on the ground, and find the mortar left over from the previous evening frozen hard in the tubs. **Mason-work in Freezing Weather.** Trying one of the small stones in the backing laid the day before, we find that instead of adhering to the mortar, and bringing away some lumps clinging to it, the stone separates easily, and comes away clean, while its surface, as well as that of the mortar in which it lay, is covered with hoar frost.

This is a plain indication that the mason-work must be suspended until the settled warm weather of spring. The film of frost which penetrates in cold weather between the mortar and the stones forms a complete separation, and adherence does not take place again after the ice is thawed, while the mortar is itself also more or less disintegrated by freezing. Mortar of pure lime, if frozen in the tubs, may be warmed and worked over so as to be used, but cement, if its first set is once broken up, either by freezing or by unnecessary stirring, will never harden again.

The foreman, mindful of his master's interests, resists the order to cover up the walls, relating anecdotes of structures built without harm in the midst of terrible frosts, and urging at great length the well-known superiority of a "frozen set" over all others; but he should be overruled. If any error is committed, it should be on the side of safety. It is true that *plastering*, of pure lime only, if frozen at the beginning of the season, and kept constantly frozen for some months, acquires at the end of that time a considerable hardness, which is not lost on thawing; while if thawed within a few days or weeks after freezing, it disintegrates and crumbles; but in this there is no question of the separation between the stones or bricks and the mortar of a piece of masonry, which takes place without regard to the duration of the freezing, and constitutes the chief danger, and moreover a "frozen set" is impossible with mortar containing cement or hydraulic lime. **Frozen Set.**

The tower walls are therefore covered with a roof of boards, and the masons sent to cut stone for the spire, or employed in building the basement piers, and as soon as the main roof is on, the lining walls in the society-room and vestibule.

Roof. The design for roof trusses should have been anxiously criticised by the superintendent, in order to detect and remedy the weaknesses, if any exist, before they cause loss and discredit through the failure of the executed structure.

If all is found satisfactory, the details are committed to the framer, who will probably need and ask for advice concerning the various joints, tenons, bolts and straps, so that the young architect will do well to refresh his knowledge on these points both from his text-books and such examples as he can reach. The accuracy of the curves, the neatness of the chamfers, the sufficiency of the bolts, should all be noted, as well as the correctness of position on the walls, and the alignment of the purlins. The rafters once on, the covering-in commences immediately, and after inspecting the quality of the boards, which should be planed on one side, smooth, straight-edged and free from loose knots, if slate are to be laid over them, it will become necessary to decide as to the best way of working the flashings and gutters. Of the former, there will be not only the valleys, but **Flashings.** the intersections of the roof with the back of the gable walls, the tower, and two chimneys, which must be protected against the entrance of water. The specification requires "all necessary flashings" to be put on "in the best manner," carelessly omitting further details. Even the material which shall be used is not specified, and the superintendent wisely resolves to come to an understanding with the contractor beforehand, and thus avoid the possibility of having to reject work after it has been put up.

Unquestionably, the "best" material for flashings is copper, which is indestructible by weather, is so tenacious as not to crack like lead or zinc when bent at a sharp angle, and so stiff that after bending it cannot be, like lead, blown up by a strong wind. Nevertheless, copper flashings are costly, and little employed, so that it would be unjust to require a contractor to use them under the circumstances, and we must content ourselves with lead and zinc. The latter metal will constitute the whole of the valley and hip flashings, and two ways of arranging these are permissible; the best practice of various localities inclining about as much to one as the other. Often the two systems are mixed, and the valleys laid by one, and the hips by the other, or *vice versa*.

By the first method, which is perhaps most popular in the large cities, the valleys are covered with a long strip of zinc, fifteen or sixteen inches wide, laid the whole length of the angle, and soldered together at the joints (Fig. 56). This is tacked at the edges, and the slates laid so as to lap over it on each side. If the metal were not subject to expansion by heat, this would perhaps be the best way; but the long strips lengthen very sensibly under a summer's sun, to contract again in winter, and the ultimate effect often is to tear them at some point, making a bad leak. The same method applied to flashing against a wall consists in covering the joint by a long strip, one edge of which is bent over, and tucked into a reglet, groove or "rag

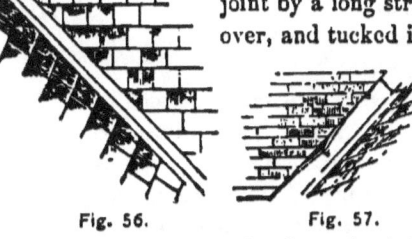

Fig. 56.   Fig. 57.

Fig. 58.

gle" cut in the stone or brick work of the wall six inches or more above the slope of the roof, and parallel with it. The efficiency of this depends on the care with which it is done. The effect of alternate heat and cold on such a flashing is to warp it until it springs out from the "raggle" either at one end or in the middle, letting a stream of water run down into the rooms below (Fig. 57), and this can only be prevented by cutting the groove quite deep, an inch or so, instead of the half-inch which is common, turning in the flashing to the very bottom of the groove, and wedging it firmly in with slate chips and cement. The wooden chips generally used for the purpose soon shrink and become loose.

Hips are covered with strips by putting a wooden "hip roll" on the boarding and laying the slates close up to it, subsequently tacking on the metal, fitting it closely around the roll, and letting it extend on each side three or four inches over the slates (Fig 58).

The essence of the rival method consists in its employment of small pieces of metal lapped over each other, without soldered joints, so that they can expand and contract freely. In forming valleys by this mode a sufficient number of trapezoidal pieces of zinc or other metal are cut out

Flashing with Small Pieces.

(Fig. 59), in length equal to one of the slates used, and in width varying from about ten inches at one end to fifteen or more at the other, according to the pitch of the roof.

These pieces are taken on the roof by the slater and "slated in," each forming a part of the two courses of slate corresponding on each side of the angle (Fig. 60), and each being laid over the course of slates next below it, while the slates themselves are laid more closely into the angle of the valley than when the other system is employed. Although more metal is used in this way, the labor is less, and the work on the whole more satisfactory, because more permanent.

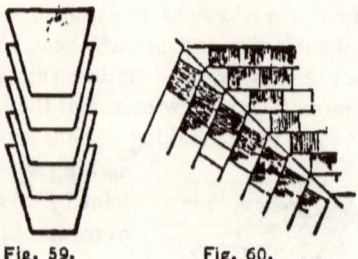

Fig. 59.   Fig. 60.

In its application to the flashing of walls and chimneys there is less to be said in favor of it, compared with a first-class job in the other style. In the "stepped flashing," as it is called, composed of small pieces, no groove is cut in the masonry, but short lengths of the horizontal mortar joints are raked out, and pieces of metal are cemented in, one above another, lapping over each other like a flight of steps. This is much more permanent than a single strip, especially if the pieces, instead of being inserted in a raked-out joint, are built into the masonry itself, as is often done; but in exposed situations the wind and rain are likely to blow into the vertical crevices which are left between the masonry and the metal when this is folded down against it, so that elastic cement, or a stopping of "paint skins" and fine sand are necessary to make them tight.

Stepped Flashing.

The principle of subdivided flashings is applied to hips by slating in pieces of metal, the slates then being laid out to the very edge. This is both tighter and neater in appearance than the hip roll with its spreading sides.

After a discussion with the roofer on all these points, we decide that the hips and valleys shall have flashings slated in, but that ordinary flashings in long strips shall be used where the roof comes against masonry. The stone-work is in places so rough that it would be impossible to turn down a sheet of stepped flashing against it so closely as to be tight, and the other mode seems preferable, but we enjoin upon the roofer the greatest care in securing the strips into the grooves.

As for the metal of the flashings, we insist that those on walls and chimneys shall be "capped," and that the capping shall be of four-pound lead. All the rest we direct to be of No. 13 (sixteen-ounce) zinc.

Capped Flashings.

The contractor demurs somewhat to capping the long flashings, but the superintendent persists, and the roofer, on being closely questioned, finally acknowledges that they cannot in any other way be made reasonably permanent, and as the contract requires that the tightness of the roof shall be guaranteed for two years, he at length, influenced by the instinct of self-protection, yields.

This capping is the best safeguard against the evil effects of expansion on long flashings, and consists in making them in a certain sense double; one strip of zinc covering the roof, and being turned up against the masonry, almost to the line of the "raggle," or groove, with a few nails to keep it in place, while a second strip of lead, thin enough to be easily dressed close against the wall, is cemented into the "raggle" just above the upper edge of the zinc strip, and turned down over it, reaching to a line an inch above the surface of the roof, so that the two pieces of metal can expand and contract independently without finally opening a joint. It must not approach nearer than an inch; if it does, it may dip into a current of water or melted snow flowing down the roof, and the capillary attraction between the two metal surfaces will draw moisture up, and over the edge of the inner strip, to find its way into the rooms below.

The mason, who is all ready to cut the reglets, asks us whether he shall make those on the rear of the copings by raking out the joint between them and the masonry below, or shall cut them in the coping itself. There is some advantage in protecting the bed of the coping, so, although we are sorry that we did not think in time to have the leaden strips laid into the mortar before setting the coping, which would have made an admirable arrangement, we order the grooves to be cut two inches above the bed joint of the coping; to be at least one inch deep, and Rosendale cement to be provided, ready mixed, for setting the lead into its place.

The gutters must next be attended to, although the specification, in calling for "gutters of No. 24 galvanized-iron, as per detail drawing (Fig. 61), running up 16 inches under slates, and to have front edge turned over a ¼ inch by ⅜-inch wrought-iron bar, with galvanized wrought-iron braces every 24 inches, running up two feet under slates, and strip of four-pound lead one inch wide soldered on under side to cover edge of stone cornice," has relieved us of the severest responsibility. But, as usual, nothing on the drawings indicates the position of the six "galvanized-iron conductors" which the specification demands, and in consequence, the direction in which the gutters shall incline must be determined by the superintendent, according to his best judgment. The task is not difficult, and the gutter, which is already on the ground, having been examined and found to be in accordance with the detailed section, well formed and straight, is passed, and immediately hoisted to the roof for putting in place.

Gutters.

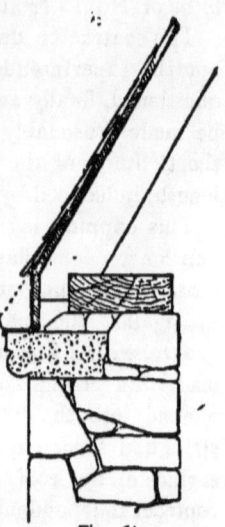

Fig. 61.

Meanwhile, we descend to the basement, to inspect the construction of the piers and the setting of the iron column in the society-room. Hardly have we turned our steps in that direction when, as we try to call to mind the basement plan, a misgiving seizes us, which is increased as the foreman mason comes out to meet us, and jerking his thumb in the direction of the society-room, begins, "About that ar column, — do you think one is enough to hold that floor?" We make no direct answer, preferring to wait until we can understand all the circumstances, which prove to be anything but reassuring. The total length of the girder in the ceiling of the society-room, from the brick wall which stands under the chancel steps to the east wall, measures forty feet, and the column is indicated on the plan as standing under its centre. The clear distance from the centre of the pillar to the wall in each direction is therefore twenty feet, spanned by a single 8' x 12" hard-pine timber. The floor-beams are all in place, but the shores set up at random under the girder to support it while the piers were being built have not yet been removed, so we are unable to judge by the deflection of the sticks whether they are overloaded or not. It is, however, very easy to determine this point by calculation. The girder F X (Fig. 62) carrying the western part of the society-room ceiling is the most strained, since it bears one-half the weight of the rectangular portion A C B D of the chancel floor, while the girder X E carries only one-half the semicircle C E D, whose area is but about four-fifths that of a rectangle of equal length and breadth, and we will therefore make our estimates for the timber F X, sure that if this is strong enough for its purpose, the one at X E will be more than sufficient. The distance F E being 40 feet, and the width of the cap of the column at X, 8 inches, being deducted from the total clear span of the two girders, leaves 19 feet 8 inches as the length of each between bearings. The distance A B is 39½ feet, so that the total floor area bearing on F X is equal to $\frac{19\frac{2}{3} \times 39\frac{1}{2}}{2} = 387$ square feet.

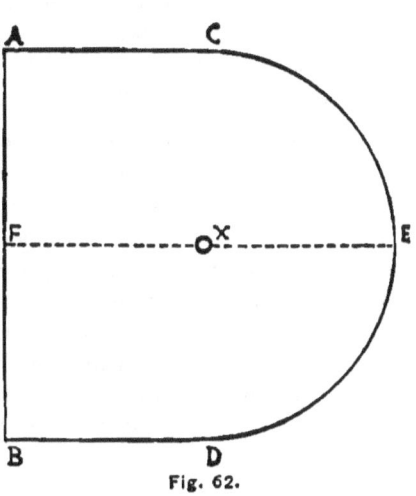

Fig. 62.

Calculation of Girder.

The chancel floor is to be tiled, and the weight of the floor, including beams and boarding, brick foundation and tiles, may be taken at seventy-one pounds per square foot, and that of the plastering on the under side at ten pounds more, making eighty-one pounds.

Besides this load, the supporting timber must be calculated to bear safely the weight of any probable crowd of persons upon it. By the New York building law, the load which must be assumed as thus liable to be brought upon the floor of a place of public assembly is estimated at one hundred and twenty pounds per square foot, independent of the weight of the structure itself, and though this is probably rather a high estimate of the weight of a compact crowd, we shall do best to adopt it as a standard. Our chancel floor must then be reckoned at $81 + 120 = 201$ pounds per square foot, and as

the girder supports 387 square feet, the load will be 201 × 387 = 77,787 pounds, equally divided by the beams over the whole length of the girder. To this must be added the weight of the girder itself, which at 45 pounds per cubic foot will be 590, making a total of 78,377 pounds. The absolute breaking weight, applied at the centre, of a Georgia-pine beam is calculated by multiplying the breadth in inches by the square of the depth in inches, and this product by 550, and dividing the result by the length in feet, or, if

b = the breadth in inches,
d = depth in inches,
l = length in feet, $\dfrac{b\, d^2 \times 550}{l}$ = breaking weight at the centre.

If the load is *equally distributed*, instead of being applied at the centre, twice as much will be required to break the beam, or $\dfrac{b\, d^2 \times 550 \times 2}{l}$ = breaking weight if uniformly distributed. As, however, timber deflects seriously and permanently long before the load on it reaches the breaking point, and as it becomes weakened by undue strain or other causes, it is considered unsafe to load it with more than one-third the calculated breaking weight, so that the result of the operation indicated in this second formula should be divided by 3 to obtain the safe distributed load; thus,

$$\dfrac{b\, d^2 \times 550 \times 2}{3\, l} = \text{safe distributed load.}$$

Applying this to our present beam, whose breadth is 8 inches and its depth 12, the length being 19¾ feet, we find that $\dfrac{8 \times 144 \times 550 \times 2}{3 \times 19\frac{3}{4}} =$ 21,478 pounds is the utmost distributed weight which can safely be put on it. As our calculations have just shown that it is liable to be called upon to bear 78,377 pounds, there is a very evident necessity for doing something to strengthen it. What steps should be taken is a matter for the architect to decide, and we notify him at once. There are several methods to choose from. Additional columns can be interposed between the central column and the walls; or the girders can be replaced by stronger ones, leaving the single column in the centre; or additional rows of girders can be put in, each supported by columns. While awaiting his reply we will inspect the brick piers which support the girders in the main

Brick Piers. cellar. Of these there are rather more than is necessary, the plan showing them spaced but six feet apart from centres. A few only have been built, of well-formed hard brick, twelve inches square, as the plan shows, but with joints of a suspicious gray-blue color, instead of brown. They have been completed some three days, but we find that a knife-blade easily penetrates the mortar after the outer crust is pierced. Calling the mason, we ask him if the piers were laid in sand and cement, only, without lime, as the specification required. He answers with considerable hesitation that "a *little* lime might perhaps have been put in, but it is mostly cement." Our suspicions are not allayed, and we ask to see the cask from which the cement was taken, and to have the mortar mixer brought before us. The foreman is about to disappear in search of these witnesses, but we detain him and send a boy in his stead, who does not return; and after a good deal of writhing, our captive confesses, being confronted with the soft mortar, that there was no cement on the ground at the time, and he had had the piers built with the best mortar he could possibly make with the materials at hand.

"What did you color the mortar with, to make it so dark?" we ask; and the foreman replies, "Well, we didn't suppose you would know the difference, so we sent over to the grocery store, and got some lampblack, and mixed it in."

We impress upon his mind our objection to such tricks by ordering all the piers demolished in our presence, and dismiss him with an admonition.

The architect is at a distance, and before his answer about the girders arrives we have an opportunity to inspect the roofing work again. The gutters are on and properly traced, the flashings finished behind one gable, and far advanced on another, and slating has begun. The slates have but just arrived, and we stop to inspect them. The specification describes them only as "good black slate," but in the lot delivered, and stacked near the hoisting tackle, we observe several different varieties. Some are thin, but with a beautifully smooth, shining surface, and very black: these are from Pennsylvania, and are of excellent quality; others from Maine are split a little thicker, and are also smooth and shining, but with a grayish lustre, like black lead; others again are thick, with a dead look, and crumble at the edges on being strongly pinched. We take one of the last-mentioned kind, and set it up on edge in a pail of water, leaving it a few minutes, when the

Slate.

moisture is seen to rise in the substance of the stone half an inch or more above the surface of the water. This slate is therefore absorbent and bad, and must be wholly rejected, while, as it would give a ragged look to see two kinds used together on a roof, we summon the contractor and request him to choose between the Maine or the Pennsylvania slate, either of which will be acceptable, and send away all others.

Ascending now to the roof, we reach the gable wall, behind which the roofer is inserting his flashings. As we approach, he hastily bundles together a quantity of pieces of zinc and throws them behind him, but the appearance of the work gives no cause for suspecting anything wrong. The lead is smoothly turned into its reglet, and the groove filled with cement; the soldering is well done, and the lower edge at the proper distance from the roof, and all closely dressed down to the stone-work. We take a convenient stick, and turn up the edge of the lead cap far enough to be able to inspect the flashing beneath. Instead of extending up the wall to within half an inch or less of the reglet in which the capping is inserted, it is turned up only about an inch and a half, so that the cap just covers the edge.

Turning to the abashed roofer for an explanation, he says that he saw no necessity for turning flashings up eight inches, which would be the distance between the roof surface and the reglet: that two inches was enough for any one, etc. We do not stop to argue the question, but simply direct the cap to be turned up throughout its length, the under flashing to be replaced in accordance with the orders first given, and the whole left exposed for a second inspection before the lead cap is turned down again into its place. Taking the roofer with us, we make our way to the other gable, where both upper and under flashing prove to be of the requisite width, but the lead, instead of being cemented into the groove, is wedged in with slips of wood. We have these replaced with slate chips, and the groove filled with fresh cement; then look to see if the tarred felt which is being spread over the roof under the slate is flexible, and the rolls perfect, instead of being, as sometimes happens, full of holes and flaws, and the material itself brittle and rotten. A glance is sufficient to satisfy us in these respects, and our duties are over for the day.

A letter received from the architect before our next visit explains the unfortunate oversight in relation to the girders in the ceiling of the society-room. It seems that three columns had been indicated in this room to support the line of girders, dividing the length into four spans, for which the timber would have been sufficiently strong, and the estimate had been made including this arrangement, and the contract signed in accordance with it. Immediately after the signing of the contract, however, the rector of the church had represented to the architect the inconvenience which would be caused by the row of pillars through the middle of a room devoted to so many uses, and had engaged him to reduce the supports to a single central column, strengthening the girders sufficiently to compensate for the increase in their length. As time pressed, the architect had hastily erased two of the columns from the tracing of the basement plan sent to the contractor for commencing operations, leaving only the single central one, fearing that any delay might cause useless foundations to be put in for the pillars first shown, and intending to arrange for the substitution of iron beams, or girders strengthened in some other way, for those specified; but this had slipped his mind until he received our letter. As a considerable saving in cost was effected by omitting the two iron columns, a corresponding allowance would have to be made from the contract price, to be offset against the extra cost of strengthening the girders. Just how this should be effected he did not wish to dictate positively. His own idea had been to substitute two rolled-iron beams, placed side by side and bolted together, for each of the wooden girders, but this would be expensive, and although he was not restricted as to the cost of making the change desired by the rector, he would like to save the church all needless expense, and if we could devise any cheaper mode of trussing the present girders, or otherwise supporting the floor, he would leave the matter wholly to our judgment, in which he had great confidence. The girders, he would remind us, were intended in any case to be plastered over.[1]

---

[1] The writer feels as if he owed an apology to his imaginary architect for exhibiting him as guilty of so many mistakes and oversights, but he is anxious to impress upon the minds of his readers the lesson that an efficient superintendent should be able to criticise and correct in good time, if necessary, the work of the architect, and to act in his stead upon occasion. His ability to do so will not be likely to make him over-forward in thrusting himself into the other's place.

In accordance with the instructions contained in the letter, we set ourselves to devise, if possible, an inexpensive mode of trussing the girders to which it relates, and thereby fulfil the architect's wishes by saving the cost of iron beams. Some further calculations will be necessary to test the expedients which occur to us. The present girders have been proved to possess only about one-fourth the requisite strength. The most natural way to increase this would be to add other timbers beside the original ones. If four girders of equal size with the one we have were placed side by side, and all bolted together, so that any burden on one would be resisted by the combined strength of the four, the problem would be solved, but in a very awkward manner, since the cap of the single column would have to be dangerously extended in order to support the ends of all the girders.

Perhaps this might be obviated in another way. Recurring to the formula for transverse strength of beams, we notice that while the resistance increases *directly* as the breadth of the timber, it is augmented in proportion to the *square* of its depth, so that although a single beam of four times the strength of one girder, and of the same depth, would need to be four times as broad, a beam equally strong would be formed by a timber of the same breadth, but of only twice the depth. As our present girder is 8 inches broad by 12 inches deep, a stick 8 inches by 24 would just meet our wants.

Unfortunately, it is impracticable to procure beams of such dimensions, and we must try some other expedient. It would be of no avail to put a second 8 x 12 girder below the first; each of them would bend independently of the other, and their united strength would be that of two beams only, instead of four.

Iron might be employed for strengthening the wooden sticks. A strip of boiler plate, 12 inches wide and half an inch thick, might be bolted on each side of the present girder, strengthening it perhaps sufficiently; or rods might be used to form a "belly-truss"

---

while it may save the latter infinite trouble and anxiety. For this reason the writer represents his architect as neglectful and ignorant, solely for the sake of showing the way in which the good superintendent can bring him out of his errors with credit, or, if both offices are united in one person, how the architect can extricate himself from them. It must be remembered that our hero, whatever he may be called, can only give proof of his varied ability by being furnished with occasions for exhibiting it.

(Fig. 63), the strength of which would depend mainly on the depth which could be given to it.

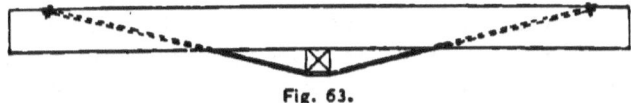

Fig. 63.

This would, however, cause an unsightly projection into the room, which must be furred over to conceal it, and the trussing, though cheaper than the flitch-plates, would still be somewhat costly.

Reflecting upon all these devices, it occurs to us that if a second beam similar to the present one could be placed below it and strongly connected with it, so as to constitute practically one timber, the desired result would be attained in a form extremely compact and simple. Turning to some text-books which we have brought with us, we learn that this can to a limited extent be accomplished by bolting or strapping the two pieces together, and either indenting them into each other as in Fig. 64, or notching them and inserting hard-wood keys, as in Fig. 65, the sole object of the indentations or keys being to prevent the slipping of the contiguous surfaces of the beams upon each other, so that in order to bend, the *whole* of the lower part of the compound beam must be stretched, and the whole

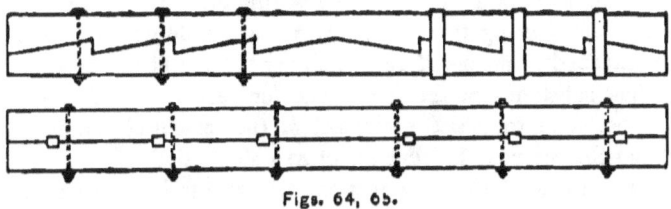

Figs. 64, 65.

of the upper part compressed, just as if the beam were in one piece (Fig. 66), instead of the upper and lower halves of each stick undergoing their own separate compression and extension, as in Fig. 67, which shows the action under stress of beams superposed, but not keyed together.

The books inform us that compound beams so keyed together and tightly bolted are nearly equal in strength to a solid stick of the same dimensions, bearing in mind that the depth of the indents or notches for keys must be *deducted* from the total depth of the two timbers in order to obtain the effective depth which can alone be used as a factor in calculating the strength.

This expedient may furnish us with a resource, but we find on careful study that if eight keys were used, which would be a sufficient number, the depth of each would have to be within a small fraction of four inches, two inches being cut out of each timber; and in order to obtain a net depth of 24 inches, which is what calculation shows to be necessary for the strength which we require, the

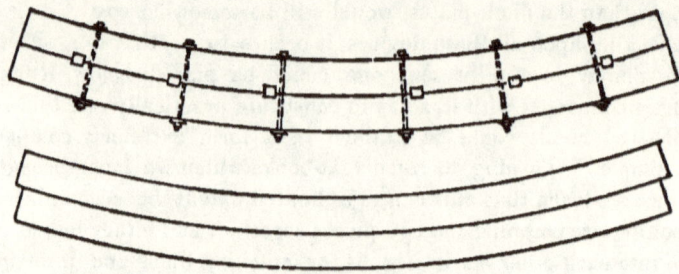

Figs. 66, 67.

aggregate depth of the two sticks before cutting must be 28 inches. Hence, as the one already in is 12 inches deep, the other must be 16 inches. A 16-inch Georgia-pine timber happens in this place to be difficult to obtain, the stock sizes running only to 14 inches; so, after considering this objection, we are induced to search for some still further means of so combining the two beams as to secure the whole substance of the lower one for resisting tension, and of the upper one to resist compression, the essential requisite for enabling the compound girder to act as a single timber.

Calling to mind the construction of certain bridge trusses, in which the upper chord is brought completely into compression and the lower one into tension by means of inclined struts, which resolve the downward pressure upon them into a push with their heads in one direction, and with their feet in the opposite direction, and being arranged so that half of them point one way and half of them the other, mutually act to compress the upper and to stretch the lower chord with their united force (Fig. 68), we reflect that if similar means could be applied to two superposed beams, the result might be just what we desire.

The action of the oblique struts in the bridge can be applied to the compound beam in the simplest possible way by nailing oblique pieces of board firmly to the timbers, reversing them on opposite sides, in imitation of the reversed struts in the bridge truss (Fig. 69). If these boards are 1¼ inches thick, their united strength will be greater than that of either of the beams, and if well nailed, so as not to spring, the lower beam will be torn apart before they will yield.

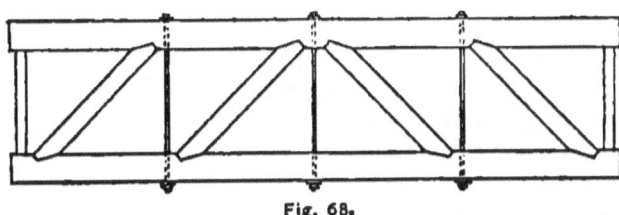

Fig. 68.

The nails, if long enough to penetrate three inches or more into the girders, can only give way by shearing, which would require a force greater than would be necessary to break a solid beam. The strength of a girder built up in this way will not be affected by shrinkage, which soon causes more or less deflection in those indented or keyed together; for its resistance is maintained by the board struts, which act only in the direction of their length; and this remains invariable whatever may be the lateral shrinkage.

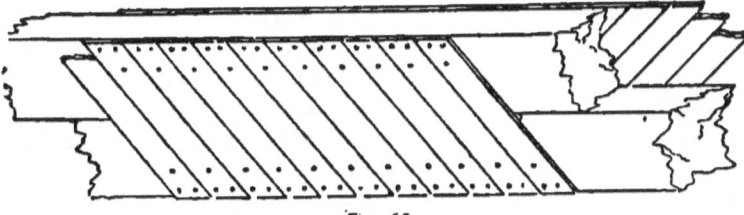

Fig. 69.

As a compound beam of this sort is a novelty in construction, we do not venture, although the theory appears satisfactory, to apply it to the important case before us without testing it by models. Two or three of these are quickly made, at one-eighth the size of the real girder, the two sticks which we propose to combine into one being accurately cut out of Georgia pine, each 1 inch wide by 1½ inches deep. At the same scale with the other dimensions, the length, to represent the length of the real girder, would be about 2½ feet, but if our theory is correct, even the small model, if cut so short as this, would require a distributed load of nearly 4,000 pounds, or a centre load of about 2,000, to break it. We can, however, reduce the transverse strength by adding to the length, and if this is made 10 feet, or four times that first proposed, the centre breaking weight will be approximately 500 pounds, — ¼ of the first result. This is quite within our reach, and we cut off the sticks at 11 feet length, set them one above the other, and put on each side slips of any wood at hand, a little more than ⅛ inch thick, set at an angle of 45° with the top of the beam, and nailed with small brads, about ⅜ inch long, three in each end of each piece. These very well represent the spikes to be used in the real beam.

Setting up two trestles, with triangular pieces put on top so as to give supports exactly ten feet apart, we hang a "scale," or platform suspended by ropes, which is used to hoist bricks or other materials to the centre of the model beam, first weighing it accurately. Before going farther, posts must be set up in pairs, enclosing the model beam between them with just enough room to allow it to deflect freely. This is to prevent the beam from turning over on its side, which so small a stick is liable to do under a centre load. Then bricks are piled on the scale, weighing each one before adding it.

The centre breaking load of a solid Georgia-pine beam 1 inch wide, 3 inches deep, and 10 feet long, is 495 pounds: that of two sticks each 1 by 1½ inches, by 10 feet, superposed, but not connected, would be 247½ pounds. The strength of our model cannot possibly exceed the former, but should, if our reasoning is correct, approach it. As the weights are placed one by one in the scale, we add up the total load, and experience a lively interest as the stress upon the beam begins to approach four hundred and fifty pounds, without any sign of giving way. The deflection is considerable, and the model is evidently under severe strain, so, to avoid the shock caused by the placing of a brick on the pile, we increase the load more gently by pouring on weighed portions of sand. Four hundred and fifty pounds are passed; then four hundred and sixty; and little by little the weight approaches four hundred and seventy. Just as this point is reached, a warning crack is heard, and we stop; but nothing further follows; we recommence pouring, but before another pound is added, the beam yields, letting the scale drop suddenly, and on examination the lower stick of the combination is found to be torn asunder. No change whatever is observed in the board struts.

The second model breaks at four hundred and seventy pounds, and a third at four hundred and seventy and a half. Comparing these with the calculated strength of a solid beam, we find that the girder built up by our simple method proves to possess a strength about equal to ninety-five per cent of that of a solid beam of the same breadth and depth; a result superior to that obtained by indentations, or keys and bolts or straps, and at a fraction of the expense. The carpenter, who had viewed our preparations with an ill-concealed scorn for "them little slivers," is profoundly impressed at the resistance which the slender model displays, and respectfully listens to our directions as to the mode of trussing the larger timbers. A very simple calculation only is needed to show that if the strength of one 8" x 12" girder of the given length is sufficient to carry safely 21,478 pounds, as previously ascertained, a solid beam of equal width, but twice the depth, would carry four times as much, namely, 85,912 pounds, and that if for the solid beam of 8 by 24 inches a compound girder were substituted, built up of two superposed 8 by 12 sticks in such a manner that it possessed ninety-five per cent of the strength of a solid beam, this built-up girder will sustain safely 81,616 pounds. As the load upon our floor with 120 possible pounds of humanity per square foot added, and including the weight both of the girder now in place, and the new one added beneath it, will be but about 79,200 pounds, this construction gives us assurance of success, and we hasten to put it into execution. Since the compound beam thus made forms virtually a lattice truss of which the upper and lower chords are in contact, its strength, like that of all trusses, could be increased by *separating* the beams, so as to make the depth of the truss greater, and in this case the side struts should be stronger in proportion to the longitudinal timbers. Our girder is, however, strong enough, and more compact than a truss. It is quite important to gain all the room possible beneath the girder, as well as to have it rest properly on the cap of the column, so we will, after shoring up the ends of the beams on each side, cut off the portion which overlaps the girder, and after this has been reinforced with the second stick, and the boards nailed on, the compound beam can be pushed up between the ends of the joists flush with their top, and these will then, instead of resting directly on the girder, be supported by strips of wood 2" x 4", spiked to the girder on each side over the board struts.

Having made careful notes of the quantity of extra timber used for the new trussed beams, and of the time occupied in cutting out, making over and replacing the work, in order to adjust the cost subsequently, we turn to look at the lining wall of the room, which is being built up inside the stone-work. As this wall is to be plastered, care must be taken not to "strike" the mortar joints off flush with the surface of the brickwork, thus taking away the projections needed for the plaster to cling to. The common way of reserving a proper "key" is to leave the mortar irregularly projecting, as it is squeezed out of the joints when the bricks are hammered into place with the trowel; but many builders find that the annoyance caused by the occasional projection of the dried mortar beyond the line of the plastering is so great that they prefer to lay the horizontal joints full of mortar, striking them off smooth, but to lay the vertical joints "slack," that is, only partly filled with mortar, so as to leave cavities into which the plaster can penetrate and obtain a firm hold.

The lining wall is often built before laying the floor-beams, which rest on this, and not on the main mass of masonry; but, although a four-inch brick wall, well tied to a thicker one behind it, possesses surprising strength, the unusual weight of the tiled chancel floor makes it desirable to support the beams which are to carry it on the heavy outside wall. In the tower vestibule, however, there is no objection to resting the timber ceiling on the lining wall, which, being of face-brick, will not be built until near the completion of the structure, to avoid marring it.

The piers are next inspected, and prove this time to be well built, of bricks properly soaked in water, and laid with cement mortar. One or two are crooked or out of plumb, and we order them to be taken down and rebuilt; and in a few instances chips of wood have been used as wedges between the top of the piers and the girders, which, as they would soon shrink and allow the floor to shake, we have replaced with stone chips or slate and cement.

The church at our next visit is a maze of scaffolding, put up for the use of the plasterers, and furring is rapidly going on. The upright studs against the walls are all in place, and spiked as strongly as may be to the masonry. Some builders rake out a joint, and drive in a wedge-shaped piece of wood for nailing the furrings to, and others build " wood bricks," or short bits of joist, at intervals into the walls, for the same purpose; but such pieces are very apt to shrink so much as to become loose, while the driving of wedges may endanger the stability of the wall, so that the practice of driving the spikes directly into the mortar is generally preferable. Very rarely the joints are raked out and slips of lead laid in, the spikes being driven between these, but the gain is hardly worth the expense. At all events, it is desirable and usual to stiffen the upright furrings by angular bridging (Fig. 70) in the same manner as a partition.

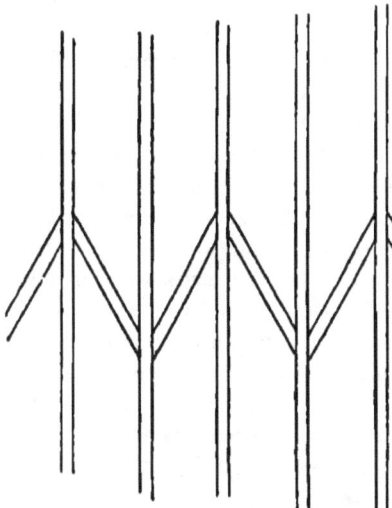

Fig. 70.

Some precautions should at this stage be taken to lessen the dangerously inflammable quality of this light construction. As all those experienced in fires know, the furring studs set against the walls of stone structures form lofty flues of inflammable material behind the plastering, up which the flames run with incredible rapidity, urged by the strong dra

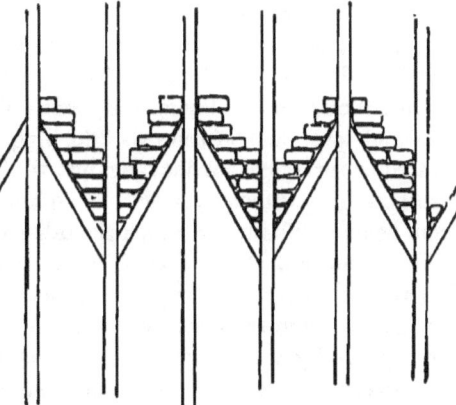

Fig. 71.

which can at all times be felt rushing up through them, and a small fire originating in the basement of such a building usually breaks out immediately in the roof, where it is uncontrollable. The means for preventing this, and confining an accidental fire within the limits where it can be reached and extinguished, are extremely simple: a single row of bricks laid in coarse mortar on the floor between the furring studs will cut off the communication between the cellar and the space behind the plastering, so that a fire catching in the basement from spontaneous combustion of coal, proximity of smoke-flues or furnace-pipes to wood-work, or any other cause, is compelled, for want of access to the concealed passages behind the furrings, to burn through into the room, where it may blaze for hours without doing much harm, and is easily quenched with a few pails of water. Two rows of bricks are still better than one, and the only precaution to be observed is to fill up the *whole* space from the stone wall to the inside finish, — wainscoting, sheathing, plaster or base-board, — with the mass, so as to leave no openings. If a second barrier of coarse mortar and chips is laid on top of the bridging (Fig. 71) all around the building, it will be doubly protected, and its destruction by fire will be rendered slow and difficult.

While the furring is going on, the plastering mortar should be mixed, and the door and window frames set. In brick buildings the latter are usually set in place at the beginning, and the walls built around them, and this is sometimes done in stone structures, but not by the best builders, as it is difficult to keep the frames from being knocked slightly out of shape by the setting of the heavy stones against them; and by laying the stone-work separately, with plumb rule and level, and afterwards trimming it, a much smoother and straighter surface is obtained, against which the frame can be fitted weather-tight, without the unreliable and often unsightly pointing from behind with mortar which the other method involves.

With the best of workmanship, however, crevices are sometimes left through which the wind can penetrate, and wherever the wind can go, rain and snow will follow, so that a certain amount of packing is generally necessary. This can best be done with cotton, driven in between the wood and the stone, where it is kept permanently in place by its own elasticity, while it checks the current of air very effectually.

The plastering is briefly specified to be the best three-coat work, sand finished throughout. At least as soon as the furring is begun, the superintendent should see that the materials are at hand for making mortar enough for the whole of the plastering. By recollecting that one hundred square yards of three-coat plastering require three casks of lime, three one-horse loads of sand, one and a half bushels of hair, and about two thousand laths, the total quantity needed of each is easily reckoned. A place should then be prepared without delay, where the whole mass requisite for the first, or scratch, coat can be mixed and allowed to cool for a week before any of it is put on the walls. In this way only can we guard against the occurrence of particles of unslaked or partially slaked lime in the mortar, which will continue to absorb moisture after being spread on the laths, and perhaps months later will cause small cracks or blisters, or throw off little chips from the plastering, disfiguring it very much when it is too late to remedy the evil.

In regard to the manner of mixing, the practice varies. Occasionally a mason is found who is willing to slake the lime by itself, and leave the paste for several days or weeks, — a year, even, in some cases, — during which it becomes somewhat more firm, and acquires a beautifully smooth, "fat" quality, something like cream cheese. This heap of paste is drawn upon as wanted, and mixed with the proper proportion of sand and hair, then put immediately on the walls. The disadvantages of this process are the difficulty of distributing the hair evenly through the stiffened paste without the help of water to loosen the tufts, and the increased labor required for working the mortar. The advantages are the perfect hydration of the lime, by which chip-cracks and blisters are wholly avoided, the smoothness and hardness of the finished plastering, and its greater tenacity, since the hair, not being added until the lime is cold, retains its full strength, instead of being burned and corroded by steeping in the hot, caustic mixture which is the first result of slaking. Few builders, however, are disposed to proceed in a way so inconvenient to them, and content themselves with spreading out the lime, pouring on water from a hose, and after a little stirring adding the hair, which is mixed into the steaming liquid, and the sand immediately thrown over it, incorporated as well as may be, and the whole mass piled up for use. The hair in this case deteriorates as fast as the lime improves, and a season of cooling which would be very beneficial to the latter ingredient will nearly destroy the former, so that a course must needs be taken midway between the two extremes.

Whatever mode is adopted, a clean floor of planks must be laid, with sides a foot high or more, to keep dirt from being mixed with the mortar. There is a process, said to be in vogue in certain country districts, of slaking the lime upon the ground, and then hoeing up grass, roots, soil and lime into a viscid mess, which is spread upon the laths, where it stays long enough for the mason to get his pay; but such methods of construction are not within the scope of this work.

It is not always easy to tell by the appearance of a heap of plastering mortar whether the lime, sand and hair are of good quality and in suitable proportion. If properly mixed, which will be shown by the absence of streaks in the mass, a small quantity of the mortar should be taken up on a trowel, slate or piece of board. If it hangs down from the edge without dropping off, the quantity of hair is sufficient; or, if it is practicable to see the mixture made, one bushel of hair to each cask of lime in the first coat will be the proper proportion. The quality of the hair can be tested in the same way. Long ox-hair is perhaps the best. It is strong, and the fibres an inch or more in length. Goat's-hair is longer, but not so strong, and short cattle and horse hair is of the least value. On drying a small quantity of mortar, an excess of sand will be shown by its being easily rubbed away with the fingers. The quality of the lime is best tested by observing the slaking. For plastering, lump-lime only is used in ordinary cases, and it should slake energetically and fall into a smooth paste, without any refractory lumps or particles of "core." If such are found, all the casks of that brand should be rejected.

Before lathing, grounds must be put on wherever necessary. These are strips of wood planed carefully to a uniform thickness, seven-eighths of an inch or more where the plastering is to be three-coat, or three-fourths for ordinary two-coat work, secured to the furrings in such a way as to give convenient nailings for the subsequent finishings; one row, for instance, being set an inch or so below the top of the future base-board, two or three in the height of a wainscoting, a border around each door and window, and so on. Being of equal thickness, and usually straightened with the straight-edge and plumb rule to correct any irregularity in the furrings or studs, they afford guides for bringing the plaster to an even surface. Further guides are formed by the angle-beads, or grounds of other shapes, which are secured to the corners of the walls before lathing.

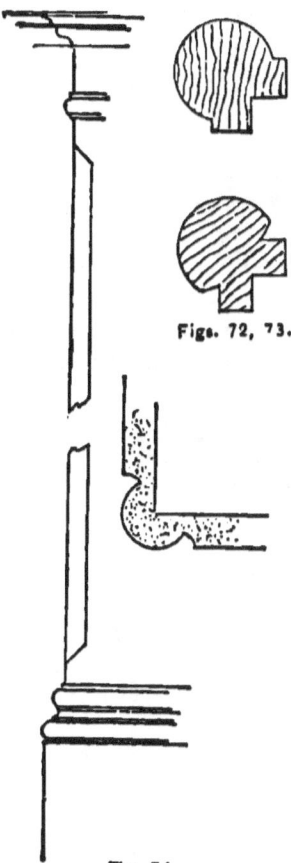

Figs. 72, 73.

Fig. 74.

The customs of different localities vary in this point, and the superintendent will probably be asked to decide on some particular mode, if the specification gives no definite instructions.

In the Eastern States, where walls are almost invariably papered or decorated as soon as finished, it is usual to nail to the studs at the angles a moulding in section like Fig. 72 or 73. This serves as a guide for putting on the mortar, and when the plastering is completed gives a finish to the corner which is not readily broken or scratched, and is easy to cover with paint or paper to good effect. In New York and other places where a pure white hard-finished surface is or has been fashionable, it is customary to turn the corners by means of a "rule-joint," worked in the plastering itself, and consisting of a vertical semi-cylindrical moulding some three-fourths of an inch in radius, which stops against a bevelled surface a little below the cornice and above the baseboard or wainscot (Fig. 74).

These rule-joints are beautifully executed by the best plasterers, but they cause a rather awkward succession of breaks in the vertical line of the angle, since the plaster must be fully brought out to the corner above and below all mouldings, as otherwise a troublesome horizontal surface would be left (Fig. 75), which with the small wooden angle-bead is not noticeable; and they are liable to scratches and abrasion during the progress of the work and afterwards. This risk of abrasion is with plaster angles a serious matter, and the difficulty of the common rule-joint is sometimes overcome by squaring out the plaster to the edge, and subsequently putting on a wooden saddle-moulding (Fig. 76) cut out of a solid piece, but this device has little to recommend it. For so large a building as a church, the rule-joints have a massive look which is pleasing, so we choose this method of finishing the corners, and leave the lathers to their task, first enjoining upon them to set the laths at least three-eighths of an inch apart, and to break joint every six courses.

The weather having now settled to steady cold, it will be necessary to heat the building by artificial means, to prevent the plaster from freezing, which will disintegrate it and cause it to crumble and fall off the laths. Some supervision should be exercised over the stoves or furnaces employed for the purpose, as workmen are incredibly careless and indifferent about the dangers to which they expose the property of other people. The windows must be well closed with boards, and temporary windows inserted.

After the first, or scratch, coat is partly on, the superintendent should endeavor to look behind the laths, to see if it has been well trowelled, so as to press the mortar through the openings and cause it to bend over by its own weight, forming a hook by which the plaster is held to the laths. As this is the only way in which the whole substance of the plastering can be kept on the walls, it is very necessary that ceilings should clinch well over every lath, and walls over every second or third lath. The scratching should be thoroughly done, as it affords the key for the

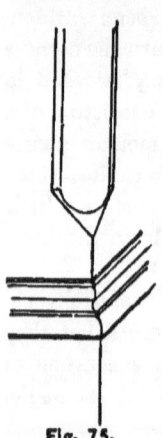

Fig. 75.

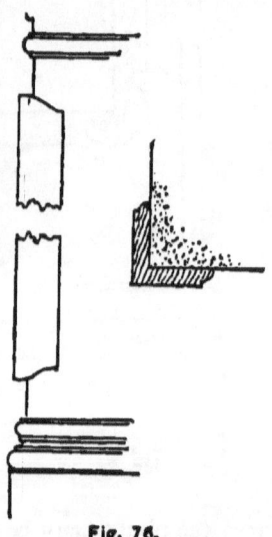

Fig. 76.

second coat, and there should be no appearance of tufts of hair, which would show that the mortar had not been thoroughly mixed.

Care should be taken to see whether the specification directs that the plastering shall be carried to the floor everywhere, or only to the grounds.

It is so habitual with builders to plaster only to the grounds that they frequently overlook the directions which require a better mode. Of course it is unnecessary to carry the hard-finish behind the wood-work, but the first two coats should be required, if any plastering in that position is specified.

By the second, or brown, coat, all the surfaces must be brought to a true plane, the angles made straight, the walls plumb and the ceilings level, since no effectual corrections can be made afterwards The wall-spaces in the interior of the church are so broken that no great care is required to obtain a true surface, and we may content ourselves with a close examination of each, to detect irregularities, and tufts of hair, which will make unsightly spots in the finished work. The brown coating must on no account be allowed to begin until the first coat is thoroughly dry. Men accustomed to two-coat work, in which the first coat is often only superficially hardened when the second is put on, frequently treat three-coat work in the same manner, thereby weakening or sometimes ruining the whole. If part of the walls are to be plastered on brickwork, and others on laths, the scratch coat is put only on the laths, and when this is dry, the brown coat is spread over the whole, including the brickwork.

After the brown coat is dry and hard, the rule-joints at the angles should first be made, and the hard-finish then applied. This, instead of being mixed with little or no sand, and with a portion of plaster of Paris, as would be proper for a smooth surface, should for our purpose contain a large proportion of rather coarse sand, sifted so as to be uniform in grain, and little or no plaster of Paris, which would set so quickly as to hinder the thorough rubbing with the float which is necessary to bring the sand evenly to the surface.

The plastering once dry, the wood finishing can proceed without hindrance. The superintendent must henceforth devote himself conscientiously to the study of the detail drawings as they arrive from the architect's office, and endeavor to forestall any slight mistakes or misfits, which are sure to happen through the unavoidable variation of the finished work from the exact dimensions shown on the plans. If a panelled wainscot is shown, he should measure on the spot the lengths of the various portions of wall to which it is to be applied, and compare them with the drawings, spacing off the panels as shown in the details, so that there may be no awkward want of continuity, or disproportionate members at the angles. The stock for the woodwork must also be looked after. A load of hard-wood lumber inevitably contains a large percentage of worm-eaten, stained or otherwise defective pieces, which must have the defects cut out in working them up, and a sharp eye is needed to see that this is done, — that a piece of black walnut streaked with white sap is not put into an out-of-the-way panel, or a knot cut out and a patch inserted in another place, where it may be unobserved while fresh from the sand-paper, but will grow more conspicuous afterwards. Any carelessness or want of decision on the part of the superintendent is apt to be taken advantage of.

Generally, where the floors are double, all bases and wainscotings and other "standing finish" are put on before the upper boarding is laid. In this way the base-boards, which extend half an inch or more below the surface of the upper boarding, which is laid up against them, can shrink, as they are certain to do more or less, without opening a crack between them and the floor, and no care is needed in fitting them to the floor (Fig. 77), while with a single flooring it is customary either to "scribe" the base to the boards, so as to fit minutely all their irregularities, which answers well enough until the shrinkage of base and floor-beams draws them apart; or to plough the base laboriously into the floor (Fig. 78). In our building we follow the former course in all parts except the chancel, whose tiled surface cannot well be fitted against the woodwork. As tiles and marble are easily injured by the operations of workmen, we will wait as long as possible before undertaking this portion, carrying the nave nearly to completion before touching the chancel at all.

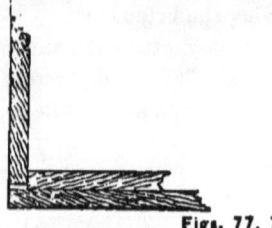

Figs. 77, 78.

When all the wood-work in the nave is finished except laying the upper floor, the marble steps at the chancel entrance may be set upon the brick wall built up to receive them.

The young superintendent should be familiar, from observation and comparison at some good marble-worker's, with the appearances which characterize the different qualities of marble: in this way only can he judge with certainty whether the proper kind of material is furnished. Often pieces of marble whose appearance is injured by obtrusive spots or streaks are sent instead of the best quality, under the pretext that there is nothing better, and sometimes even a coarse, soft Vermont marble, streaked with a blue somewhat resembling the Italian, is palmed off on a contractor in place of it, but the inferior kind can easily be distinguished by its coarse grain, and the yellowish cast of the white portions.

As soon as the steps are accurately set, they should be protected with boards, and the laying of the tiles may begin. For convenience in building, an under floor has been laid over the chancel, as well as the nave and aisles, and this rough flooring should now be taken up, sawed into short pieces so as to fit between the beams, and these pieces laid in, on strips previously nailed to the sides of the beams.

The best, and the only durable, way of laying encaustic tiles on a floor framed with wooden beams is to make the foundation for them of bricks set edgeways on the short pieces of board between the beams, but to save material the bricks are sometimes laid flat, though the result is much inferior. In our case the proper mode is specified, and it is only necessary for us to make sure that the strips or "fillets" are nailed on at the proper distance from the top of the beams, a matter about which workmen are very careless.

The tiles are from $\frac{1}{2}$ to $\frac{5}{8}$ of an inch thick, and to ensure a continuous but thin bed of cement between them and the brick, the top of the latter should be $\frac{3}{4}$ of an inch below the line of the finished floor. The brick to be used may measure from 3 to $4\frac{1}{2}$ inches in width, according to locality, so we try those on the ground, and find them to vary from $3\frac{1}{2}$ to $3\frac{3}{4}$ inches. The maximum width must be taken which added to the $\frac{3}{4}$ of an inch for tiles and cement gives $4\frac{1}{2}$ inches depth from the finished floor to the boarding, or, as the boarding is $\frac{7}{8}$ of an inch thick, $5\frac{3}{8}$ inches from the finished floor to the top of the fillet. The beams should stand in the same relation to the tiling as they would to a wooden floor of double boarding, in order to avoid a disagreeable break between the chancel floor and that of the adjacent robing-room, and as each portion of the double flooring is $\frac{7}{8}$ of an inch thick, the whole distance from the finished floor to the beams will be $1\frac{3}{4}$ inches, which being deducted from the distance last found will give $3\frac{5}{8}$ inches as the proper gauge from the top of the beams to the top of the fillet. A thorough understanding of this matter will save much subsequent annoyance and expense in cutting off brick which are too high, or concreting up from a surface set too low.

It is better and much more economical to have tiles laid by the parties who furnish them, and contracts are generally made for the floor complete, but cases may occur where the local masons will be called upon to do the work. In such cases the brick surface must be swept clean and *thoroughly wet*, and the tiles must be soaked in water for some time before they are used. Without these precautions, either the brick or the tile will absorb water from the thin layer of cement between them, making it powdery and useless. The best Portland cement only is suitable for use — the American brands, if fresh, being quite equal to most of the English as generally found in our markets — and is to be mixed rather thin, without any addition of sand. The pattern must be commenced *from the centre*, which is to be very exactly ascertained by previous measurement, and straight-edged strips of board should be put down as guides for each day's work, not only for regulating the lines of the pattern, but also for securing a uniform surface, which is done by first levelling them carefully, and setting the tiles by means of a straight-edge resting on the strips. Each tile is set in a bed of cement spread for it, and beaten down to the proper point with the wooden handle of the trowel. After a sufficient number have been laid, the joints may be grouted with liquid cement, which must, however, be immediately wiped off the surface of the tiles, since it is difficult to remove it when dry.

If the cement is good, the tiles cannot, after a few days, be removed without breaking, so that too much care cannot be exercised in placing them properly at first. When the pattern reaches the edge it is usually necessary to cut many of the tiles. This can be done by soaking them well in water, and then scoring a line with a sharp chisel where the separation is to be made; then placing the chisel exactly on the line, a sharp blow will divide the tile neatly. Wide chisels should be used, and unless the tile is well soaked, it is apt to fly into fragments.

After the floor is done, it is covered with sawdust an inch or two deep, and planks laid over it to walk on. The base-boards and wainscoting are then fitted down upon it. The marble tiling is laid in the same way, on bricks set on edge, but the marble is thicker, usually varying from seven-eighths to one and a quarter inches, the under side being quite rough; and the fillets should be set accordingly. The laying is much easier than that of clay tiles, and mortar of cement, lime and sand in equal parts may be used.

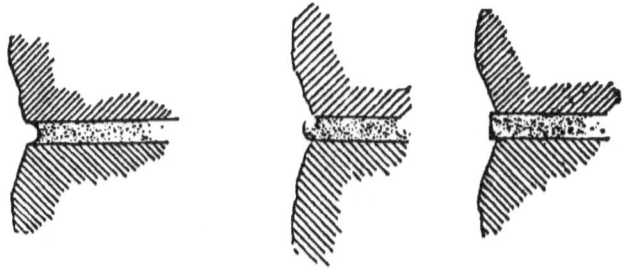

Fig. 79.   Fig. 80.   Fig. 81.

By the time our building has reached this point, spring has advanced, and the warm, moist days of May present the best possible opportunity for pointing and cleaning down the exterior of the stone-work. The brownstone ashlar will look best if pointed with mortar of nearly the same color as the stone, which can be made by mixing burnt umber with the cement used. The best Portland cement is preferable to any other, and is to be thoroughly mixed with an equal bulk of sand, and such coloring matter as may be required, but with only just water enough to give the compound a mealy consistency. The old mortar is raked out of the joints to a depth of an inch, or if, as is likely, that in the upper part is found to be frozen and powdery for a greater distance inward, it should be completely removed as far as the freezing has extended; then the pointing mortar is inserted and strongly driven in with a steel jointer or S-shaped instrument, rubbing it until the moisture is squeezed out upon the surface. The tool is formed to mould the edge of the joint in various ways as it is rubbed. The most durable form is the hollow (Fig. 79), but the half-round (Fig. 80) is often used, as well as the fillet (Fig. 81). In the two latter the mortar is less thoroughly compressed, and the projecting part may fall off.

The cleaning down is done with muriatic acid and water, applied with a sponge, and followed with pure water. This removes the lime stains, and leaves all neat.

The subsequent operations in the church, such as painting, glazing, and decoration, can best be studied in connection with other constructions in which they play a more important part.

BUILDING SUPERINTENDENCE.

CHAPTER II.

WOODEN DWELLING-HOUSES.

ONE of the most difficult portions of an architect's business is precisely that which amateurs usually imagine to be the easiest, — the superintendence of the work connected with dwelling-houses. It is natural to suppose that an intelligent householder, who has spent a large part of his time for years in observing the defects in his own habitation, and comparing it with those of his neighbors, would find no difficulty in directing the construction of a similar building, but experience soon shows that the knowledge which most persons have of the structures they live in is a very superficial one, consisting in the observation of results, rather than of the processes by which the results are obtained; so that the amateur house-builder is apt to find himself quite at fault in endeavoring to give the necessary preliminary directions for securing the particular objects of strength, durability, healthfulness, appearance or finish which he has most at heart. Nevertheless, it is often necessary, and always desirable, that unprofessional persons should be able to direct the operations of mechanics, and make contracts for various kinds of work, and it is hoped that the suggestions contained in the following pages, although intended primarily for young architects and superintendents, will not be found too technical for the ordinary reader, and that the explanations given with regard to the objects which it is desirable to seek, and the means by which they can be attained, will be found serviceable to the large class of persons who are interested in building, either for themselves or others, as well as to those occupants of houses already built who would be glad to understand more clearly the structure of their dwellings, with a view either to the correction of defects, or the planning of improvements.

The point in which amateurs are particularly liable to fail is the choosing of a proper site, and as even experienced architects are not always successful in this respect, a few directions in regard to the placing of houses upon the ground should not be omitted.

In all cases it is essential to determine the position approximately before the plans are begun, in order that the building may be so arranged as to present an agreeable appearance from **Determining Location.** the neighboring streets, securing at the same time the greatest pleasantness of prospect from the windows, with the most cheerful light and sunshine in the rooms, that the situation can be made to yield. In our climate, the best aspect for the **Aspect.** windows of living-rooms, particularly of bed-chambers, is south or south-east. Such rooms are warm in winter and cool in summer, and cheerful at all seasons. Next to a southern exposure, the eastern is the pleasantest, and may be appropriated for dining-rooms, which will thus enjoy the advantage of the early morning sunshine, with coolness during the rest of the day. Between the western and the northern aspect there is little to choose; the cheerlessness of the one is hardly more objectionable than the heat, on summer evenings, of the other, so that these sides of a dwelling-house should, as far as possible, be given up to inferior rooms, halls and stairways. Of course, considerations of prospect and position with respect to the approaches must affect the plan more or less, but the skill of the designer will be shown by the success with which he contrives to satisfy all the requirements at the same time.

Other points will, however, claim the attention of the careful architect, besides the more obvious ones of situation and exposure, and the final staking out of the building should never be at- **Character of Ground.** tempted until they have been thoroughly considered. Foremost among these is the character of the ground, whether wet or dry, springy or well-drained. The only certain test of this consists in sinking pits at different places, to the depth proposed for the future cellar, or a little below, if the trials are made in the dry season; but indications may be found in the conformation of the surface. Depressions, or level spots hemmed in by ledges of rock or elevations, even slight ones, generally retain water near the surface. The ridges which enclose the basin may be at some distance, but the effect will be the same. The upper part of hill slopes, also, contrary to the common notion, is very apt to be wet and springy, while rocky regions seldom furnish dry cellars unless unusual precautions are taken.

Such precautions should, however, be taken wherever there is occasion for them, since a house with a wet cellar is, to speak briefly, unfit for habitation. The main principle to be borne in mind in ordinary soils is that the ground-water stands nearly at a level (Fig. 82), varying, like the tides, with the season; and that the smaller elevations in gravelly soil form islands in the **Ground-Water.**

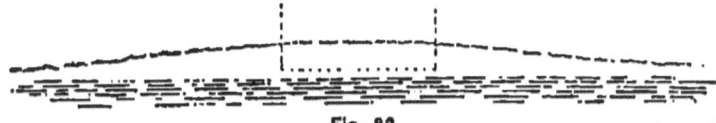

Fig. 82.

subterranean lake, upon which a house may be built with perfect safety, while another, a few rods distant, would have its basement perpetually steeped in moisture. The existence of ground-water a foot or two below the cellar bottom is of comparatively little consequence, provided that this is well concreted, and that the water never rises any higher.

More important elevations generally contain a substructure of rock, above which the rain collects, and flows downward, coming to the surface at intervals in the form of springs. As the soil is always thinner near the top of the hill, such springs are more common there, and are the more annoying because their existence often cannot be detected until the progress of the excavation brings them to light. These streams can, however, in ordinary soils be intercepted without much difficulty, and their current turned harmlessly toward one side; but it is otherwise with cellars cut in rock. These are almost invariably infested with small veins of water, which run along the seams of the ledge and collect in the basin formed by the excavation, soaking through the house walls, and saturating the concrete floors; and the extra cost of intercepting such veins and of cutting a channel to convey the water to a proper outfall is usually very serious, so that locations of this kind should be avoided; or if that is impossible, the level of the cellar bottom should be set so high that little or no excavation in the rock will be needed, and the downward course of the water will not be interrupted. This may always be done, without detriment to **Cellars on Hills.** **Rocky Soil.**

the appearance of the building, by means of suitable grading or terracing.

Clay. Clayey soils are also unfavorable. Being impervious, they retain the water which may settle into the new excavation just outside the cellar walls until it finds an escape for itself, very probably into the building. Moreover, they expand greatly in wet seasons, or in frosty weather, to contract again in summer; while the tenacity with which frozen clay clings to stone or brick work often causes the dislocation or derangement of cellar walls and piers.

Most house-lots in the country or suburban towns offer at least some choice of location, and as a few inches difference in the level of the cellar bottom, or a few feet difference in the distance of the building from a ledge, may be quite sufficient, without special precautions, to determine the wetness or dryness of the cellar, and therefore the healthfulness or unhealthfulness of the house, a judicious study of the site is of great importance.

Where the limits or the character of the plot admit of no choice, it will in very many cases be necessary to incur extra and unforeseen expense in draining the excavation thoroughly. Some of the expedients to be adopted have been described in the first part, and others will be mentioned below; so that with the help of these hints the architect or private owner will be able to deal with any difficulties which he is likely to meet.

Description of House. The house whose construction we have to follow is situated on the side of a rather steep hill, sloping toward the north. The lot on which the house stands comprises about six acres, and the avenue leading to the building passes for some distance through a cutting, seven or eight feet deep. Rocks appear above the surface in various portions of the lot, and several springs ooze through the sides of the cutting made for the driveway, indicating that the ledge is not far beneath. The house is to be of modest dimensions, comprising a parlor, dining-room, hall, kitchen, staircase-hall and back staircase on the ground floor, and four chambers, dressing-room and bath-room on the second story, with four finished attics above. The cellar contains a laundry, furnace-room, vegetable-cellar, and open place for storage, and a servants' water-closet. The hall is finished in oak, with oak floor; the principal staircase is of cherry, with mahogany posts and rail; the parlor is finished in maple, with pine floor, bordered with maple and cherry parquetry work; and the dining-room is in ash. The kitchen and laundry are finished in hard pine, with floors of the same.

All the rooms in the second story are in whitewood, except the bath-room, which, in order to give a pleasant liveliness of effect, is finished with alternate black walnut and maple.

The attics are in pine throughout. The hall and one large attic room are to be finished so as to show the natural color of the wood; the other rooms will be painted.

The plans and specifications having been carefully drawn, the owner, if unacquainted with building matters, should take pains to understand them fully. While still in the architect's hands, he should with his help consider the various points in regard to which he has any particular theory or preference, and by measuring for himself in the houses of his friends the dimensions of doors, windows, stairs, closets, sinks, baths and other details, and comparing them with those shown on his plans, he can form a clear notion of what is intended by them, and satisfy himself that they indicate just what he wishes his house to be. The specifications, also, he should study carefully at home with his family, and may take advantage of the suggestions which they furnish to modify certain points, if he wishes, in a way to please the fancy of the persons who will occupy the various rooms. Such study as this is of much value to the owner, who is thereby often saved from expense in altering work already done in the ordinary way in order to gain some object desirable to himself, but about which he had forgotten to inform the architect; while the architect is always glad to furnish all the assistance in his power, knowing that a little time spent in promoting the thorough comprehension by all parties of the structure indicated in the drawings will enable him to do his work with much more satisfaction to himself as well as to his employer. *The Owner and the Plans.*

In regard to the contracts, the question always arises, whether the whole work shall be entrusted to one man, or two or more separate contracts made; and it is not always easy to answer it. The practice of various localities has much to do with the matter. In some places it is rare for mechanics to make sub-contracts, and therefore each trade must be dealt with *Contractors.*

separately, while in others the best contractors prefer to have the sole control of their buildings, and endeavor to keep the work entirely in their hands.

Where a building must be speedily completed, it is generally easier to attain that object by putting the whole contract into the hands of one man. Two contractors, responsible only to the owner, and jealous or indifferent in regard to each other's interests, always charge each other with the responsibility for the delays which usually occur under such circumstances, and the owner finds it difficult, if not impossible, either to enforce his contract as to time of completion, or to collect indemnity for the delay without doing injustice. Where, however, the time is not restricted to the shortest possible space, most architects will agree that the best results are obtained by making at least four separate agreements; the cellar-work and grading forming the subject of one; the carpenter-work, including painting and glazing, of a second; the brickwork and plastering of a third; and the plumbing of the fourth. It is often desirable to make a fifth agreement for the painting and glazing, but if the carpenter is trustworthy, there is generally some advantage in allowing these to be included in his contract.

By the system of separate contracts better work is usually to be obtained in each branch, and, considering its quality, at a cheaper rate; although speculative builders have ways of making sub-contracts at prices which seem incredibly low to those who are not familiar with the difference between the good and the "jerry" style of work. The best mechanics always prefer to treat directly with the owner; they are in this way sure of their pay, and can therefore afford to work at a lower rate; while the owner saves the percentage of profit which the principal contractor feels himself entitled to charge upon the tenders made to him by his sub-contractors. Whatever mode is adopted, too much care cannot be taken to have the plans and specifications as full and explicit as possible. If these are what they should be, a building so simple as a dwelling-house can be, and generally is, where the owner knows his own mind in regard to the kind of house he wishes, and takes the trouble to see that the plans express it, carried out to completion without any "extras" whatever; generally to the great surprise of the proprietor, who is sure to be informed by volunteer counsellors before he begins operations that his extra bill will inevitably be "at least as large as the contract price;" that he "ought to restrict the architect to half the sum that he intends to spend," and so on. As an example of what is desirable in such documents, forms of specifications and contracts are subjoined, such as have been used, with the necessary variations, for a considerable number of houses, all of which have been finished complete at the contract price, without a dollar of extra charge, except in case of unexpected difficulties of ground, or unless the owner has desired to make alterations as the work went on, or to add to the contract the execution of some parts of the furniture, as mantels, fixed book-shelves, seats, and the like.

Armed with such instruments as these, we enter upon the execution of the work. We have made separate contracts for the cellarwork, the carpentry, the brickwork and plastering, and the plumbing, and have also selected a good furnace, and arranged with the makers to put in the requisite pipes and registers in the best manner when the proper time comes, under a guaranty that the apparatus shall heat a given number of rooms to a temperature of 70° when the thermometer outside stands at 0°, without taking air from the cellar or any other part of the house, and without regard to the direction of the wind.

Occasionally, the heating apparatus is included in the principal contract, but this is most unwise. As with plumbing, the work to be done is so difficult for any one but an expert to understand or criticise, and the difference between good and inferior work is so great, in value, even more than in cost, that it should never be made the interest of any man to get it done as cheaply as possible. Explanations of these points will be given further on. Meanwhile, we hasten to get the cellar under way.

Although no rock appears in the immediate vicinity of the site selected for the house, it is not improbable that it will be found somewhere in the excavation, and the contract with the cellar mason provides a certain price which shall be paid for whatever blasting may be necessary, stipulating at the same time that the rock taken out shall be used in the cellar walls, and an allowance made for it. If there is other stone to be had near at hand, the cost of taking out rock from the cellar is fairly offset by its value for walling material.

The occurrence of ledge in the excavation will usually be accompanied by small springs, but it would be inexpedient to burden the contract with an allowance for draining them away properly, so, if they occur, the operations which they may render necessary will be best treated as extra work.

The contractor and the superintendent stake out the ground together, the latter checking the rectangularity of the lines by measuring the diagonals; and batter-boards are set up; and a bench-mark, showing the level of the top of the cellar wall, is made on one of the batter-boards. The house should always be set high enough to give good cellar windows, with a sufficient fall to the surface of the ground away from the building on all sides. Three feet distance from the highest point of the natural surface in the perimeter of the building to the top of the cellar wall is none too much. This will give two and a half feet of underpinning all around, and insure a light, well-ventilated cellar.

Staking out.

The excavation is required to be eight inches wider on all sides than the outer line of the walls, in order that the latter may be carried up smooth and strong, outside as well as inside. The whole is laid in mortar containing equal parts of lime and cement, and both the outer and inner faces are to be pointed neatly. The trenches are to be dug two feet below the proposed cellar bottom, and eighteen inches of dry stone chips are first put in, before starting the cement wall. All this is expensive, and the cellar will cost in this way at least twice as much as if constructed in the usual country fashion; but it will be more than ten times as good, and nothing short of this fulfils the conditions which modern ideas regard as essential to a wholesome dwelling.

Cellar-Building.

The clear height of the cellar should be eight feet in the smallest house intended for winter occupancy, and more than this in larger mansions, in order to give sufficient height above the furnace to allow of a proper ascent in the tin hot-air pipes, without which the heat cannot be successfully distributed to the various rooms. This, with the two feet additional below the cellar bottom, will make ten feet of stone-work, one and one-half feet of which will be laid dry, and the remainder in cement. Whether the material of the cellar wall shall be brick or stone may depend upon the local custom Hard stone makes the best wall. It is non-absorbent, and as the frozen earth adheres but slightly to its surface, a wall built of it is not subject, as brickwork is, to gradual loosening and decay at the surface of the ground. Where brick must be used, they should be of the hardest quality.

The first operation is the stripping of the surface loam from the whole area covered by the building, and about eight feet additional on all sides, and the stacking of this in some convenient place for use in the subsequent grading. The excavation of the cellar then proceeds in conformity with the lines given by the batter-boards. As the top of the wall is fixed by the bench-mark at 3 feet above the highest point of the ground, and the clear height of the cellar is 8 feet, the main part of the excavation, exclusive of the trenches for the walls, will be nowhere more than $5\frac{1}{4}$ feet below the ground; the extra 3 inches being allowed for the thickness of the concrete.

We have chosen our site well, and are fortunate enough to find no rock in the excavation, and no wet places in the cellar itself, but in the trenches, two feet or so below, the water stands in several places. This is an indication of a moisture in the subsoil, which will increase after spring rains so as to fill the trenches, **Draining** and these will overflow into the cellar unless some **the Soil.** other outlet is provided for the water. The slope of the ground, and still more the comparative shallowness of the cellar, make this a simple matter: all that is necessary is to continue the trench beyond the line of the house, giving it a slight fall to some point where the descending ground will allow it to reach the surface.

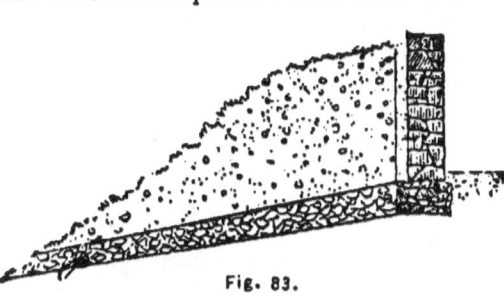

Fig. 83.

The trench may be filled nearly to the top with loose stones or broken brick, and then covered with straw and loam, or a pipe-drain may be laid through it (Fig. 83). All the water that collects under the walls will then be immediately drained away, and no matter what may be the level of the ground-water outside the house, the trench with its outlet forms a barrier which will prevent the moisture from ever making its appearance above the cellar floor (Fig. 84). It is true that in some clayey grounds, especially if traversed by veins of sand, water may rise through the cellar bottom in some places, but such soils are rare, and moisture so introduced can easily be carried off by small "French drains" of broken stone, or lines of agricultural tiles, leading from the wet spot to the main drain under the walls, or, still better, extending from wall to wall across the place to be drained.

Where the ground is very soft or sandy, the outlet drain should be laid at a gentle pitch, in order that the current toward it may not be swift enough to scour out the soil beneath the foundations and cause settlement. Some architects, for fear of this, prefer to make

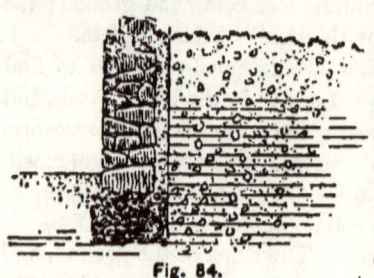

Fig. 84.

stone or tile drains entirely outside the walls (Fig. 85, or Fig. 86), but this is somewhat more expensive, as the excavation must be made proportionally larger, while the former method is more effectual in keeping the cellar walls and floor dry, and if carefully carried out should be no more liable to cause settlements than the other.

If small stones are used under the walls, they should be compacted with a rammer, and thus form an incompressible mass.

The building of the masonry upon this foundation is a simple matter, but must be sharply watched, for in no detail of construction is the common practice so vicious as in the laying of cellar walls. In

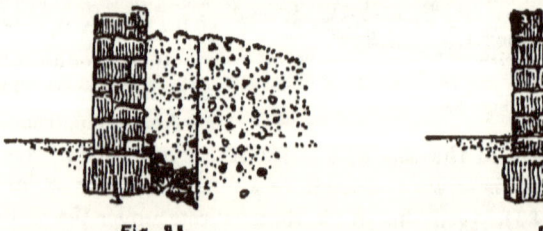

Fig. 85.   Fig. 86.

stone districts, the majority of houses stand upon basements built with blocks of the most irregular shapes, laid "dry," that is, with out mortar, and depending partly for support upon the earth outside of them. The smoothest face of the stones is inside, while the outside presents a ragged, bristling mass of projections; and to improve the visible face, the crevices between the stones are "chinked" with small chips, and the joints are "pointed" by rubbing mortar over them with the point of a trowel. A coat of whitewash completes a work which, while it is new, presents on the inside precisely the same aspect as the best sort of wall. But a brief period only is needed to make its defects manifest. Streams of water, after heavy rains, pour through the loose structure, followed by rats, which burrow down next the outside of the walls until they find a wide joint from which they can easily push out the pointing mortar and obtain access to the interior; and little by little the earth washes into the crevices between the bank and the stone-work, until the latter yields to the pressure, presenting the characteristic inward convexity of country cellar-walls.

Every feature of these constructions must be avoided. In place of a dry wall, furnished only with a miserable pointing of mortar on the inside, the whole thickness must be solidly filled with cement; in place of a rough outside surface, the exterior must be the smoothest face, that water may not collect upon projections and be conducted into the wall; instead of leaning against the bank, the masonry must stand at least eight inches away from it, and the intervening space must be filled with porous gravel or sand, in order that the subterranean water-courses may be intercepted and conducted away harmlessly into the drain beneath; while all the details of bonding and proper jointing of the masonry should be as carefully attended to as in the case of a wall above ground. Further details on these points have been given in Chapter I., and need not be repeated here.

Without assiduous watching, most country masons lapse continually into the wretched workmanship which has become habitual to them: long stones, which they will not take the trouble to break, are set in the wall with a fair inner face, but with long "tails" projecting from the exterior into the bank, whereas the reverse would be the preferable way; and short stones are similarly set, leaving a cavity on the outside to gather water and conduct it into the wall. As the earth is usually filled in behind the stone-work as fast as it is laid, such faults are not generally detected until heavy rains reveal them, too late to apply a remedy; but something may be ascertained by using the steel rod spoken of in Chapter I., while the mortar and the filling outside are yet soft.

It is essential that the material next the wall should be permanently porous. If sand or gravel cannot be had, unsifted coal-ashes form a good substitute, and broken bricks, stone or slate chips may be used.

Gravel outside of Wall.

The underpinning, or portion of the wall above ground, is very commonly made different from the rest, in order to obtain a smoother face. Long slabs of split granite or freestone are often used, and for cheaper work an eight-inch brick wall is sometimes built on top of the stone-work, from the grade line upward. Of these, the granite underpinning is much the best: sandstone and brick absorb moisture from the ground, as well as from snow lying against them in winter, and communicate it to the interior, besides being themselves subject to exfoliation and decay at the ground line.

Underpinning.

Independent of appearance, the best construction is to carry the cellar wall of the full thickness to the very top, trusting to careful "drawing" of the joints for giving the exposed portions a satisfactory finish. Care should be taken that the wall is not thinned at the top (Fig. 87), as is very commonly done in country work, to give opportunity for the vicious form of floor framing by which the joists are set flush with the upper surface of the sill, their lower portion hanging down inside. Whatever the thickness,— twelve inches for brick, sixteen, eighteen, or twenty for stone, according to the character of the material,— it should continue to the under side of the sill, leaving only the necessary places for inserting the girders, and levelling off the top carefully. Frames of basement windows should, if possible, be built into the wall. The method of spiking them to the under side of the sill, leaving large, irregular holes in the stone-work for their reception, is objectionable, as the subsequent filling up is apt to be less solid than the surrounding masonry.

Fig. 87.

While the wall is in process of construction, the framing of the timber, that is, the cutting into lengths, fitting and mortising, will have been going on. Usually this is done upon the ground, from material selected in some neighboring yard, as near the required dimensions as may be; but it is not uncommon for contractors to procure an "ordered frame," by sending framing plans and elevations, with a proper specification, to some saw-mill in the timber region, where the pieces are cut from the logs of the exact sizes required. One or two establishments do more than this, and ship the frame ready mortised and fitted for putting together, including when desired the boarding, shingles, clapboards, doors, windows, and other simple wood-work. There is an economy in the use of such frames, as waste is avoided, and more perfect timber is obtained, but the yard timber is usually better seasoned, and some contractors think that the time spent in overhauling after delivery the innumerable pieces of an ordered frame, except of the simplest kind, in order to select the sticks that are needed, quite offsets the waste and extra expense of framing from yard timber in the usual way.

Framing.

The inspection of the rough lumber is not difficult. White pine, spruce, and hemlock are the woods most commonly used. In the far West, cottonwood is sometimes employed, and redwood is the ordinary framing timber of the Pacific coast. Of these, redwood is much the best, being strong, straight-grained, obtainable in any dimensions, and less subject to shrinkage or movement than any other. White pine is next in value, for similar reasons. Cottonwood is soft, and shrinks very much. Spruce resembles pine, but shrinks more, and is apt to warp and twist with great force during the drying process, and to "check," or crack open near the middle of the stick. The dimensions run rather small, compared with white pine or the enormous sizes of redwood lumber, but pieces up to 12" x 12", and 25 or 30 feet long are always to be had, and wooden dwelling-houses rarely require anything more. Hemlock is a harder wood than either spruce or pine, and can be had in large sizes, but its strength is injured by the want of adhesion between its annual rings, which disposes the timber to crack very badly in drying. The hemlock trees are very tall, and their swaying in the forest often "shakes" or separates the rings of the heart-wood, so that when sawn into scantlings or boards the in-

Inspection of Timber.

terior is little better than a mass of splinters. Hard pine, or, as it is sometimes called, Georgia pine, is much used in city buildings, where its great stiffness is of advantage in enabling floors of wide span to be covered without employing timbers of inconvenient size, but it is rarely necessary in country houses to incur the additional expense of using it for this purpose, although it is employed in other ways. It is a very good timber, though shrinking considerably in drying. The principal point in examining the lumber will be to ascertain whether the sizes are according to the specifications, for which a few measurements will suffice. It should be borne in mind that only green lumber will show the full dimensions; and seasoning reduces them somewhat. A white-pine plank originally cut 12 inches wide will when dry measure about $11\frac{3}{4}$ inches; spruce, hemlock, and hard pine somewhat less; cottonwood will shrink a whole inch in the same width, while redwood scarcely shows any change; so that suitable allowances should be made for every case. Crooked and "wancy" pieces (see Figs. 46, 47, page 65) should be condemned, as well as those affected with serious shakes. Longitudinal cracks in the middle of a *thick* piece of spruce need not condemn it: they are almost inevitable if the timber is dry, and do not detract much from its strength; but if they occur in a thin piece, as a floor joist, and extend entirely through it, they constitute serious defects. Sticks of which portions, especially at the ends, appear livid and friable should be totally condemned. They are infected with dry-rot, and will communicate the infection to others. Amputation of the diseased part is not sufficient, for the threads of the fungus may extend a long distance into the sound part of the wood.

The first timber set in place is the sill, and this should have a thick bed of soft cement mortar prepared for it by the mason, and be hammered firmly down into it. This closes up the crevice between the top of the stone-work and the timber, through which, in badly built houses, much cold air finds its way into the hollow floors. For additional protection, after the adjustment which is generally necessary to get the sill into position is over, it is best to point up with similar mortar along its outer edge (Fig. 88). The inside might be similarly treated, but this would interfere with the still more effectual process of lining with bricks and mortar, as described below. It is now

Bedding Sill in Mortar.

usual to make the sill of pine or spruce, like the other portions of the frame. Our forefathers, who had no furnaces, set their dwellings very low, and banked them up with earth in winter so as to cover the lower portions of the wood-work, in order to keep the cold air from the cellar and the floors; and they found chestnut or cedar to be the only material which would resist dry-rot under such circumstances. We, however, prefer high, light cellars, with cement walls, kept dry by furnaces, and little dampness reaches the sills, but there is, as in all cases where wood comes in contact with masonry, a possibility of its being affected by moisture in the pores of the stone and mortar, so that it is best to provide a repellent coating by painting the sill on the under side. The other sides should be left untouched, so as not to impede the drying action of the air. The ends of girders also, where they enter the wall, may be similarly painted.

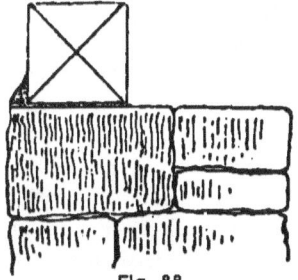

Fig. 88.

In most cases, the sill is the subject of more notching, mortising, and cutting than any other timber in the building, and must be of sufficient dimensions to allow for this. Six by six inches is a common size, or six by eight, where the basement openings are large, or there is danger of decay affecting the under side. The angles are halved together (Fig. 89), and pinned or strongly spiked. Many builders secure the sill to the foundation-walls by means of vertical bolts, about two feet long, built for the greater part of their length into the walls, at intervals of eight or ten feet. Corresponding holes are bored in the sill, and this is slipped over the bolts, and secured by nuts and washers. With light structures in exposed situations this forms a valuable safeguard against storms.

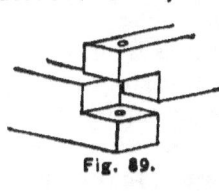

Fig. 89.

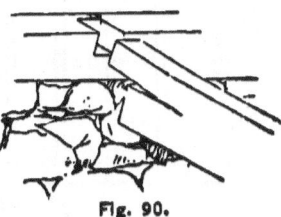

Fig. 90.

Into the sill the floor beams are framed in various ways. A very common and bad mode is to notch it some three inches deep, cutting a corresponding tenon on the upper corner of the beam, so that the upper surface of this, when in place, is flush with the top of the sill,

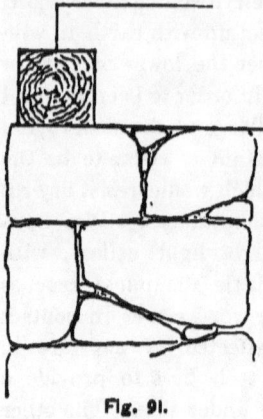

Fig. 91.

the lower part projecting below. This necessitates the thinning off of the upper part of the wall, which would otherwise come in the way of the beams, while the tenon, from which hangs the whole weight of the beam and its load, often splits off (Fig. 90). The beams should, instead of this, be cut so as to bring their lower edges flush with the bottom of the sill (Fig. 91), with a notch, perhaps two inches deep, to hold them in place. Then the tenon will be deep enough to hold safely, and the wall can be made of the full thickness to the very top.

Mortises must also be made for the corner posts, and for those to be set at the intersection of interior partitions with the outside walls. Usually, each of the "filling-in" studs (Fig. 92) has also its appropriate mortise, even in "balloon" framing, but occasionally a cheap builder contents himself with simply setting the end of the stud down on the sill, and securing it by nails driven diagonally through the foot. Of course, the position of all the mortises is taken from the framing plans and elevations; which cannot be too carefully made, or the execution of them too closely watched, as a window, a chimney-opening, or a stairway, once framed in the wrong place, cannot be altered subsequently without injury to

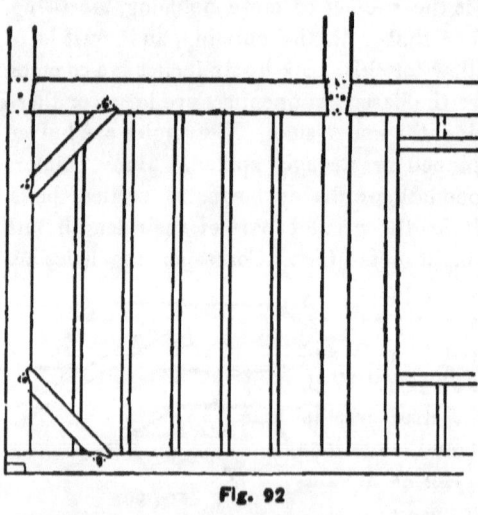

Fig. 92

the solidity of the building. Workmen are very careless about such matters; we have known a foreman to use a framing plan traced on transparent cloth wrong side up, and regardless of the careful lettering and figures on the right side, to cut the mortises for the whole side of a house by scaling with a foot-rule the dimensions as seen inverted through the waxed linen; and it is rare, even with the most carefully drawn plans before them, for framers to complete the mortising of a sill without gross mistakes. The young architect or superintendent should therefore, as soon as the sills are set in place, verify every measurement. The figures on the plans for a wooden building should always give the distances of the *centres* of openings from the corners of the building and from each other; then the middle point between the mortises made in the sill for the studs which form the jambs of the opening can be readily found, and compared directly with the figure. If the plans are figured, as is sometimes done, in the way appropriate to stone or brick buildings, by giving the distances to the *jambs* of the openings, the verification is much more difficult, since the rough studs are always set two inches or more wider on each side than the finished opening is intended to be, to allow for the weights and lines in the case of windows, and in doors to admit of a little play in setting the frames.

After all the mortises for the studs have been examined, and such corrections as are found necessary made on the spot under the superintendent's eye, those intended for the floor timbers should undergo an equally rigid inspection. The openings for staircases and chimneys will almost always be found misplaced, or made either too large or too small. The latter is much the worse fault: it is possible to fill up an excess of space, but too small an opening, which can only be made available by cutting away and weakening the trimmer-beams, or by constricting the flues, is a serious misfortune. When these tests have been thoroughly applied, the beams may be set in place. In most cases the floor rests partly on girders, which are larger sticks, generally from 6" x 10" to 8" x 12" for the light strains of country houses, running through the cellar under the "fore-and-aft" partitions, or those which carry the floors above, and supported by brick piers in the cellar.

Occasionally, brick walls, eight inches thick, and pierced with arches for communication, take the place of the girders, but without any material advantage, unless they are carried up to the under side of the floor boards, in which case they serve to keep the floor warm and diminish the danger from fire, by intercepting the spaces between the beams.

If girders are used, special attention should be given to contriving the framing of the beams into girders and sills so that the shrinkage shall be the same at each end. This point is almost always neglected, to the detriment of the work, which begins, a year or two after the completion of the building, to undergo settlements and deformations, which instead of being inevitable, as is usually supposed, might easily have been avoided by a little care at the commencement. If, for example (Fig. 93), a ten-inch beam is framed at one end by a three-inch tenon into a six-inch sill, all flush on top, and at the other end is "sized" down one inch, without mortising, upon a 6 x 10 girder, the total height of shrinkable timber between the floor boards and the unyielding masonry will be, at the sill end, 6 inches, and at the other end 10 + 9 = 19 inches.

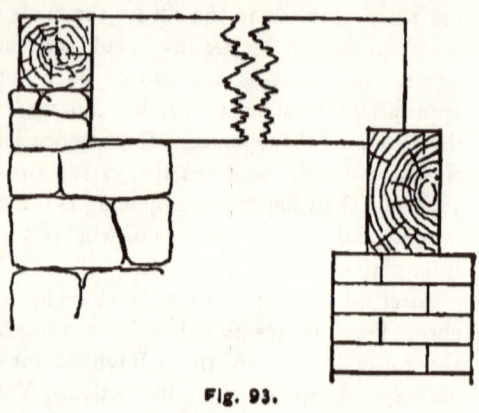

Fig. 93.

Now a six-inch timber will shrink perhaps ¼ of an inch in drying, while at the same rate, 19 inches will shrink some ¾ of an inch, and after a year's seasoning the inner ends of the beams will thus be half an inch lower than the outer, and the floor to the same extent out of level, cracking the plastering of the walls above, distorting the door-frames, so that the doors no longer fit their places, and causing ugly dislocations in base-boards and wainscotings.

To avoid these evils, an equal height of timber should be left at each end between the flooring boards and the masonry. If the proper mode of framing into the sill is adopted (see Fig. 91), a ten-inch beam will have a six-inch tenon resting upon the bottom of the notch, with four inches of the wood of the sill between it and the cellar wall; in all, ten inches of wood. We need, therefore, ten inches of wood, and no more, between the brick piers and the floor boards at the other end. But if the girder is ten inches high, this will furnish the whole, with none to spare for projection of the beam above it, so the latter must be framed into the girder flush with its top. This is for various reasons the best way of framing into girders. Not only is it advantageous to get rid of their projection below the cellar ceiling, but the circulation between the beams is effectually cut off. If the girder is of ample strength, it may be notched, say five inches deep, to receive a

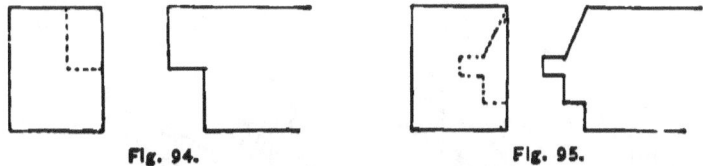

Fig. 94.   Fig. 95.

five-inch tenon on the beam (Fig. 94), but the best mode, preserving most effectually the strength both of girder and beam, is the joint with "tenon and tusk" (Fig. 95) by which the cutting is brought nearer the neutral axis of the girder, while the tusk tenon allows the joint to be bored and pinned from above.

As it would be difficult to lay heavy timbers on a row of isolated piers without overturning them, the girders are generally held in place by shores set beneath them until the floor beams are all on, or sometimes even longer; and the piers are then built up beneath them. These should never be less than 12" x 12", of hard brick, laid in cement mortar. Piers 8" x 8", as often seen, soon bend, while those of soft or "pier" brick are liable to be worn and kicked away at the foot. The proper spacing for piers depends on the size of the girders, and the load upon them, but should not be over 8 feet, for fear of deflection, even with strong timbers.

In the Eastern states, the next step always is to lay an under-floor of planed hemlock or spruce boards, over which the men move freely, while it forms a roof to the cellar, which can immediately be used for storage of tools and materials. **Under-Floor.**
Whether this is done or not, after the floor is made practicable for the

passage of workmen across it the large posts should be set up at the angles, and at the intersection of interior partitions with the outside walls. These should be 4" x 8", at least, even in a "balloon" frame, not so much for strength as to give good nailings for the angles of interior furrings, wainscot and base-boards (Fig. 96).

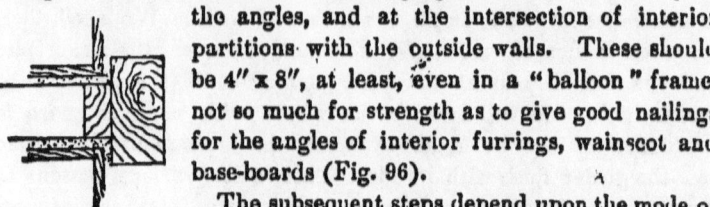

Fig. 96.

The subsequent steps depend upon the mode of construction adopted,—whether a "balloon" or a "braced" frame is specified; and it is of importance to the young

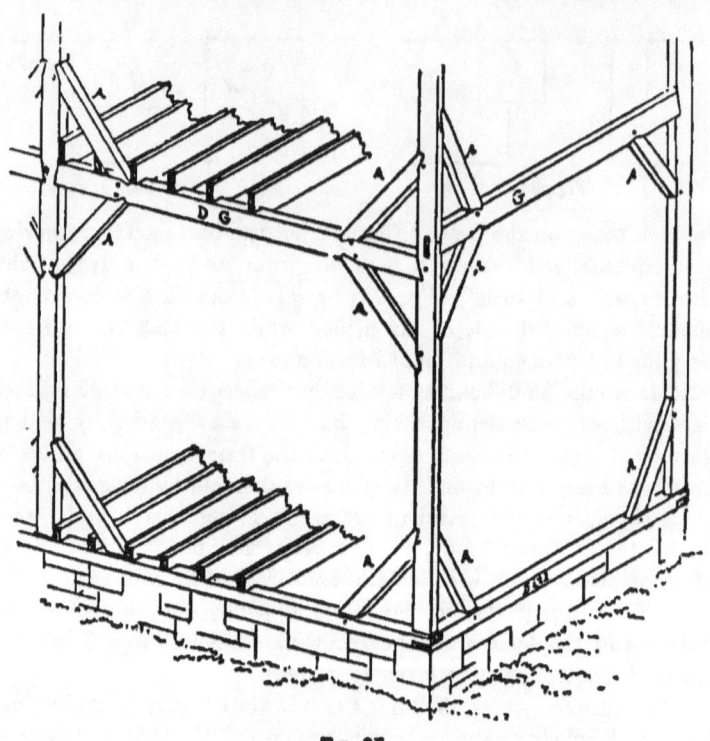

Fig. 97.

architect to understand thoroughly the nature and advantages of each.

Supposing a "braced" or "old-fashioned" frame to be called for, the next step after setting the corner posts firmly into their mortises will be to secure them in their upright position by means of the braces (Fig. 97), which have been pre-**Braced Frame.** viously fitted to mortises cut in the sill and post, and on being inserted and a hard-wood pin or trenail driven through the hole, hold the post fast. The shape of the tenons is shown in Figure 98.

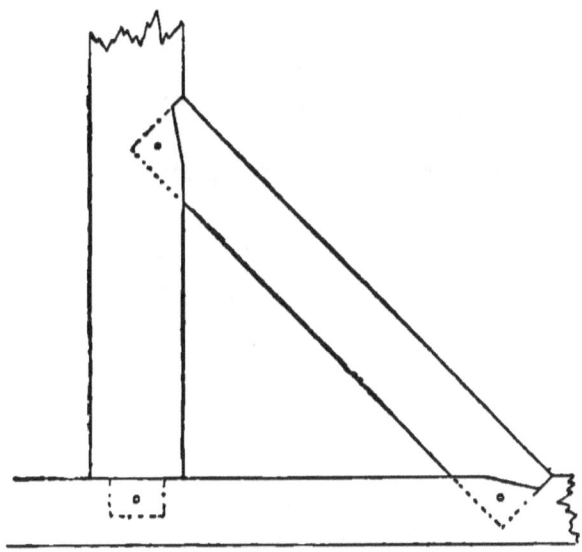

Fig. 98

While setting and bracing the corner posts, which always extend the whole height from the sill to the plate supporting the roof, the girts (G and D G, Fig. 97), or horizontal timbers which tie the frame at each floor-level, must be put in place. In a simple rectangular frame, like the one shown on the drawing, two **Girts.** of these will run parallel to the floor beams, and are for convenience generally set at the same level with them. The other two, marked D G in the figure, cross the ends of the beams, and are utilized to support them. If it were possible to continue the girts all around at the same level, the beams might with advantage be framed with tenon and tusk into those which run transversely to them, in the same manner as into the girders in the first floor, but as the tenons by which the girts are framed into the posts would, if these were set at the same level, intersect and cut each other off, it is necessary to place one pair entirely below the others. The latter are called the "dropped girts," and are

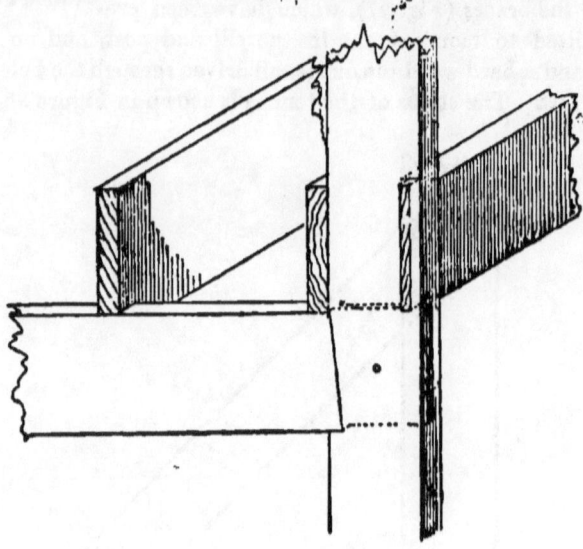

Fig. 99.

generally arranged so that the floor-beams may be notched or "sized" one or two inches down upon them. (Fig. 99.)

The proper joint for posts and girts is shown in Figure 100. After setting these and pinning them securely, a second set of braces should be put in to tie the angle between them and the posts, as shown in Figure 97, using the same form of mortises and tenons as before. A third set is then placed in the angles between the posts and the upper side of the girts. All these pieces are accurately cut and fitted before any of the work is set up, so that when inserted in their mortises, and the pins driven home, the posts are drawn into their proper position, exactly at right angles with the girts, notwithstanding the bending or twisting which the heat of the sun often causes in them when first set up;

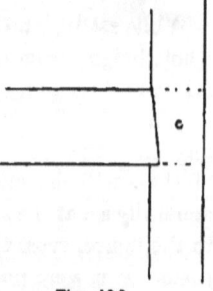

Fig. 100

and are thus prepared to receive the plate, which will not fit unless the posts are accurately parallel. The angles of the plates are halved together and mortised entirely through, so as to receive a long tenon left on top of the posts; and to facilitate this, as well as to give a certain lateral stiffness to re- **Plate.** sist the thrust of the rafters, it is common to make the plate 4" x 6" or 4" x 8", and lay it flatways. If the roof is to have gables, a somewhat simple construction is used, the plate which receives the rafter first being framed on top of the posts, while the gable is supported by a sort of dropped girt, tenoned into the posts below the plate proper. The angles of posts and plate may finally be braced, making a strong and rigid skeleton, as shown in Figure 97, upon which the rafters can be set at leisure, and the framework of the side completed with "filling-in" studs, set at a suitable distance for nailing the laths inside, and the clapboards or other covering outside. (Fig. 101.)

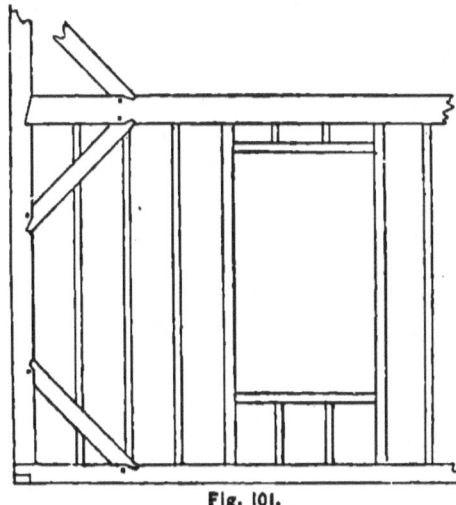

Fig. 101.

The principle of the balloon frame is totally different, and although it may be, as is claimed for it, more philosophical, it is far inferior to the braced frame in many important **Balloon Frames.** respects.

If balloon framing were specified for our building, the next step after setting the corner posts would be to secure them temporarily in place by means of "stay-laths," or pieces of board nailed diagonally to post and sill. The "filling-in" studs would then be set all around the building, each stud in this system extending the whole height from sill to plate. The best carpenters mortise the feet of the studs into the sill, but this is frequently omitted, nails being simply driven diagonally through. No attempt is made to cut the pieces to the right length, and their upper ends present for a time an appearance like Fig. 102. To straighten them boards are temporarily nailed on outside, and more "stay-laths" brace the studs inward to the floor. As soon as the building is so far advanced as to admit of climbing safely to the top of the studs, a line is marked with a chalked string at the proper height for the underside of the plate, and the studs are cut off at that level. If any prove too short, an additional piece is set on top, and "fished" by nailing a bit of board on each side. (Fig. 103.) When all are brought to the line, a 2" x 4" or 2" x 6" timber of random length is laid on top, and spikes driven through it into the top of each stud. Other similar sticks are laid in the same manner until the circuit of the building is completed, when a second row is laid on top of the first, breaking joint with them, and overlapping at the angles. (Fig. 104.)

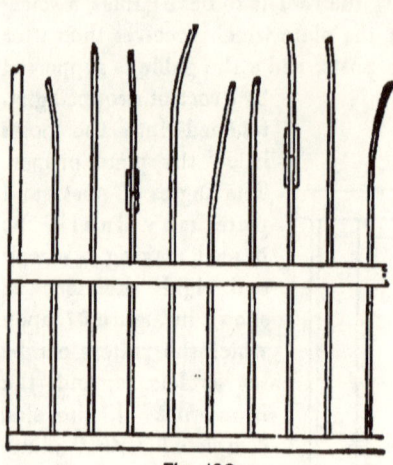

Fig. 102.   Fig. 103.

This operation brings the studs to a vertical and parallel position, but provision is needed for supporting the floor-beams. As before, no notching or mortising is done before setting up the frame, but when all the studs are in place, the chalked string is again brought into requisition to mark upon them two lines, 4 inches apart,

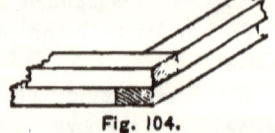

Fig. 104.

at such a height that the upper one will be an inch above the proposed level of the underside of the floor beams. Each stud is then notched one inch deep between the lines, and a "ledger-board" or "false girt," consisting of a strip of board an inch thick and 4 inches wide, is inserted and nailed in place. (Fig. 105).

This gives a support which is strong enough for the work required of it, but excessively slender in appearance, and liable, if fire should get into the spaces between the studs, to be quickly burned off, perhaps letting the floors fall.

The bracing of the angles, which forms an important part of the old-fashioned framing,

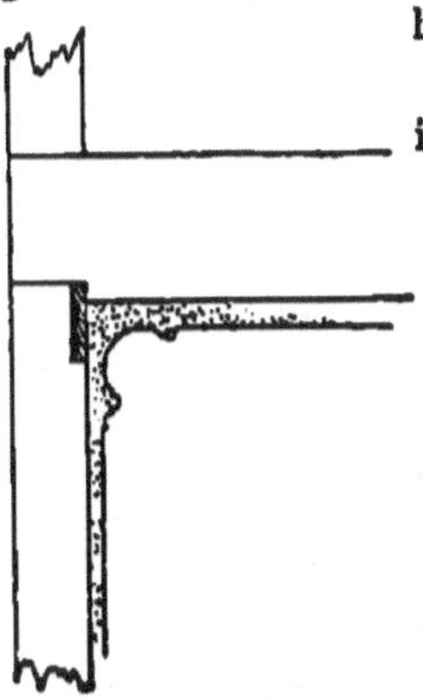

Fig. 105. 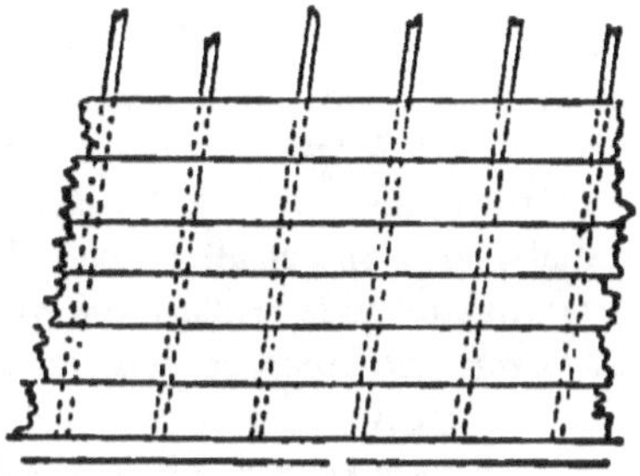   Fig. 106.

is entirely omitted by most carpenters in setting balloon frames, so that nothing but the resistance of a few nails prevents the building, as the outside boarding shrinks, from leaning gradually in one direction or another, according to the prevailing winds. (Fig. 106).

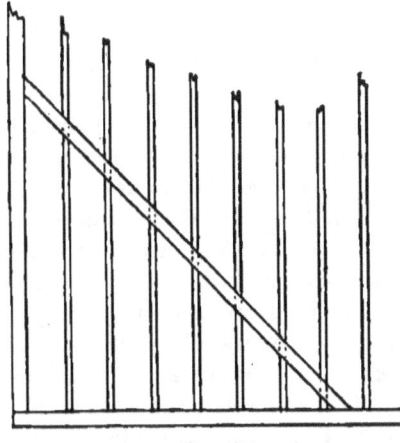

Fig. 107.

This may, however, be prevented by what is called long bracing (Fig. 107). consisting of stout strips set flatways their whole depth into notches cut in the studs to receive them. The notches are made either on the outside or inside of the studs, but better bracing can be had by placing them outside. A spike is driven through the braces into each stud, and the angles are thus very strongly tied, but at the expense of strength in the vertical studding.

Long Bracing.

By these very different modes the balloon and the braced fram ing accomplish the same result, the construction of a timber skeleton bounded by sill, plate and corner posts, and included, with all bracings, tenons and fastenings, between two planes, 4 or 5 inches apart, as the case may be, so that laths can be nailed uniformly all over the inner surface, and boards over the outside. All the subsequent steps, until the completion of the building, are the same for either mode of construction. The design of the edifice is, however, somewhat dependent upon the mode employed. In both cases, heavy studs are used to form the sides of window and door openings, in order to give rigid support to the casings, and as in balloon framing all the studs, large and small, extend the whole height of the building, it is quite desirable to place the windows in the different stories vertically over each other, so that one pair of "window studs" may serve for two openings; and any variation from this direct superposition involves expense or weakening of the portion of the structure involved. With a braced frame, the main posts only extend to the plate, all the filling-in studs, large and small, terminating at the girts; so that the studding of one story is completely independent of that above or below; an important consideration where any picturesque irregularity of fenestration is to be attempted.

As soon as the studding of the walls is in place, the outside boarding is begun. Hemlock or inferior spruce is used for this purpose, and no great care is taken to lay the joints close, but the boards must be mill-planed on one side to reduce them to an even thickness, or the subsequent shingling or clapboards will not lie evenly. While this is going on, the beams of the second floor are to be set. As it rarely happens that the beams extend in one span from the girts in one wall to those opposite, an intermediate support must be provided for them, consisting generally of the head of some interior partition, on which their inner ends are notched or "sized" down, just as their outer ends are upon the dropped girts. As the 3" x 3" or 3" x 4" piece which forms the partition-head is capable of withstanding a considerable cross-strain, it is unnecessary to set all the studs under it at first, and isolated ones are usually put in, some 3 or 4 feet apart, in order to allow free passing between them. The second-floor beams can then be immediately laid in place, and another partition in the story above, set in the same way, serves for supporting the third floor.

Boarding.

The notching or "sizing" of all beams upon their horizontal supports is made necessary by their inequality in size. Ordinary timbers often vary one-fourth to one-half an inch from their specified dimensions, but by notching them to a uniform distance *from the top* they will, when laid in place, have their upper sides level ready to receive the floor. (Fig. 108.) The undersides will be uneven, but the subsequent cross-furring will conceal this.

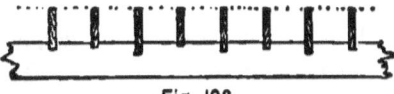

Fig. 108.

How to Provide against Unequal Settlement.

It is very common, but not judicious, to set the studs of all interior partitions either on the under-floor or on a horizontal piece resting upon the beams. If the inner ends of a tier of floor-beams are supported, as is usually the case, by a partition extending from the basement, these ends will be subject to a settlement equivalent to the sum of the shrinkage of all the horizontal pieces interposed between the underside of the beams in question and the immovable supports at the bottom of the partition; in this instance, the brick piers in the basement. If the partition studs stand on the beams, with a 2" x 4' "sole" interposed, a partition extending from the cellar piers to the floor of the third story will be interrupted by the basement girder, first-story beams, sole of partition, cap of the same, second-story beams, and another sole and cap, in all, from 30 to 40 inches of horizontal timber, the shrinkage of which in such a position would be from one to two inches. The outer ends of the same third-story beams will rest upon the framing of the outside wall, which would with a balloon frame, in which the studs are continuous, present only a six-inch sill and one four-inch ledger-board of shrinkable timber between them and the immovable basement wall. The shrinkage of this 10 inches of horizontal wood would amount to less than half an inch, so that when the wood-work became fully dried, which in our furnace-heated houses is in a year or two, the inner ends of the third-story beams would be an inch to an inch and a half below their outer ends. Such a difference in level is quite sufficient to cause cracking of the plastered walls in the second and third stories, and to distort the openings in the cross partitions (Fig. 109), so as to make the doors fit badly, and "bind," or require to be trimmed, or the hinges "set up" to adapt them to the altered shape.

These phenomena, which every one has observed in city as well as country houses, depend solely upon the unequal shrinkage and consequent settlement of the two vertical structures by which the opposite ends of the beams are supported, and can be avoided by any device which shall make the settlement the same at both ends. In wooden houses this may be approximately accomplished by setting the studs,— not on the floor or on the beams, but *between* the latter and on the same support, so that the beams cease to form a part of the vertical frame. The studs of the first-story partitions will then stand *on the girders*, extending thence to the underside of the second-story beams which rest upon their cap. The studs of the second-story partitions again, instead of standing on the floor, or on a sole-piece, will extend down between the beams to the cap of the first-story partition (Fig. 110). By this arrangement there will be, supposing the height of the girder to be 10 inches, and that of the partition caps 3 inches, 16 inches only of shrinkable wood in the partitions between the basement piers and the underside of the third-story beams, and the difference of level after drying between their inner and their outer ends, supposing these to be supported by balloon frame, would be about ⅛ of an inch instead of three or four times that amount, as in the case previously described.

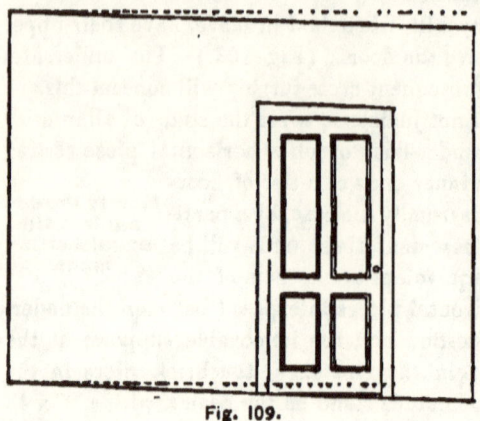

Fig. 109.

Settlement in Balloon Frames.

It is desirable to avoid even this inequality if possible, and as it is evidently impracticable to diminish the amount of horizontal timber, and consequent settlement, in the interior partitions, it will be advantageous to *increase* that in the outer walls. As the height of the sill is fixed, the aggregate shrinkage can only be increased by adding to the width of the horizontal timbers on which the upper beams rest. With a balloon frame nothing can be gained in this way, since the ledger-board must be nailed to the studs, and the free portion between the nails and the upper edge of the board is alone capable of affecting the beams by its shrinkage; and with such frames an unequal settlement is practically inevitable. The braced frame, however, sustains the beams upon a wide girt, resting on the corner posts by its lower edge, so that the whole effect of the shrinkage tends to make the upper edge descend, and with it the beams which may rest upon it.

Settlement in Braced Frames.

A braced frame, therefore, with a six-inch sill and a ten-inch dropped-girt, upon which the beams of the third story are "sized down in the same manner as on the partition-heads supporting their other ends, will give sixteen inches in verti-

Fig. 110.

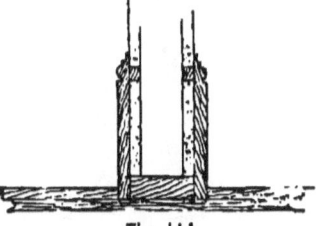
Fig. 111.

cal height of shrinkable timber, and supposing the girders and partition-heads to be as before, the settlement at both ends of the beams will be the same, and the floor will remain perfectly level, the doorframes square, and the plaster probably unbroken, for an indefinite period.

The partitions which extend from the first floor through two or more stories, even though no beams rest upon them, as in the case of those running parallel with the beams, should be set in the same way, the studs in the upper story resting on the cap of the partition below; not for the sake of lessening the shrinkage, which would in this case do no harm, but to relieve the floor-beams from the weight of the partition by making the support continuous from the basement girder upward; and partitions enclosing stairs should be similarly constructed.

Besides these, there will usually be some partitions, especially in the second story, which have no corresponding partition below. These must be supported on the beams. The simplest way of setting them is to lay the sole directly upon the under-flooring, where its position can be accurately marked; and the sole may with advantage he 5¼ or 5⅜ inches wide, so as to project beyond the stud on each side by an amount equal to the thickness of the plastering. In the subsequent finishing this projection will be of great service for keeping the base-boards firmly in their proper position, and for nailing them at the lower edge if required (Fig. 111.) Where the partition runs parallel with the beams, it is common to provide in the framing plans for a timber of extra size, or two timbers spiked together, under it, to give the extra support required. A better way is to set two

Beams under Partitions. beams equidistant from the centre line of the partition, seven or eight inches apart from centres, instead of close together (Fig. 112.) The same strength will be obtained,

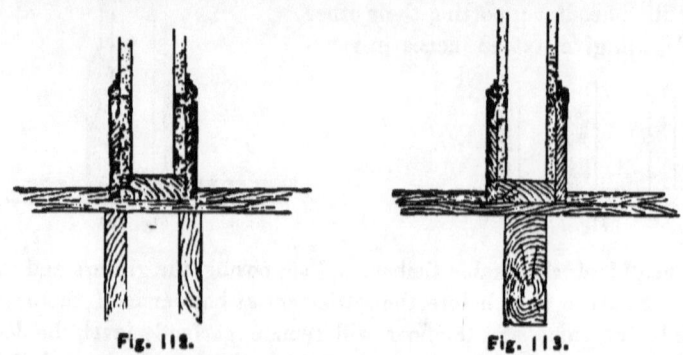

Fig. 112.   Fig. 113.

and there will be opportunity for a solid nailing at the ends of the floor-boards that abut against the partition, which cannot be had by the other method (Fig. 113.) For the same reason, the framing-plans should always show a beam placed close against outer walls and partitions extending from below. A floor where the ends of the boards, for want of these precautions, are secured only to the thin under-flooring soon acquires an uneven and slovenly look.

The first studs of the partitions are usually set, and the floors bridged, before the roof is begun. In the short spans usual in country houses, this construction is a matter of little diffi-

Roofs. culty. Where support is needed, it is generally obtained by carrying up the partitions which extend from the firm foundation in the basement, and heavy trusses and purlins are rarely necessary, the weight being equally distributed over all the rafters, which may be tied with "collars" of plank where required. At the same time, the form of such roofs is often very complex, and the framing-plans should be carefully and clearly drawn. Every ridge, valley and hip must be marked in plain letters, and the lengths of hip, valley, common and jack rafters should be calculated and written on the drawings. Without these precautions, the architect is very likely, during the framing, to find a hip substituted for a valley, or *vice versa*, and not infrequently, either by accident or design, the height of a picturesque roof will be materially lessened without consulting the designer, who does not discover until too late the reason why its appearance in execution is so disappointing.

The covering-in of the building gives the signal for a multitude of minor operations, the principal among which is the construction of the chimneys, which should be commenced at the earliest practicable moment, in order to avoid delay in finishing the roof. The bricks furnished for this work should be rigidly inspected. As the chimneys in frame houses are usually plastered outside as fast as built, in order to lessen the danger of sparks passing through the joints of the masonry among the furrings, the opportunity for using soft, half-burnt bricks without detection is unusually favorable, and the young architect should look sharply to see that none of that kind are allowed to be delivered on the ground. For the purpose of aiding the meaner builders to impose bad materials upon their employers, it is common at the brick-yards to denominate the half-burnt material from the outside of the kilns "chimney brick," "pier brick," or "place brick." The name, however, does not change the quality, and any work containing bricks whose edges can be crumbled by the fingers should be pulled down at once, and rebuilt with better materials. Unless this is done, no reliance can be placed upon the masonry; the piers are liable to be broken away and bend, and chimneys may crack open at any moment after being enclosed by furring.

Chimneys.

Bricks.

We suppose that the position and size of all the openings made in the floors for the passage of the chimneys have been carefully verified long before. If not, this should be done without delay. Masons rarely think of questioning the accuracy of the carpenter's work, and whenever they find an opening framed, they suspend plumb-lines from its four corners and commence laying bricks between them; and to the endless mistakes made by the inferior workmen who are employed in framing they add others of their own. One fertile source of errors is a want of some common understanding in regard to the system of figuring plans. Most architects, unless very experienced, figure all horizontal dimensions in wooden buildings from the nearest surface of the studs; thus, a fire-place in the middle of one side of a room 16' long in the clear would generally be figured as 8' 1" from the inside of the studs to its centre. Nearly all framers, however, measure to the *outside* face of *outside* studding, although interior dimensions are taken to the *nearest* face; and the workman will probably set his trimmer-beams, or lay out his chimney, by measuring the figured distance on a ten-foot pole thrust between the studs against the outside boarding, the point thus falling four inches short of the place intended; and the mistake, if discovered, is very likely to be rectified by shifting the chimney over bodily, and resting it upon the trimmer-beam. It is safest in any case to figure the openings in floors two or three inches wider on the framing-plans than they are actually intended to be. This gives a little lee-way for contingencies, and it is always easy to fill out an excess of room by nailing pieces to the timbers, while the cutting away of beams to gain necessary space should be avoided.

All flue-doors, ash-doors, stove-rings and ventilating registers should be marked on the plans, and inserted as the work goes on. If left for subsequent cutting they are sure to be forgotten. Rings for furnace smoke-pipes should never come within sixteen inches of the cellar ceiling. All flues must be closed at the bottom, and kept separate to the top. Bad workmen often leave them open at the bottom into the ash-pit, or, where two flues run side by side, omit the partition, or "with," in the lower part. In either case the draught is spoiled.

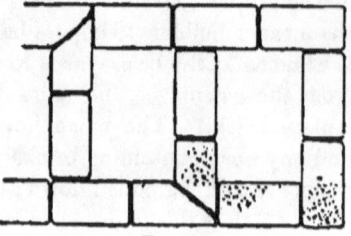

Fig. 114.

Bonding of Chimneys. The withs should be four inches thick, and at least once in every eight courses in ordinary chimneys they should be bonded by two bricks roughly mitred with the stretchers of the walls. (Fig. 114.) Without this precaution, which it is not easy to enforce, the with forms a mere tongue of superposed bricks, standing upright in the rectangular shaft of the chimney, and held in place only by the feeble adhesion of the mortar, so that it not unfrequently loses its balance and bends over, stopping up the adjoining flue. Ties of tin or hoop-iron laid in the joints are sometimes used to sustain the withs, but the other bond is better, particularly in tall chimneys, where a thorough interlocking of the withs with the walls adds very greatly to the strength of the shaft.

Stacks of irregular plan, (Fig. 115), can be better bonded than those of the common form, and are much stronger. If carried up smooth and nearly straight, without twisting or constriction in any part, an 8" x 8" flue is ample for any stove or ordinary hot-air furnace, and is sufficient for an open fireplace of moderate size; but the danger of some obstruction is so great that it is prudent to provide the latter with 8" x 12" flues where practicable.

Flues.

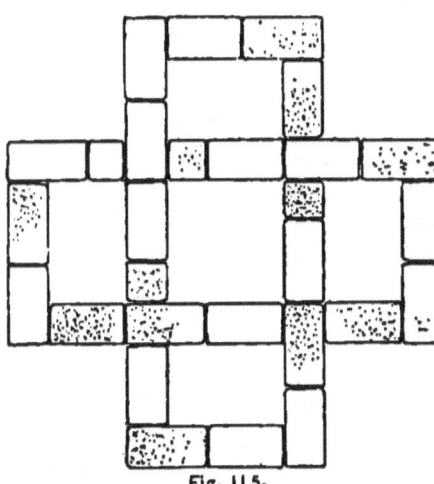

Fig. 115.

Fireplaces are usually roughly formed during the construction of the chimney, to be subsequently lined with soapstone or brick, but if the latter material is to be used, it is better to finish the whole at once and cover it up with boards to prevent injury during the progress of the work. By this method there will be no danger of settlements and open joints between the rough work and the lining, through which sparks may reach the space behind the furrings. Wrought-iron chimney-bars must be used to support the brickwork above each fireplace opening. Two inches by half an inch is the usual size, and two bars should be used. The whole support should be given by the bars, without any assistance from arches, which are liable to spread and split the masonry. The depth and form of fireplaces depend on

Fireplaces.

Chimney-Bars.

the use to which they are to be put. Small hard-coal grates are often set with only a four-inch recess in the masonry, the front of the grate projecting three or four inches beyond the face of the wall, and work very well so if the draught is good, but eight inches is better, and soft coal or wood need at least twelve inches depth. The "splay"

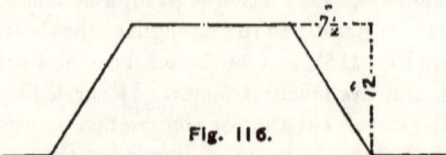

Fig. 116.

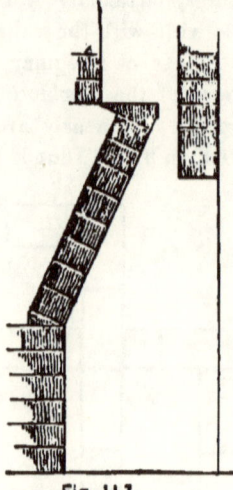

Fig. 117.

or bevel to be given to the sides should conform to that of the grate, if one is to be subsequently inserted, and as they are made with various angles, the choice should be made before the fireplace is begun. If this has not been done, a splay of seven and a half in twelve does very well, and will fit many grates. (Fig. 116.) In vertical section, the back of the fireplace should be built up plumb about six courses, and then inclined forward, making the throat of the chimney about two inches wide (Fig. 117), finishing with a *level* surface, of cut bricks, about six inches above the line of the chimney bars. By this narrowing of the throat the hot gases are concentrated, and the draught much improved, while the level surface at the foot of the flue checks and repels any downward current, instead of deflecting it forward into the room. In laying out the fire-

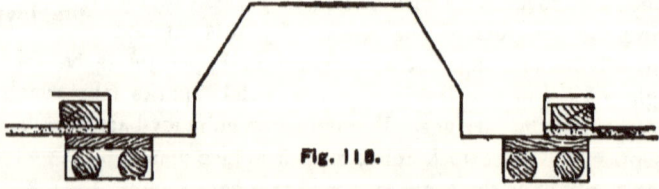

Fig. 118.

places, it should not be forgotten that they must project from the general surface of the chimney at least as far as the line of the plastering. If the chimney is furred with 2" x 4" studs, set flatways, and one inch clear of the masonry, and then lathed and plastered, the plaster surface will be four inches in front of the brickwork (Fig. 118), and if the facings of the fireplace are brought out to this point, a mantel which is flat on the back can be used. Most marble man-

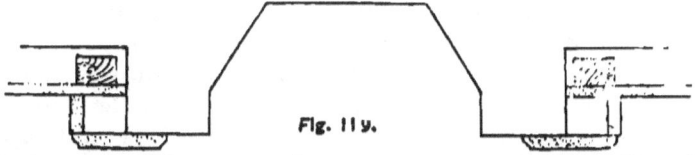

Fig. 119.

tels, however, and some wooden ones, are constructed to allow of a further projection of four inches beyond the plaster line (Fig. 119), so that a choice should be made as early as possible. The superintendent must watch the construction of the flues assiduously, as the only means of making sure that they are smooth and uniform in size. Whether they shall be "pargeted," or plastered inside with mortar, depends upon circumstances as well as on local custom. There is some danger that the pargeting may scale off and fall into the flue, dragging with it the mortar from the joints of the brickwork, so as to open a passage for sparks, and for this reason the practice is forbidden in some places, but if the mortar contains, as is advisable, half as much cement as lime, and the brickwork is kept wet, pargeting may be safely used, and certainly assists in giving smoothness and continuity of surface to the flue.

In regard to plastering the outside of the chimney, there is no difference of opinion, and where it is to be subsequently concealed by furring, the superintendent must insist upon its being thoroughly covered from the basement floor to the underside of the roof boarding.

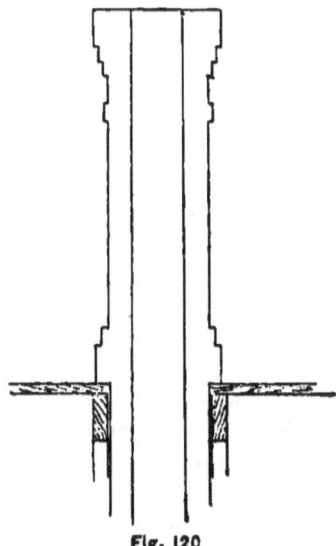

Fig. 120.

The "topping out" of the chimney, above the roof, should be done with mortar containing equal parts of lime and cement. Unless thus made waterproof, every rain will saturate the mortar, dissolving and loosening it until the whole stack begins to lean toward the windward side, and then speedily decays. For the same reason the four upper courses should be laid in clear cement, unless a stone or iron cap is used. Nothing else will long withstand the disintegrating action of rain, added to that of the acid vapors from the burning fuel. No overhanging projection in the shape of a base should be allowed where the chimney leaves the roof. (Fig. 120). In the inevitable settlement of the whole stack the upper portion will be caught upon the rafters, and the remainder sinking away from it, a dangerous seam will be opened just above the boarding.

Even before the chimneys are started, the cross-furring of the ceilings will begin. For this planed strips are used 12 inches apart from centres, two inches wide, and $\frac{7}{8}$ or $1\frac{1}{4}$ inches thick, the latter for three coat plastering. It is of great importance to get these truly level, to prevent inequalities in the finished ceiling. Ordinary carpenters try the strips with a straight edge as they nail them to the beams, hacking away a little from one

Furring.

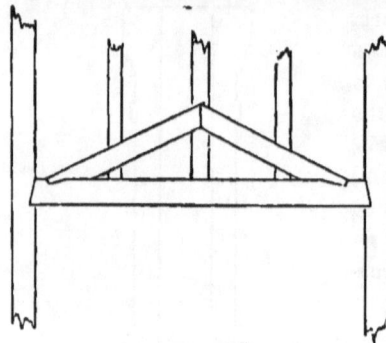

Fig. 121.

beam and filling up a deficiency in another by means of a chip, until an approximately even surface is obtained; but a much better way is to notch all the beams for the furring strips before putting them on, gauging from the upper side in the same way as in sizing upon the partition boards or girts. After cross-furring, the setting of the partitions is finished, those that need it are trussed, so as to throw the weight upon firm points of support, and small trusses are put over all openings in the partitions.

The trussing of partitions should be studied beforehand, and indicated on the framing plans, so that doors can be arranged without cutting off the braces. Trusses over openings should be framed like Figure 121. The dimensions and position of all the doors and partitions should now be thoroughly verified. No dependence whatever can be placed on the care of workmen in these respects, and the proportions of the plan are very likely to be hopelessly mangled unless a rigid watch is kept. The door openings must be framed about 5 inches wider and two inches higher than the finished door, to allow of proper blocking; and the distance from the angles of the room to the openings must be verified to insure symmetry, if that is intended; and sufficient space should be allowed for the architraves.

After all is made correct, the partitions may be bridged. This is often done by nailing in short horizontal pieces between the studs; a process which has its use, but is valueless for the present purpose. The proper way is to cut in diagonal pieces (Fig. 122), which present a considerable resistance to the sagging of the partition. **Bridging Partitions.** The final operation will be to try the partitions on each side with a straight edge, and correct the crooked studs by sawing half through them on the concave side, forcing them into place, and driving wedges into the incision.

The chimneys are next enclosed with furring, consisting of a cage of studs, supported by posts at the angles of the breast. These furrings should be measured to see that they are accurately placed in the room, that they are of the proper dimensions, and that the fireplace comes accurately in the middle of them; such details being little regarded by the average framer. Nothing now remains but the fixing of the grounds to prepare

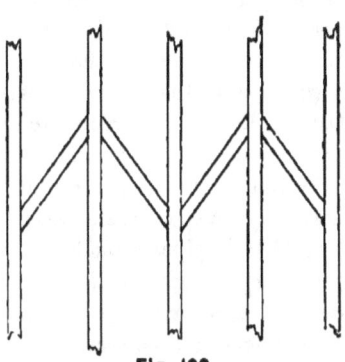

Fig. 122.

the inside of the house for lathing and plastering. Meanwhile, however, various other mechanics have been busy inside the building, and the work of finishing the outside has gone continuously on. Every hot-air pipe, gas, drain and water-pipe, bell-tube, speaking-tube and electric wire which is not intended to appear outside the plastering must be fixed in place during the short interval which **Hot-Air Pipes.** elapses between the completion of the studding and the commencement of the lathing. If the plans are prepared with any care, the position and size of all hot-air pipes will be indicated on them, but so much thought is required in arranging them to the best advantage that the superintendent should study the plans for himself, in season to suggest changes if circumstances render them desirable.

The proper size of the pipes depends upon the form of heating-apparatus employed, and this should therefore be selected as early as possible. Most persons make contracts for furnace pipes and registers complete, but if the pipes are put in separately it must be remembered that heaters with small radiating surface, like most of the wrought-iron furnaces, deliver air at a very high temperature, but can deal only with a small current, and therefore work better with few and small pipes and registers, while the best cast-iron furnaces as well as a few forms in wrought-iron, and all indirect steam and hot-water heating apparatus, possess a large radiating surface, and therefore warm a much greater volume of air in a given time, but to a lower temperature, so that larger pipes are necessary to convey an adequate supply of heat to the rooms. As the only fresh air supply in winter is derived from the registers, it follows that better ventilation is obtained by the introduction of a copious, but moderately warm current than from a small admixture of very hot air, but the latter is the cheaper mode, in consumption of coal as well as in the first cost of the apparatus.

With heaters of large radiating surface, the pipes to the principal first-story rooms should not be less than 12 inches in diameter, and 9 or 10 inch pipes should supply the larger chambers; but for those consisting of a mere cylinder of wrought or cast iron, 9-inch pipes are large enough for the ground-floor, and 6 or 7 inch for upper rooms. Inexperienced architects and house-owners often make the mistake of putting in too many pipes and registers. Every unnecessary one, by the leakage at the dampers and the chilling of the exposed pipe, interferes with the working of all the others, and many houses are almost uninhabitable in cold weather, with a register in every room, which would be quite comfortable with half the number judiciously arranged.

Wherever possible, the hot-air pipes should run through closets or inferior rooms, exposed to view, but it is usually necessary to carry up some behind the furring of chimney-breasts, or even in partitions. This involves danger of fire, which must be guarded against. Where the pipes pass through the floors, a tin ring, flanged over floor and ceiling, is usually inserted, through which the hot-air pipe runs, leaving half an inch air-space all around; and through the whole of its course, all wood-work exposed to the radiation from it should be covered with pieces of *bright* tin. Specifications for furnace-work generally require all wood within two inches of the pipe to be so protected, but this will rarely be faithfully done in contract work unless the supervision is very close. A much better way is to have all the pipes which are to be concealed behind wood-work made double, a space of half an inch or so being left between the outer and inner cylinders, through which air circulates freely. Where the pipes are carried up in partitions without such protection, heavy wire lath should replace wood over them.

Of the plumber's work, only a small part will generally need to be done before plastering. Although supply-pipes are now generally planned to run outside the walls, in positions where they can be easily reached, iron waste and ventilation pipes are often best carried up in partitions, and these must be set in place in time to avoid subsequent cutting of the plaster, so that the plumber should be notified in season. **Plumber's Work.**

Gas-piping must be all completed before the lathing can commence, and will need vigilant watching. Careless and stupid in regard to the work of others as plumbers are apt to be, the cheap gas-fitters far surpass them in capacity for botching their own work and destroying that of others. **Gas-Piping.** If the plan shows two or three rooms *en suite*, with a centre light in the ceiling of each, the workman will generally, if the superintendent is out of sight, proceed to take up a board through the middle of the floor above, and notch the entire tier of beams from one to six inches deep in the middle of their span, weakening the floor dangerously and irreparably; or short cuts will be taken in a similar manner across girders, braces, or anything else that may happen to be in the way. All good specifications stipulate that no beam shall be notched at a greater distance than two feet from the bearing, where the cutting down does not materially affect the strength, and that all centre-lights shall be supplied by branch pipes at the proper points, but irresponsible journeymen, unless watched, pay little regard to such documents.

Care must also be taken to see that all pipes are laid with a continuous fall toward the meter, as otherwise they may be choked by liquid condensed from the gas; and the position of all centre-lights and brackets should be verified. No pains will generally be taken to have them accurately placed, so long as they come within a few inches of their proper position, unless the men are strictly looked after, and annoying difficulties are likely to be encountered in consequence when the ceiling is decorated. The height of the bracket outlets from the floor will also generally be varied to save the workman the trouble of cutting his pipe, rather than with a view to their appearance or use. The proper height for bracket outlets in chambers is 4 feet 10 inches from the top of the under floor; in halls and first-story rooms, 5 feet 7 inches is the rule. Mirror lights should have outlets 8 feet from the floor.

Even if the outlets are properly placed, it is usually too much trouble for the workman to cut the nipples of the right length, or to make their direction normal to the surface from which they project, and consequently the fixtures, after the house is completed and paid for, will be found to take all sorts of unexpected angles with walls and ceilings, which can only be remedied by hanging the chandeliers with ball-and-socket joints, or bending the nipples, at the imminent risk of causing a bad leak. The proper time to make sure of this point is while the work is going on. A straight piece of pipe, a foot or more in length, and of the proper calibre, should be screwed upon each nipple, as soon as set, and carefully levelled and plumbed, and tested with a carpenter's square. If any deviation from the true position is observed, it should be rectified before the inspection of the pipes by the mercury gauge, so that any leaks caused by the correction may be detected. All bends, tees, and other fittings for gas-pipe under two inches in diameter should be specified to be of malleable iron, which admits of a little bending, and this is often the easiest way to bring the pipes back to place. The nipples, or short pieces of pipe which project through the plastering and receive the fixtures, should all be cut to the right length, so that they may project not more than $1\frac{1}{4}$ or less than $\frac{3}{4}$ of an inch beyond the finished plastering, allowing for the projection of the "bud," or middle portion of the plaster centre-piece. It saves trouble — to the gas-fitter, — to utilize waste bits of miscellaneous lengths for the purpose, so that it is generally necessary, after the house is done, to dig some of the nipples out of the plaster in which they are buried, and to replace others of inordinate length by pieces of proper size; at the risk in both cases of detaching the centre-pieces or causing leaks within the walls or ceilings.

After the pipes are all in place, their whole extent should be examined, to make sure that no split or defective pipe, made temporarily tight with putty or red lead, forms a portion of the system. This once ascertained, the caps are to be screwed on, and the pipes inspected by the mercury gauge, which will always be done on a request left at the office of the company which will supply gas to the house. To one of the outlets is attached a manometer tube, or inverted siphon, with the short leg closed, filled with mercury. Air is then pumped into the pipes until the mercury stands at a given height in the tube, and the whole left over night. The next morning, if the mercury column remains at the same height, the piping is pronounced tight. The superintendent should witness this process, and satisfy himself of the result, instead of taking the word of the gas-fitter in regard to it.

Bell-wires should always run in zinc tubes, which are sometimes fastened to the outside of the laths, and buried in the plastering, but are much better run between the studs. Their position should be verified, or, if not marked on the plans, determined with reference to the furniture which will occupy the rooms. **Bell-Wires.** Nothing, in a layman's eyes, does more discredit to an architect than to find gas outlets situated behind doors, or bell-pulls so placed that a bed or dressing-case must inevitably come in front of them.

Electric bells, which are very generally used instead of the ordinary kind, require only wires, which need not run through tubing, but are simply fixed to the studs by means of staples. With them should be put any other wires which may subsequently be needed to conduct electrical currents, such as those for burglar alarms, electric gas-lighting, or the electric light itself.

While these operations are going on inside, the outside finish has been rapidly advancing, so that the whole building may be so far as possible tight against rain by the time the interior is given up to the plasterers. The gutters are first put on, and the shingling or slating, beginning at the eaves, is carried to the top. Care should be taken that the gutter is so placed as to catch rainwater, but allow snow to slide over it. (Fig. 123.) Young architects often find their detail drawings for cornices defective in this particular. For additional protection against snow-water backing up under the shingles, or the overflow of the gutter dribbling through behind the cornice, the gutter should be ploughed at top and bottom, and "facias" inserted as shown in the figure. The shingling or slating begins with a double course, and the gauge is then marked off regularly to the top. Ordinary shingles, sixteen inches long, should not show more than four-and-a-half inches to the weather, unless on very steep roofs. The thick Michigan pine shingles, eighteen and twenty inches long, can be laid with much more projection without fear of breaking or curling. Each must be nailed with two nails, which should be galvanized if a very permanent roof is desired. Common nails rust out long before good shingles, well painted, become unserviceable.

Gutters.

Shingles.

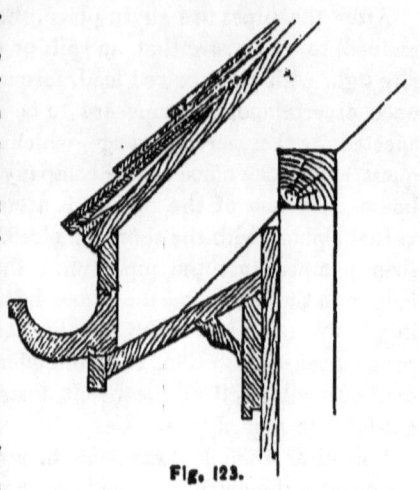

Fig. 123.

It is easy to judge of the quality of shingles. Freedom from knots and cross grain, and an approximation to uniform width, are the principal requisites. Spruce shingles, which are unfit to use in any but inferior buildings, are easily distinguished by their appearance and smell, which differ completely from the aromatic odor and silky grain of the "white cedar," or arbor-vitæ wood, from which those in ordinary use are made. Pine shingles are of a special size, and those of the Virginia cypress, which have the reputation of being everlasting, are somewhat costly, and have but a limited market. The choice between sawed and shaved shingles depends upon circumstances. The latter allow water to run off more freely, and are to be preferred if unpainted, while the former hold paint better, and are therefore generally used by architects.

In laying, the widest shingles are selected for the hips and valleys, where cutting is necessary, in order to give room for two nails. (Fig. 124.) Many of the best carpenters lay hips like Fig. 125, so that the cut shingle will not

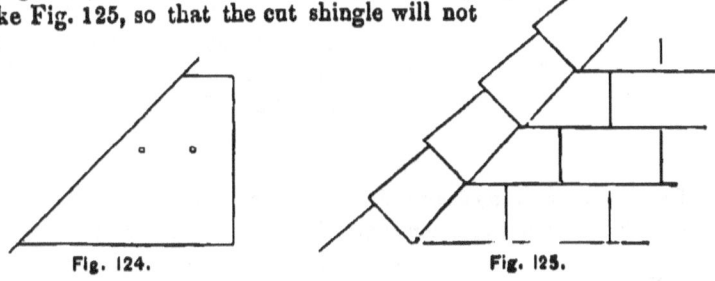

Fig. 124.   Fig. 125.

come at the extreme edge, and the effect is picturesque, while the durability of the roof is improved.

The painting of a shingle roof is important. Many architects specify that each shingle shall be dipped in paint, some even requiring the paint to be hot; but this is tedious and expensive. A simpler and very good way is to paint each course as it is laid; and the cheapest is to do it all at once after the roofers are out of the way. The last process rather hastens the decay of the shingles, by forming little dams of paint which hold back the rain-water against the unprotected portions, but is usually adopted.

Painting Shingles.

Where shingles are to be used, the roof boarding is generally of hemlock or inferior spruce, planed to an even thickness, and one or two "plies" of tarred felt are laid under the shingles, to prevent fine snow in heavy storms from finding its way into the rooms. With slates still greater precaution is necessary, and the tarred felt should not only be double with all joints well broken, but matched pine boards should be used underneath. The "lap" of the slates will vary according to the size. Ten inches by twenty, or eight by sixteen are generally used, and for the best work should be laid with "three-inch lap"; that is, each slate should lap three inches beyond the head of the second slate below, (Fig. 126), and the length of the exposed portion will be found by deducting the lap from the whole length of the slate, and dividing the remainder by

Slating.

two. Two inches lap, is however, common. The slates should be put on with galvanized nails, not driven too hard, for fear of breaking them, nor too little, lest they should rattle and blow off in high winds, and the cut slates at the hips should be watched to see that they are not hung by one nail. Patent slating nails, which have a circle of japanned tin around the head, are useful. The flashings should last be looked after, and the roof will then be complete.

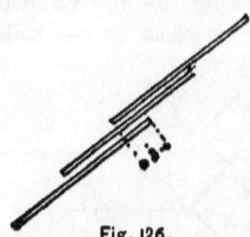

Fig. 126.

Leaks in Roofs. If the young architect is ambitious of being able to say that no roof built under his superintendence ever leaked, he will need to exercise both thoroughness in inspection and skill in providing for various contingencies. The worst leaks come from improper position of the gutters, by which wet snow sliding from the roof is caught and held back. It soon freezes to the roof along the lower edge, the upper portion remaining free, and the water subsequently running down the slope is caught as in a long, deep pocket, in which it rises rapidly until its level reaches that of the upper edge of a course of slates or shingles, over which it pours in a sheet, to find its way into the rooms below. Next to this defect insufficient flashing in valleys is perhaps the worst. As metal is expensive, the roofer's interest is to save as much of it as possible, and the superintendent must consider the circumstances of pitch and extent of roof surface draining into the valley, and the slope of the valley itself which should determine the depth which the water will probably obtain in it. In certain cases, where the roofs are large, this may be eighteen inches or more in summer showers, and the only security is to make the valley flashings of corresponding size.

A very common place for a small leak is around the chimneys, where rain or snow often blow through between the bricks and the flaps of a "stepped flashing." The remedy is a liberal application of elastic or "Boston" cement between the brickwork and the metal. The same cement is also needed to prevent water from getting in at the angle of a hip in a slated roof, unless the hip is protected by flashings (See Part I, page 80.) This should always be the case where a permanent construction is intended, since the cement soon

BUILDING SUPERINTENDENCE.   151

" burns out " and crumbles by the heat of the sun on the slates. Shingles fit much more closely, and will generally make a tight hip without cement or metal.

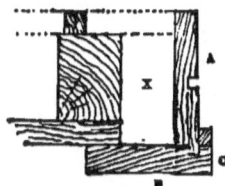

Fig. 127.

While the roofers are at work, the window frames are being rapidly set in place. These, in wooden houses, are mere fronts, in plan like Fig. 127, A being the "pulley-style," grooved for the parting head, while B is the outside casing. When this is put in place and nailed to the rough-boarding, the space X, between the pulley-style and the stud, forms the pocket for the weights.

Window Frames.

The casing B is often moulded or ornamented. If plain, it should be 1¼ inches thick, to prevent curling under the heat of the sun. Its inner edge, projecting half an inch beyond the face of the pulley-style, is usually made to form one side of the channel in which the upper sash slides, but it is much better to increase the depth of the reveal by inserting a slip C, some five-eighths or three-quarters of an inch wide, changing the position of the parting-bead to correspond. Independent of the improved appearance of such a frame, room is

   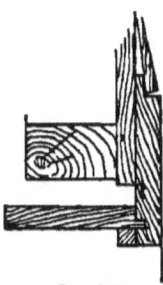   

Fig. 128.   Fig. 129.   Fig. 130.

thus given for mosquito-nets and blinds. The shingles or clapboards are laid close up to the other edge of the casing, but when this shrinks a vertical opening is left, through which rain penetrates, and tarred felt, or still better, strips of zinc, must be laid in behind the casing and the adjacent work. The junction of the clapboards or shingles with the top of the casing must also be protected. Some carpenters do this by tacking a strip of lead to the boards just over the casing, and turning it down over the edge (Fig. 128), but it is neater and tighter to rebate the top of the casing. (Fig. 129.) If this is done, and the vertical sides of the casing are grooved into the head (Fig. 130), the sills set to a sharp pitch, one-and-a-half inches or so, and grooved underneath for inserting the shingles or clapboards which come below them, the superintendent may be tolerably sure that his building will not show that most annoying of defects, leakage around the edges of the openings, and the shingling or clapboarding may be commenced at once.

Before this, however, the back-plastering, if any is specified, should have been completed, in order that its drying may be favored by the circulation of air through the open joints of the boarding;

Back-Plastering. and notice must be given to the plasterers in ample season. The mode usually considered best is to nail fillets to the sides of the studs, and to lath on these, so that in theory a double air-space is formed by the outside boarding, the sheet of back-plastering, and the inner plaster; but in practice it is inconvenient to nail fillets in the narrow space between the ledger boards and the outer boarding of a balloon frame, or just above the dropped-girt of a braced frame, and still more so to nail laths to them, so that these spaces are usually neglected, and the wind which blows in under the clapboards and paper of even the best built house finds an issue at such points into the interior. The real object should be to spread a continuous sheet of mortar from sill to plate, and this can be much better accomplished by omitting the fillets, and nailing two lines of laths vertically between each pair of studs, then lathing on these, and plastering with a thick coat of mortar, well pressed on, so as to fill in between the laths and the boarding. By this method room is left for the mason to insert his trowel and hand behind the ledger-boards or flooring-strips and reach every inch of the space between the studs, closing it against the wind, if care is taken to bring the mortar well up on the studs, with almost the imperviousness of a solid brick wall.

The meaner contractors, not in consequence of scientific deduction, but from desire of gain, often caricature this mode, by lathing vertically directly on the boarding, without the interposition of the first laths, which are essential to give a key to the mortar, and prevent it from cracking off. Some even content themselves, if not observed, with simply spreading the mortar on the boards, without any lathing whatever. It stays in place just about long enough for the contractor to get his pay.

As soon as the back-plastering is done, it is usual in good houses to apply some precautions against the spread of fire and the passage of rats and mice from room to room. In the cellar, **Fire-Stops.** the whole space between the beams, from the sill to the inner surface of the wall, should be filled in solid with brickwork and mortar (Fig. 131), up to the under side of the floor-boards. This not only keeps out the cold wind which would otherwise blow freely through the chinks about the sill and the base-mouldings of the house, but renders it impossible for rats and mice to climb on top of the wall and gnaw their way through the floor above. If the cellar ceiling is not plastered, the space above the girders should be bricked up, for the same reason; and in any case a brick partition, at least four inches thick, should be built, in mortar, on top of every girder, between the studs of the partition which rest on it, and on the heads of all partitions above which cross the beams, to a height of two or three inches above the floor. By this means the hollow spaces between the ceilings and floors are divided up into compartments, effectually cutting off the circulation of cold air through the floors which is so much to be feared in country houses, while fire is prevented from spreading laterally between the beams. It is very desirable also to fill in with brickbats and mortar between the studs of all partitions, at about half the height of the room. This "fire-stop" belt may be laid on top of the bridging, and will check the upward course of fire, forcing it to burn out into the open room, where it will be discovered and means applied for extinguishing it. The principle should never be lost sight of, that the chief danger from fire in dwelling-houses comes from allowing it to find its way into the wooden tubes, lined with bristling splinters, nearly as inflammable as tinder, which are

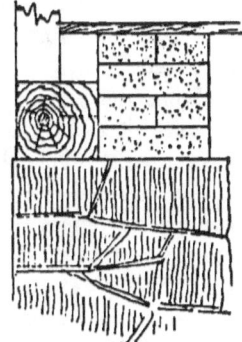

Fig. 131.

formed by the interior surfaces of the laths, studs, beams and flooring-boards. In these the flames creep, undiscovered and inaccessible, through and through a house in which no device is adopted for intercepting the communication between such air-spaces, until they gain strength enough to burst out fiercely in a dozen places at once, and the building, already permeated in all directions by the fire, is quickly destroyed.

To carry out the system of protection with any thoroughness, the spaces between the outside studs must also be divided into sections by incombustible material. With braced frames this can very well be done by means of brickwork laid upon the girts, but for balloon frames it is necessary to nail in short pieces of timber between the studs at the level of the ledger boards, to build upon. A good body of masonry should then be formed on top of the plate, filling the whole space between the rafters to the under side of the roof-boards. This is an invaluable defence against cold air as well as fire, both of which usually find the cornice a very vulnerable feature.

One other point remains to be guarded: the hollow space around the chimneys, behind the furring of the breasts. The easiest way of obtaining a partial protection is to lath and plaster the ceiling from the chimney outward as far as the breast will extend, before setting the furrings. A single rough coat only need be used, and it can be put on at the same time as the back-plastering, so as not to cause any delay. If stout wire-lathing is used instead of the ordinary kind, and especially if the mortar is "gauged" with a liberal dose of plaster-of-Paris, a very efficient protection is obtained for the portion of the building most exposed to danger.

Wire-Lath.

Where the owner is disposed to incur a small extra expense for the sake of additional safety, wire-lath will be found generally of great service. Only the heaviest kind should be used, as the numerous furrings necessary to secure the more flexible varieties detract very much from its value as a protection against fire. By applying it to the surface of chimney-breasts and the under side of the stairs, great security is gained for the building, at a trifling expense, and it may be employed also in the ceilings and walls about furnaces and stoves. Where it is necessary, as it will sometimes be, to carry hot-air pipes in partitions, the lathing over them should always be of wire, unless the pipes are made double, or a casing of bright tin put all around them on the wood-work and under the laths before plastering. By first nailing laths vertically on the studs, the wire can be brought to the same plane as the wooden laths on each side, and the plastering will form an unbroken surface. Even the cost of using wire-lath throughout a dwelling-house is in many ways repaid. Its power of holding the mortar so firmly that no amount of force will detach it, to which, far more than to its incombustible material, it owes its efficiency in resisting fire, is equally valuable in maintaining a perfect surface for walls and ceilings. Such plastering does not crack or sag; no violence can shake it down, and the most prolonged water-soaking fails to detach it from the lathing.

Whether a larger or smaller number of these precautions shall be adopted will depend on circumstances. If wooden laths are to be used for the plastering, the superintendent should see that they have the requisite number of nailings, and are not placed too near together. Three-eighths of an inch is the proper distance: if nearer together the mortar will not be effectually pressed through the intervals, and its hold will be feeble: if farther apart, it will not, while soft, sustain its own weight. It is usual to specify that the joints shall be broken every six courses; but it is much better and not much more troublesome to break joints at every course: and care should be taken that the laths above the door and window heads extend at least to the next stud beyond the jamb (Fig. 132), so as to

Wooden Laths.

Breaking Joint.

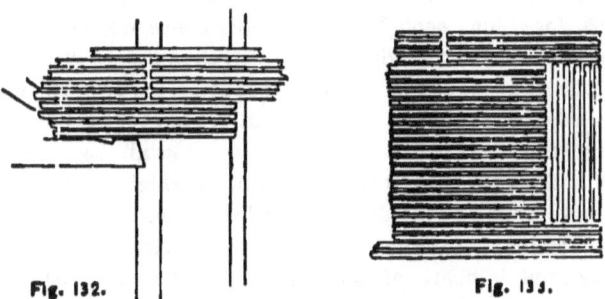

Fig. 132.   Fig. 133.

prevent the radiating cracks which are apt to appear at that point. Where the men meet with a small space, whose longest direction is perpendicular to that of the other lathing, they are apt to fill it up with laths set at right angles with the rest. (Fig. 133.) This should never be permitted, as cracks are sure to appear afterward where the change of direction takes place.

After the application of the laths, if not before, the furrings should be verified. The soffits of dormer windows, the under side of stairs, and all angles of walls and ceilings should be carefully observed to see that their surfaces are plane. Chimney-breasts should be tried with a carpenter's square, to make sure that their external and internal angles are right angles, and the position of the chimney-breasts in the room should be finally examined. All laths which show knots, portions of bark or stains should be pulled off on the spot and thrown away, and their places supplied with fresh ones, as such imperfections in the laths are liable to cause discoloration in the plaster over them; and for the same reason all brickwork which is to be plastered should be cleared from soot, tobacco juice or soluble defilements.

At least seven days before the lathing is finished all the mortar for the first coat of plastering should have been mixed and stacked; if possible in some place outside the house, since the evaporation from so large a mass of wet material, if stored, as is often done, in the cellar, causes the timbers to swell through dampness. The lime must be of the best quality, free from every trace of underburnt "core." If the lumps, on being covered with water, slake for a time with considerable violence, but leave a residue which must be crushed by the hoe, the whole should be rejected. Such lime, unless it can be stored long enough to insure this conversion of every portion not absolutely inert, will surely cause the plastering made from it to blister and "chip-crack," sometimes after the lapse of months; and with the best lime it is unsafe to put the mortar on the wall until it has "cooled" for a week or more, to allow all the particles to become hydrated. The ancient method, by which the slaked lime was stored in pits and not drawn upon for use until after one or two years' seasoning, unfortunately has become obsolete.

Loamy sand is nearly as much to be dreaded in plastering as underburnt lime. Some varieties contain particles of compact clay or soil which will, after a season's drying, assume a powdery condition, expanding as they do so, and throwing off the mortar in hundreds of little pits, like the scars of small-pox. Moistening the sand and rubbing it on the hands will

Sand.

usually give sufficient evidence of the presence of loam or clay by the stain which it leaves on the skin.

After seeing that the mortar is well pushed through the laths with the trowels, so as to bend over on the inside, the straightness of the angles, both vertical and horizontal, between walls and ceilings should be assiduously criticised. In two-coat work, such as is used throughout the Eastern States for dwelling-houses of moderate cost, there is no opportunity for bringing the surfaces to a true plane after the first coat is on, since the second or "skim coat" is a mere varnish, less than ⅛ of an inch in thickness; and the plasterer must use his judgment in laying on just mortar enough to fill out to the line after it has been trowelled down enough to force it well through the laths. This can seldom be done with any accuracy over the larger surfaces, but by applying the long, thin-edged board which is used to finish against the angle beads as a ruler, any reasonably careful man can insure the straightness and accuracy of the corners, and this should be insisted upon, as the eye immediately detects any irregularity in the angle between walls, or between the wall and ceiling, while inequalities of the intermediate portions are unnoticed. With three coats it is easy to obtain surfaces absolutely plane, by using the proper means. The scratch-coat is to be very strongly trowelled and well scratched up with a sort of comb, made of sharpened laths, nailed in a row on a stick. After this is *thoroughly dry*, "screeds" should, for a first-class job, be run all around the margin of the ceiling, consisting of strips of mortar, carefully put on, and brought to a perfectly plane and horizontal surface by means of the spirit level and a long straight edge, applied diagonally across the corners as well as along each strip. For a small ceiling this will be sufficient, but a large one requires intermediate screeds, brought accurately to the plane of the first ones. When the screeds have hardened a little, the space between them is filled with "brown" mortar, which is easily made perfectly even by means of the straight-edge. Similar screeds should be formed in the vertical angles of the room, plumbed, and the intermediate spaces filled up to a plane surface. If cornices are to be run, which is always done before the last coat of plastering the angles should be as rough as possible, to give them sufficient "key." The superintendent should study the profile

Cornices. of the proposed mouldings, and if a large mass of mortar will be left in the angle, he should order nails to be driven to hold up the coarse mortar which is used for "dubbing out" the cornice before the finer material is applied.

Fig. 134.

(Fig. 134.) Some care is necessary to see that the final coat of plastering is not injured by freezing in winter or by too rapid drying in summer. From the latter cause the finished work near the windows is often found covered with a network of minute cracks, particularly on the side which the wind strikes, while a breeze barely at the freezing point will cover the surface with radiating crystals, disintegrating it so that on thawing again the mortar will scale off in patches. The remedy for this is to keep all openings protected by

Screens. temporary windows or screens, consisting of wooden frames covered with cotton cloth, well fitted to the openings. Whether the plasterer or the carpenter shall provide these screens or temporary windows depends on the terms of their respective specifications. Perhaps the best way is to require the carpenter to supply and fit them, and the plasterer to shift them in such a way as to secure his work against freezing or unequal drying.

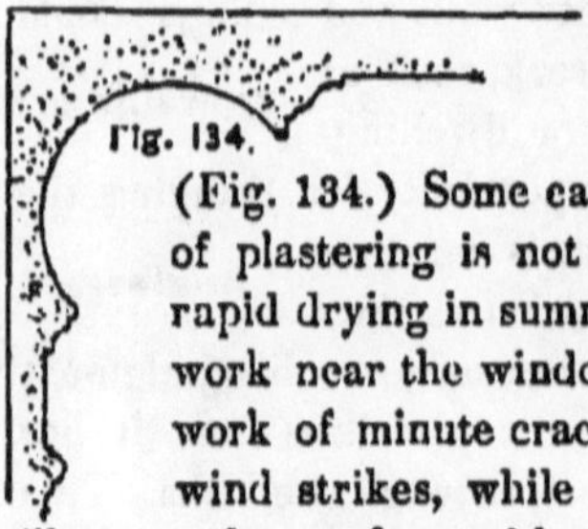

As soon as the plastering is completed, the plumber must be summoned to finish his work, so as not to delay the joinery. The pipes will be first put up, and the superintendent must thor-

Plumbing. oughly understand the purposes and requirements of each. Cast-iron pipes should be carefully scrutinized, especially where cut or broken. The metal, unless double-thick pipes are specified, will be very light, and in the poorer makes it is apt from careless casting to be much thinner on one side than the other. If any particular kind is called for, or known to be good, the shape of the "hub" will serve to distinguish it, if the name of the

Soil Pipes. maker is not cast upon the pieces. In some places iron pipes are coated with asphaltum at the factory, for the use of the best plumbers; the inside as well as the outside being treated. Elsewhere, painting with red lead is customary; and this is generally confined to the outside, as the inside would soon lose its coating. The asphaltum forms the best covering, but whatever is used, the exterior of the pipes must be completely coated before they are brought to the building. The joints should be made with oakum, not paper or shavings, driven in tight, and finished with melted lead, which, after the pipes are fixed in position, is to be thoroughly calked all around. It is often much more convenient to calk the joint before securing the pipes in place; but the jarring so occasioned may loosen the lead, and where the joint will be accessible after fixing its completion should be deferred. It must not be forgotten that the melted lead by itself will not make a tight joint, since the shrinkage of the metal in cooling draws it away from the iron, and it must be forced again into contact with the calking iron, applied at every point of the circumference. A first-class workman will use three or four pounds of lead for each joint, filling the hub completely, and showing the marks of the tool all around. Inferior plumbers leave a little space above the lead, which they afterwards fill up with putty, smoothing it neatly, and, if possible, getting a coat of paint over it before the superintendent comes. Such joints will pass the test, but are not durable if there has been any carelessness in calking the lead. Intentional swindlers fill the joint with shavings, paper, mortar or anything else which happens to be at hand, and daub the top over with putty, or perhaps with a little lead, ladled out of the pot.

Rust joints, of sal-ammoniac and iron turnings, are sometimes, though rarely, used. They are said to be tight and durable, but likely, if unskilfully made, to burst the hub by the expansion of the mass. All iron pipe should be very strongly supported, by iron straps and hooks, never by wires.

Lead pipes should be examined as delivered. The weight per foot, or the letters denoting the same thing, are stamped on the ends of the coils: after the lengths are cut off, it is more difficult to ascertain whether they comply in weight with the requirements of the specification. Most lead manufacturers furnish cards showing the thickness of metal corresponding to a given weight for each calibre, but the saw used to cut it spreads out the lead, increasing the apparent thickness. In general, lead supply-pipes, unless for a tank or other very light pressure, should not be *less* than $\frac{3}{16}$ of an inch in thickness of metal. Waste and air pipes will be little more than half of this. There is some difference in the quality of lead pipes, but it is not easy to detect it except by analysis or the test of use. Honey-combed or corroded pipe, and any which shows unequal thickness of metal, should however be at once rejected.

In certain localities the seamless brass tubing, drawn over a mandril in the same way as lead pipe, is much in favor for plumbing work, and where the pressure is very heavy or the water is so soft as to attack lead, it is well worth the additional cost, which is not usually more than ten to fifteen per cent on the whole amount of the plumber's contract. They can be had either plain, or coated inside and outside with tin, or, for use in conspicuous situations, plated with nickel and polished; and couplings, unions, bends, tees, and all varieties of cocks and fittings are furnished to correspond with each kind of pipe. Where the brass tubing is employed throughout a house, it is common to have the cold-water supply tinned inside, while the hot-water, which is not likely to be used for drinking, is conveyed in the ordinary kind. If there is no danger that either brass or lead will be corroded by the water, it is not unusual in the best work to make the hot-water pipes only of brass, using lead for the others. In this case the harder metal possesses the advantage that if properly put up, with the angles left free to move a little back and forth to accommodate the expansion and contraction of the pipe between them, it is not injuriously affected by repeated alternations of heat and cold, which with the inelastic and ductile lead first stretch the pipe by contraction, and then, as it does not possess elasticity enough to recover its shape, cause it, on being again extended by the passage of warm water through it, to sag down between its supports, this effect increasing by repetitions of the cause until the undulations of the pipe become sufficiently pronounced to retard or stop altogether the flow of water through it. Where, however, the brass tubing is tightly confined at the ends of a long line, the joints and fittings often become strained and leaky by the contraction of the pipe in cold weather, since the very rigidity of the metal prevents it from accommodating itself, like ductile lead, to the force exerted upon it; so that it should be used intelligently to obtain the full benefit of its good qualities. One last precaution should be observed: at the completion of a piece of brass-pipe plumbing all exposed portions of the

Brass Pipes.

metal must be varnished with a good coat of shellac, or it will soon become corroded and unsightly. This may be made a portion either of the plumber's or the painter's contract, but unless the duty is distinctly imposed upon one or the other, it will be neglected. Paint, which was once generally applied both to brass and lead pipes, is best omitted unless required for appearance.

As a cheap substitute for brass tubing, where there is reason to fear the corrosion of lead pipe from the softness of the water, or its bursting from the heavy pressure under which it is delivered, iron is often used, galvanized or enamelled in various ways. It may be obtained with a lining of pure block tin, **Wrought-Iron Pipe.** forming a very strong and pure channel for water, and glass-lined iron pipe is sometimes used; but the ordinary coating is one of coal-tar or paraffine enamel, giving a shining black surface. A smooth, red covering is sometimes seen, which is said to have a base of vulcanized rubber. Whatever may be the protecting medium employed, the unions, bends, and other fittings are treated with the same, and if well put together the water nowhere comes in contact with the metal. The galvanized or zinc-coated pipes are more expensive than those merely enamelled, but more durable, unless in acid waters.

The same precautions against the straining of the joints which are necessary with brass pipes should be observed with those of iron. Moreover, iron being a very rapid conductor of heat, cold-water supply-pipes of that metal will in warm, sultry weather condense a great deal of water upon their surfaces, which trickles down them and may in time cause serious injury to paper or other decorations beneath. Where there is any risk of this the pipes should be encased in a tubing of zinc, which will catch the con- **Zinc Casings.** densed drops and conduct them to a place of safety. This is even done with lead and brass pipes in city work of the best character. Where costly decorations or papers are in danger of being injured by a possible leak, it is always advisable and is generally required of the plumber by the best architects, to enclose all supply-pipes in zinc tubes, which will retain the jet from a lead pipe burst by freezing or water-hammer, or the drops of condensed water, and conduct them to a safe outlet.

The course joints and fastening of all the pipes, whatever their material, should be carefully observed, and the hand of a skilful and conscientious plumber will be more quickly recognized in this than in any other detail of the work. Such a man can always arrange his pipes so that they will fall naturally into their proper places, without that dodging over or under each other which characterizes the "botch's" work; his lines will be perfectly straight, and all hot-water pipes separated by a small distance from the cold, to avoid loss of heat from one to the other; the supports will be neatly put up, at equal and small intervals, so that no sagging of the pipes will be possible between them; and all will be laid with a continuous fall toward some faucet by means of which the water can be thoroughly drained from them.

Boards should be put up by the carpenter, well secured to the walls or ceilings, wherever pipes are to run, and similar ones fitted in between the beams where it is necessary to conceal them between floor and ceiling. To these boards the pipes must be attached at intervals of about four feet where they run vertically, two feet where they follow the underside of a ceiling, and six or eight feet where they simply lie on a horizontal surface. Horizontal pipes should be secured by stout brass bands screwed to the boards; and where several pipes run side by side, a first-rate plumber will separate them far enough to allow screws to be put into the band between them. Vertical pipes, in order to prevent them from creeping downward by alternate expansion and contraction, must be fixed in place by hard-metal "tacks" soldered to the pipes and screwed to the boards. All joints in lead pipes should be "wiped joints," excepting only those connecting the couplings of basin, sink, or other wastes to the pipes, which cannot easily be wiped, and do not require to resist a pulling strain, so that a "cup joint" is sufficient.

Brass and iron pipes are connected by means of unions, or short pieces of cast-brass, tapped at each end to receive the screw thread cut on the lengths of tubing. In the case of brass, the joint is covered with red lead before screwing up. Enamelled or galvanized iron fittings may be put together without such luting.

If the workman shows a disposition to neglect any of these niceties of his art, he should be always under suspicion of greater errors, and should be watched to see that he does not fasten the pipes in out-of-the-way corners by means of hooks, or leave them suspended from a ceiling, or between beams, by attachments so far apart that alternate expansion and contraction will in time cause them to sag down, forming a hollow from which the water cannot be drawn off, so that the pipe is likely to burst there if the house should be left vacant in freezing weather. The opposite fault, of allowing the pipe to take an upward bend in any part of its course, is still more to be avoided. Such a bend, whether accidental or made by ignorant intention, soon becomes filled with the air always carried in bubbles through the water, forming an "air trap," which may stop the passage of water entirely unless relieved by opening the pipe at the highest point of the curve.

The number and courses of all the pipes should be minutely described in the specification. If, however, no details are given, as sometimes happens, the superintendent will have a little difficulty in deciding upon the arrangement which he can require with justice both to the plumber and the owner. Much depends upon the custom of the locality, and something also upon the price for which the work is to be done. In Massachusetts or New York, for instance, under a mere agreement for a "first-class job," *with an adequate consideration*, the plumber would be expected, without special orders, to provide stop-cocks in convenient positions for shutting off both hot and cold water from any part of the house at pleasure, and draining the pipes; and in Massachusetts he should arrange a tank in the roof, itself supplied by a rising main and ball-cock from the regular house service, to contain water for the copper bath-boiler below, and carry up an "expansion pipe" from the highest part of the hot-water system to this tank, turning over the edge just above the water-line, so as to allow steam and froth to escape freely; and in any large city he would be expected to put separate traps under all fixtures, and provide air-pipes to the same, and to carry up such air-pipes, as well as the soil-pipes, above the roof; and finally, to fit up separate cisterns over each water-closet, for their exclusive supply, and place "safes" of sheet-lead turned up around the edge under all plumbing fixtures above the first story, or over furnished rooms in the basement, with waste-pipes of their own, emptying over a sink or bath.

In States which have statutes for the regulation of plumbing work, the provisions of the constituted authority must be taken, in the absence of specific details, as forming a part of the plumber's contract.

In New York and elsewhere, however, it is customary to fit up bath-boilers without the tank in the upper story, rising main and expansion pipe, and in some small country towns the old-fashioned pan water-closet, supplied without the intervention of a service cistern by a valve with branch from the main pipe, is still used in the best ordinary practice.

In cases where a first-rate job is clearly not intended, the superintendent may still do much by attentive study of the subject to secure, if not the best possible arrangement, at least a safe and strong one. When the time comes for setting in place the sinks, baths, wash-bowls and water-closets, this special knowledge will be still more necessary, for without an adequate and constantly advancing acquaintance with the improved apparatus continually placed upon the market, he can neither criticise justly that which may appear to him valuable or the reverse in the construction or fitting up of new appliances, nor make suggestions to meet peculiar needs.

Such knowledge must, however, be gained from technical books and journals; a general treatise can do little more than indicate a few main principles; as for instance, that waste and supply pipes should, for any kind of apparatus, be large, to insure speedy filling and discharge, and the construction as simple as possible; that putty joints, rubber washers, and inaccessible floats, check-valves or other moving parts are to be avoided: and that the ideal of every plumbing appliance would be a solid glazed stoneware basin, in one piece, supplied by a quick and copious flow of water through a large pipe, and discharged through the shortest possible waste-pipe and ventilated trap, all connected with calked lead joints, into a thoroughly aërated, porcelain-lined soil-pipe.

At present this ideal is far from being fully realized in any form of apparatus. The nearest approach to it is perhaps the ordinary wash-bowl well fitted up, supplied through half-inch basin-cocks, and drained through 1¼-inch pipe, with a 1¼-inch S-trap, ventilated by a pipe of equal calibre carried to a main air-pipe. In such an apparatus there is no cavity to collect foulness; the supply is sufficiently copious to wash the sides of the basin by the force of its flow, and the calibre of the waste-pipe, including the trap, being uniform, the impetus of the discharge through it is nowhere checked, so that the friction of the swiftly passing water scours away the slime which tends to collect upon the inner surface of the pipes. Next to the wash-basin comes the pantry-sink, or "dish-washer," which is hardly more than a large basin lined with tinned copper instead of porcelain, and drained through the same size of pipe and ventilated S-trap.

The ventilation of these traps, as well as of all similar ones, whether used under bowls, sinks, baths, water-closets or other apparatus, although absolutely essential to their security as a seal, is not yet so universally demanded by architects, or practised by plumbers, as it should be. Many and varied experiments with glass and other traps have conclusively shown that the effect of the discharge of a volume of water through those of the ordinary form, even when ventilated, is not, as apparently it would be, to leave a residue of liquid in the bend, standing at the level of the bottom of the outlet-pipe (Fig. 135) but is rather, by the impetus given to the whole moving mass, to throw a considerable portion of the residuary water beyond the bend, a portion running off through the outlet into the drain, while the remainder sinks back into the trap, partially filling it. (Fig. 136.) When the trap is unventilated it very frequently happens that the column of water passing through it, and over the outer bend, sets up a siphon-like action, the rising portion of the bend forming the short leg and the discharge-pipe the long leg of the siphon. When this happens, all the water in the trap is drawn over after the main stream, leaving it empty (Fig. 137), and of course destroying its efficacy as a check to the ascent of foul air. Another kind of siphon action is produced in the same traps whenever a considerable quantity of water is thrown down any waste connected with the drainage system. The main waste being filled, or nearly so, by the charge, is partially exhausted of air by the passage of the water through it, and unless the deficiency is supplied through ventilation-pipes the external atmosphere will force its way in through the traps, pushing the sealing

Ventilation of Traps.

Siphonage.

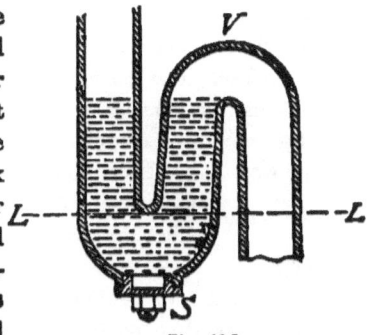
Fig. 135.

water before it, and leaving them open. To guard against the splashing of water over the bend, it is now common to make the trap very deep; and siphonage of all kinds is prevented by ventilation pipes attached to every trap at the outlet bend, and carried to some main air-pipe. It was once usual to make these ventilation-pipes of small calibre, but experience has shown that air

Fig. 136.

Fig. 137.

passes so slowly through small tubes that nothing less than 1¼ inches is now used for small traps and 2 inches for larger ones.

If nothing but clear water passed through house-traps, the science of plumbing would be an easy matter, but it is not so; and with the kitchen sink serious difficulties begin to appear.

Kitchen Sinks. The supply, by means of two "bibb-cocks" for hot and cold water respectively, is simple enough, but the waste liquid from the kitchen inevitably contains more or less fatty matter, generally in a melted state, and suspended in small globules in the water, which during its passage through the cold pipes is chilled, and the floating particles striking the metal walls are congealed, covering them with a tallow-like coating, which gradually thickens from accretions of similar matter until the pipe or siphon trap is nearly or wholly choked. The coating is very hard, so that no flushing, however vigorous, can do more than hasten the passage of the liquid and remove the inevitable congelation to a point farther down the pipe, and the only way to avoid certain ultimate stoppage at some point is to provide a reservoir or catch-basin on the line of discharge,

Grease-Traps. which may retain the waste waters long enough to allow the grease to congeal and separate, and shall at the same time be large enough to contain a certain accumulation of solid matter without obstructing the water-way, and in a position where it can be readily reached and removed upon occasion. Years ago this was done by placing at some distance beyond the trap of the sink, either in the cellar or out of doors, a "grease-trap," consisting of a small tight brick cesspool, into which the waste-pipe from the sink emptied, while a second pipe, inserted in the wall of the grease-trap with its mouth turned downward (Fig. 138), so as to prevent the entrance of floating cakes of fat, carried the overflow to the drain. This form of trap would often hold the accumulated grease of many months, but offensive decomposition was apt to appear in the accompanying liquid, and it is generally preferable to discard in such a situation both the S-trap and the little cesspool, substituting a device which combines the advantages of both in the shape of a " round trap," shown in section in Figure 139, and provided with a large brass trap-screw at the top, which can be opened whenever necessary and

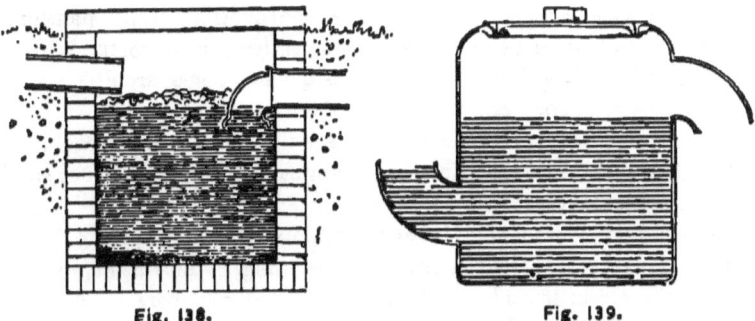

Fig. 138.   Fig. 139.

the fatty coating removed in a few moments. The round traps commonly used for sinks are of lead, six inches in diameter, with a 4-inch brass trap-screw inserted in the top. Until nearly filled with grease they are not easily siphoned out, and for this reason it is not customary to provide them with air-pipes, but the advantages of ventilating all wastes are so great that no exception should be made in their favor.

Wash-trays are usually drained through a round trap of the same size and construction in order to prevent any possible stoppage of the pipes by soap, which will in time form a greasy coating upon them. One trap is enough for a set of three or four trays.

Wash-Trays.

For the kitchen sinks themselves no perfect material has yet been introduced. Small ones are made of white earthenware, which is all that could be desired, but specimens up to 24″ × 48″, the average size for good houses, are hardly yet attempted, and soapstone forms perhaps the best substitute. Earthenware wash-trays are howeve.

already in use, and though they are more expensive than soapstone the forty or fifty dollars of extra cost is more than repaid by their beauty and cleanliness. Iron sinks are much used and are made either plain, galvanized, or enamelled; the galvanizing being, perhaps, the most durable finish.

Bath-Tubs. Bath-tubs can be had of earthenware, but at considerable additional expense, the ordinary material being tinned and planished copper of suitable thickness. Fourteen ounces to the square foot is the weight which will generally be used unless the specifications direct otherwise, but copper of this thickness will soon "cockle" and become uneven from the expansion caused by hot water flowing over it, and sixteen-ounce metal is the lightest which is suitable for a first-rate job. Some architects require eighteen-ounce copper, which gives excellent results. The best manufacturers of baths stamp the weight of metal on each, so that there is no difficulty in discovering whether the contract has been complied with in this respect.

Every bath should have its own ventilated S-trap, separate from all other fixtures. The old custom of running bath or basin wastes into water-closet traps is obsolete among good plumbers. The supply may be brought through plain bibb-cocks (Fig. 140), bath-bibbs (Fig. 141), or by means of various forms of combination-cocks or concealed valves. It is quite common to arrange the cocks for hot and cold water behind the end of the tub, operating them by means of small handles above the cap-

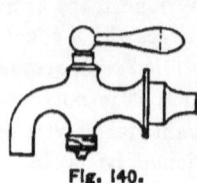
Fig. 140.

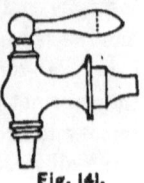
Fig. 141.

ing. Inside the bath is a single small mouth-piece through which the hot and cold streams issue, mixed, at the temperature desired.

The supply, may be placed close to the bottom of the tub, so that it fills noiselessly, but in this case, unless the ordinary pressure is very considerable, a check-valve should be put in the cold water supply, or the opening of a cold-water cock in a basin or sink on a story below may, if the main supply should happen to be cut off, draw water from the bath.

For sinks, wash-trays, basins, baths, slop-hoppers and similar apparatus, the supply of water is drawn from the pipes by means of cocks varying much in form and shape, and still more in construction. As a rule, the particular variety to be employed is mentioned in the specification, but there are some architects, and many more builders, who are content to regard the whole science of plumbing as a mystery beyond their comprehension, and either pass over such details in silence, or specify the cheapest varieties without regard to the conditions under which they are to be used.

Ground-Cocks.

All the forms of cocks, however varied in appearance and use, belong to one or the other of two great divisions, the "ground" or the "compression" cocks. The ground-cocks operate by means of a plug, which is inserted into the bore of the pipe, and is itself pierced with a hole of nearly similar size, so that when the plug is turned in such a way as to bring the hole into the axis of the bore of the pipe, the water runs freely through it; but by turning it at an angle with its previous direction the solid part of the plug is brought across the water-way and the flow cut off, either partially or wholly, according to the length of the arc through which it is turned. Figures 140 and 141 show two common forms of ground-cocks, and Figures 142, 143, and 144 explain the action of the pierced plug, Figure 140 showing a horizontal section of the cock when open, fully Figure 143 when partially closed, and Figure 144 when wholly closed. Of course the efficiency of such a cock, which is exactly the same in principle as the key of an ordinary gas-fixture, depends upon the accuracy with which the plug is ground into its seat; and to provide for tightening it after the surfaces have become abraded by the friction of use, it is customary to taper the plug and to insert a set-screw in the bottom, acting upon a strong spring interposed between it and the main body of the faucet, so that a turn of this will draw the plug down further into its seat.

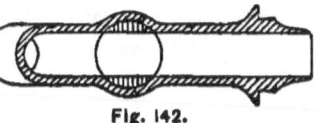

Fig. 142.

Fig. 143.

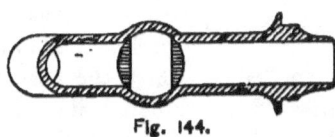

Fig. 144.

This answers very well with clean water, but many public and private wells, reservoirs, pumps or other sources of supply deliver water containing small particles of sand, which are drawn into the faucets and get between the plug and its seat, where they soon cut small grooves which allow water to pass around the plug at all times, and the faucet drips persistently, notwithstanding the tightening of the set-screw. Such a leak cannot be remedied, except by putting in a new faucet, which involves a considerable expense. This difficulty is obviated by employing the compression or screw-down cocks, through which the water flows in a devious course, passing at one point through a strong metallic diaphragm, pierced with an opening of the requisite size, but capable of being closed by the application of a piston, armed with a leather or rubber washer, which is brought down upon it by a screw operated by a cross-handle from the outside. There are many varieties, some employing a lever instead of a screw, and closing with the flow of water instead of against it. but the principle is the same in all: the current being controlled wholly by the movement of the soft washer, the grains of sand which may be present can do no worse than lie on the top of the diaphragm, and indent themselves in the leather as it descends upon them, without affecting the metallic portions of the cock in the least; and when the washer is worn away, as it will be in course of time, a fresh one can be inserted by any person in a few minutes.

Compression-Cocks.

The only common form of apparatus which remains to be described is the water-closet, of which many varieties are used. The worst of these in principle is the ordinary pan-closet (Fig. 145), in which a pan, so arranged with valves and cistern or other supply as to be kept full of water when at rest, is held up by a counter-balance weight against the lower orifice of a stoneware basin. The pan, counter-balance and other working parts are enclosed in an iron "container," or "pot," above which the basin is set, while itself discharges through a lead S-trap into the main soil-pipe. Upon occasion the pan is tilted by pulling the handle beside the seat, as shown by the dotted lines, and throws its contents into the "pot," whence they find their way into the trap, and thence to the drain. The dashing of the soil and water against the rough inner surface of the cast-iron container

Water-Closet.

Pan-Closet.

soon smears it with filth, which decomposes, evolving much foul vapor, and although the water-seal of the trap below prevents the gases of the sewer from rising through the closet, the same barrier serves to hold within the container the effluvium which proceeds from the matter adhering to its walls till the unclosing of the mouth of the basin above by the raising of the handle, and consequent tilting of the pan, allows it to escape upward, which it does in a nauseating whiff, familiar to every one.

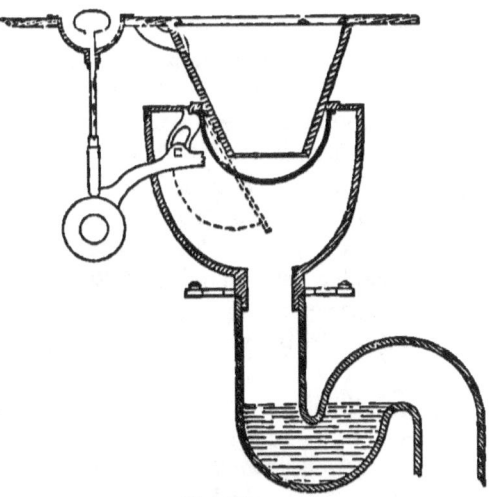

Fig. 145.

Besides this avenue of occasional exit a smaller one is always open, through the journal in which the pivot works which serves to tilt the pan. This might be made gas-tight, but in practice never is, and after the putty which is put around it on first setting is broken away by the movement of the pivot a continual leakage of effluvium takes place. Various devices have been employed to lessen this annoyance: the best closets of the kind, in places where they are not yet superseded by very different apparatus, have the containers enamelled with smooth white porcelain, which is more easily washed by the flushing water; and others are fitted with ventilating tubes, to give a safe outlet to the air of the container; but the amelioration of the evil is only partial, and good plumbing work now admits no form of water-closet which gives any lodgment for filth, or opportunity for the generation of gas, on the house side of the trap, this, if ventilated as it should be, forming an effectual barrier against the return of any kind of vapor.

One variety of improved closet which is now popular, and if well made is very good, substitutes for the bulky pan under the bowl a small valve, which fits tightly against the mouth of the bowl, and holds the water in at the height determined by an overflow until the lifting of the handle drops it, and allows the contents of the bowl to escape into the trap. As in the case of the pan-closet, a receiver must be arranged under the bowl, to give room for the movements of the valve, but this is very small compared with the container of the pan-closet, so that it is well washed by the passage of the large body of water discharged from the bowl, and its surface being enamelled, little or nothing remains in it to decompose. Nevertheless, it is necessary for a first-class job to ventilate even this small receiver by a separate air-pipe. If this is done, not only is the vapor rising from a possible contamination of its walls prevented from issuing into the house on the opening of the valve, but that from the water of the trap below is withdrawn safely. It is now usual, in closets of this kind, to make the bowl with a "flushing-rim," or **Flushing-rim Closets.** pierced pipe formed in the earthenware itself around the upper edge. The flushing water enters this and is carried entirely around the rim, descending on all sides and washing every portion of the bowl.

This flushing-rim is an essential requisite of a good closet. Without it the best patterns fail to give perfect satisfaction, and with it a very simple closet can be kept thoroughly clean. In fact, the practice of the best architects is gradually inclining to the **Hoppers.** use under all circumstances of a closet consisting merely of a well-made stoneware bowl or "hopper," with either a stoneware trap made in one piece with it or a separate lead trap, as circumstances may decide, and furnished with a flushing-rim, without valves, pan, or moving parts of any kind. The supply is so arranged as to deliver a considerable quantity of water with a sudden rush all around the rim, and the effect is not only to wash the bowl thoroughly at each discharge, but to urge the contents of the trap forward with such force as to insure their passage into the soil-pipe, leaving only clean water to form a seal. Without the sudden and copious downward supply, filth is sure to be left floating in the water of the trap, exposed to sight and smell, so that hoppers should only be used where this can be insured.

A fourth variety of closet should be mentioned. This is the plunger-closet, of which the earliest form was the Jennings patent (Fig. 146), consisting in an earthenware bowl with side outlet and trap beneath in one piece with the bowl. The water is retained in the bowl by a plunger, as shown in the figure, fitting against a rubber seating. When the plunger is lifted by means of the handle, the contents of the bowl flow out beneath it, into and over the trap. The overflow in the original Jennings closet was arranged to take place through the plunger, which was made hollow for the purpose, but effluvium from matters which might be floating in the trap passed readily up through the same avenue, and escaped into the room around the handle by which the plunger was moved.

Plunger-Closets.

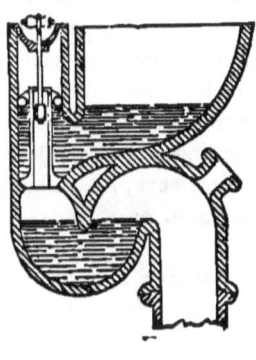
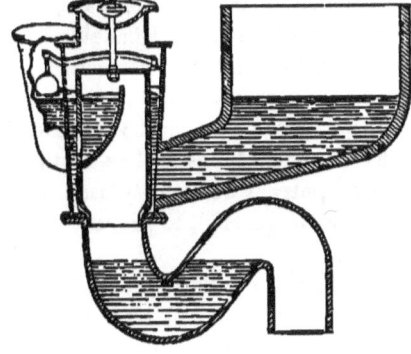

Fig. 146.   Fig. 147.

The supply was delivered through a delicate valve, which required to be adjusted to the pressure for each closet, and where this was variable, often allowed water at times to run to waste through the closet and the overflow. Both these defects have been remedied in various ways. In the Demarest pattern (Fig. 147), a small cistern is attached to the bowl, communicating with it, and a ball-cock in this affords a reliable means of shutting off the supply when the basin is full. The overflow takes place through the plunger, which is considerably larger than in the Jennings apparatus, but is trapped in its passage

All the closets of this description, as indeed of any other, require careful use, to prevent filth from lodging upon or about the plunger, where it decomposes, sending up faint odors around the handle. If this should happen, the cup can be unscrewed and the plunger lifted out and cleaned in a few minutes, but even this trouble is to be avoided if possible.

The main point to be observed in setting closets is the security of the connections. The old-fashioned bowls have earthenware "horns" to receive the supply, and the only joint possible between these and the lead pipe is made with a mass of putty, tied with a rag; but all modern apparatus has brass ferrules or couplings baked into the earthenware horn, and a permanent soldered joint is easily made. In the same way, the connection between the outlet of the container, hopper, or trap, as the case may be, whether of iron or stoneware, and the lead pipe or trap beyond, is still usually made by "flanging" the end of the lead pipe out on the floor, inserting the projecting mouth of the closet, screwing it down, through the lead, to the floor, and daubing the whole with putty; but the very best apparatus is now provided with brass ferrules at the outlet as well as at the inlet, securing a perfect joint. Where putty is necessarily used about plumbing work, it should be mixed with red lead, to prevent rats from eating it, or it will soon disappear.

Connections.

At the end of this Part will be found a plumber's specification, which will serve to call attention to certain points which need not here be mentioned.

The plumber's work continues, simultaneously with the other building operations, nearly until the completion of the house, so that in point of time the description of a considerable portion of it belongs with work mentioned hereafter, but it is more convenient to gather together in one place whatever needs to be said upon the subject, and one more matter remains to be explained before we dismiss it entirely from our minds.

After the contract is completed, and the connections made between the bath-boiler and the water-front of the range or cooking-stove, the water should be turned into *all* the pipes, in order that imperfect joints, or the holes frequently caused by the careless driving of nails, may be detected and remedied. At some subsequent time, when the traps have been filled by use, the tightness of the drain-pipes should be tested with oil of peppermint. The oil is sold expressly for the purpose, in two-ounce vials, hermetically sealed by melting the glass together over the

Peppermint Test.

mouth. A man is sent up to the roof with the vial of oil, and after stopping up temporarily all ventilation or air pipes connected with any part of the drainage system, he breaks off the top of the vial, and pours the contents down into the soil-pipe, which will as a matter of course in any modern house project above the roof. A pitcher of hot water is immediately handed up to him, and he pours this down after the peppermint, and closes the mouth of the soil-pipe by stuffing in paper or rags. The peppermint is volatilized by the heat of the water, and the vapor, unable to escape, penetrates by diffusion every part of the system. Meanwhile, another man examines all the drain, soil, and waste pipes in the house, and if the operation has been properly conducted, the slightest odor of peppermint in the building will be conclusive evidence of some defect, either in a joint or pipe, which must be at once remedied. It is important that the man who carries and applies the peppermint should not be allowed to enter the house, as he is sure to carry with him some trace of the powerful scent, which will make the test useless. After the trial is over, the pipes above the roof may be unclosed, and if no leak has been detected, the plumbing can be pronounced safe. Plumbers often profess to apply this test, but do so in a manner which makes the result unreliable. Unless the apertures of vent and soil pipes are closed, a circulation is very apt to exist between the upper portion of the soil-pipe and the nearest air-pipe, which will, especially if no water is used to help the diffusion of the oil, carry off the fragrant vapor before it can penetrate into the comparatively stagnant atmosphere which fills the lower portions of the system.

The fumes of burning sulphur are sometimes substituted for the peppermint vapor, but the application is more troublesome, and the result no more satisfactory.

At or before this stage in the construction the furnace should be put in, and the cellar floor concreted. If left, as is often the case, till a later period, when the kiln-dried finishings or floors are in place, these are very apt to absorb dampness from the mass of wet cement in the basement, and lose their shape or their glossy surface.

Furnace.

We have to deal with a client who insists upon an ample supply of fresh air at all seasons, and have therefore advised him to select a furnace possessing as large a radiating surface as possible, in order to secure the delivery into his rooms in cold weather of an abundant supply of moderately warmed air; and in accordance with this intention we have provided for large hot-air pipes and registers everywhere. That this will involve greater expense, both in the original cost of the apparatus and in consumption of fuel, than would be necessary for obtaining the same amount of warmth by means of a smaller volume of hotter air from a furnace with less radiating surface, we have frankly told him, but he is wise enough to think that true economy lies in sacrificing something for the sake of the health and good spirits which only fresh air can give.

He is, indeed, so bent upon securing a perfectly pure atmosphere in his house in winter as to be quite alarmed when we propose to him the purchase of a cast-iron apparatus, and reads to us extracts from the circulars of various manufacturers of wrought-iron furnaces, which, as he says, "prove" that carbonic acid, carbonic oxide, and other deleterious gases "pass freely through the pores of cast-iron," and escape into the house. We assure him that this danger is greatly exaggerated, if not entirely imaginary; while the large radiating surface, which is absolutely essential to the effect which he desires, can be had only in one or two costly forms of wrought-iron furnace, most of these consisting simply of a short cylinder of sheet metal, inverted over the fire-pot, and presenting a very limited, but very hot surface to the air flowing past it. As air can only be warmed by *actual contact* with a heated body, such a furnace, if set in a large casing, with ample supply of air, instead of warming the whole to a moderate degree would heat a small portion intensely, leaving the remainder as cold as ever, and the registers would either deliver into the rooms alternate puffs of very hot and very cold air, or certain rooms only could be heated, at the expense of the others. To be operated successfully, this sort of apparatus must be fitted with a *small* air-chamber, and small pipes and regis-

Small Radiating Surface. ters. The casing of the air-chamber is then heated by radiation from the dome of the furnace, close by, and the small volume of air which passes between the two surfaces is thoroughly and strongly warmed, acquiring thereby a powerful ascensive force, which throws it easily in any required direction through pipes of appropriate size, and heats the rooms above, not by introducing a full volume of warm air, but by means of a small current of very hot air; which mixes with that already in the apartment, so as to raise the whole to the required temperature.

To obtain an abundant supply of moderately warm air, it is essential, as furnaces are now constructed, to provide a large air-chamber, and distribute the smoke-tubes and other radiating surfaces in it in such a way that the air cannot pass through without striking one or more of them. The air issuing from such a furnace will all be warm, instead of partly cold and partly hot, as it would be without this division of the heating surface; and the quantity being greater, a much lower temperature will suffice to produce the same effect in warming the rooms above. *Large Radiating Surface.*

Most cast-iron furnaces are designed with special reference to this end, which has been recognized as desirable ever since heating apparatus first came into use; and many of them secure it tolerably well. Unfortunately, the castings are sometimes defective, and the joints are subject in several forms to separate or break by the effect of expansion and contraction, with the result of allowing smoke and gas to escape and mingle with the fresh air in the pipes. We have, however, selected a pattern in which the castings appear on close examination smooth and sound, and the joints, while occurring at the most favorable points, are all put together with short sleeves, which allow of expansion and contraction without harm. Unlike most furnaces, which receive the air from the cold-air box a little above the level of the ash reservoir, ours is intended to stand over a pit dug in the cellar floor, into which the cold air is brought by an underground conduit, to circulate first beneath the pan of ashes before it ascends among the hotter surfaces above. Unless the ground below the cellar bottom is well drained, such subterranean conduits are liable to infiltrations of unwholesome moisture, and this point should be determined before the choice of apparatus is made. The air trunk is made of brick, with brick bottom, plastered with cement, and covered with flag-stones. The pit into which it opens is walled with brick; the same wall being extended upward if the furnace is to be "brick-set," or forming merely the foundation for the sheet-iron casing if the "portable" variety is used. A brick pier in the centre serves to support the heavy castings above. Into the further end of the brick trunk is cemented the cold-air box, of iron or wood, which brings air to the furnace from a window or other opening.

178   BUILDING SUPERINTENDENCE.

The superintendent must see that the cold-air box is not made too small. The obvious rule for determining the proper size is that it should be capable of conveying into the furnace-chamber as much air as is to be drawn out by means of the registers; or, to put it in another way, that the capacity of the cold-air box should be equal to that of all the hot-air pipes which will ever be in use at one time, less one-sixth, which represents the gain in volume which the air acquires by expansion in passing through the furnace. In our present example, the registers in the parlor, dining-room, hall, and staircase-hall are, for the sake of insuring an abundant ventilation, supplied through circular tin pipes, twelve inches in diameter. Two chambers in the second story have ten-inch pipes, two have eight-inch, and the bath-room has a six-inch supply. The aggregate sectional area of these will be, expressed in square feet, $(1^2 + 1^2 + 1^2 + 1^2 + (\tfrac{10}{12})^2 + (\tfrac{10}{12})^2 + (\tfrac{8}{12})^2 + (\tfrac{8}{12})^2 + (\tfrac{1}{2})^2) \times .7854 = 5.13$ square feet. Six-sevenths of 5.13 will be 4.39 square feet, and this will be the necessary minimum sectional area of the cold-air conduit to insure a supply of warm air at each of the nine registers, in case they are all open at once, as they should generally be. If the cold-air box is made smaller than this calculation would require, the flow of warm air at the registers will be feeble and uncertain, or "wiredrawn," or perhaps at some of them it may cease entirely, or even be reversed by the draught of the longer pipes, which, unable to obtain through the contracted cold-air box the quantity which they require, draw down through the registers nearest the furnace an additional supply.

Capacity of Cold-Air Box.

Where so liberal a provision of fresh air is to be made, it is particularly necessary to see that the outer opening of the supply-conduit is not so situated as to be unfavorably acted upon by the wind. It is usual to place the opening toward the north or west, as the coldest winds come from those points, and while they blow the air is drawn through the furnace with greater rapidity than usual, and if fire enough is kept up, the supply of warm air at the registers is correspondingly increased. There is, however, under these circumstances, some danger that a high wind may drive the air through the conduit so rapidly that it cannot stay long enough to get warmed on the passage, and blows out from the registers in a chilly stream; and to guard against this a slide-damper

Air-Supply.

is usually inserted, which can be partially closed to temper the force of the incoming blast; but if a change then takes place in the direction of the breeze, the furnace is left without its needful supply of air. Occasionally the single inlet proves to be the source of still worse troubles. If, while it remains open, a violent wind should spring up from the quarter opposite to that toward which it faces, the partial vacuum which always exists on the lee side of a building may become so decided as to cause air from within the house to flow toward it by the most direct channel, which will be downward through the registers, into the air-chamber of the furnace, and thence by means of the cold-air box to the outside. By this reversal of the ordinary course, not only are the rooms deprived of heat, but the air drawn from them at a comparatively high temperature becomes intensely hot in passing again over the radiating surfaces of the furnace, and may even, if the cold-air box is of wood, be the means of setting this on fire, and with it the house itself. This is a much more common accident than most persons imagine, and safety as well as comfort make it important to guard against the causes which may occasion it.

The simplest way of preventing reversed currents in the cold-air box is to give it *two* openings to the outer air, as nearly as possible opposite to each other; then, whatever may be the direction of the wind, the air cannot be drawn out of the furnace. It may, however, still blow through the registers, and a still better mode is to carry the cold-air box entirely across the building, at a little distance from the furnace, opening to the outside at each end, drawing from this the supply to the furnace by means of a short, but sufficiently capacious pipe, opening into the main conduit, at right angles to it. Then the wind may blow at will through the main trunk, without affecting the current in the short pipe, which will continue to draw at all times just the supply that the furnace needs, and no more.

Whether the cold-air box shall be made of wood or metal is a question to be decided according to the circumstances of each case. Galvanized iron has the great advantages of being impermeable, so that no cellar air can be mixed with the pure current from out of doors or its passage through the furnace and the pipes, and of being fire-proof, so that there will be no danger, however hard the wind may blow, of having

Material of Cold-Air Box.

the building set on fire by an unexpected back draught; but it is very expensive, and those who wish to secure its advantages must pay for them. On explaining the matter to our client, we find that even his enthusiasm for fresh air is a little damped at learning that a galvanized-iron air-box for his furnace would cost more than the furnace itself, and he takes the question under consideration for a few days; but an inspection of the wooden air-boxes in the houses of his friends shows them to be full of crevices, sometimes large enough to admit the hand, and in all cases quite capable of allowing an unlimited amount of cellar air and dust to be drawn into the furnace and discharged into the rooms above, so that he finally declares in favor of an impervious conduit at any cost.

Concrete. The concreting of the cellar floor is generally done before the furnace is set, to avoid spattering the iron or brickwork with mortar, but all air conduits, ash pits, and other work below the floor level should be completed, to avoid breaking into the concrete coating subsequently. The thickness of the stratum should be determined by circumstances. On very soft soil four to six inches may be necessary to prevent settlement and cracks, but under ordinary conditions it may with safety be made three inches thick. If less than this the falling of heavy weights upon it, or long-continued movement on it, may break it, and when once fractured it crumbles and deteriorates rapidly. Only the very best fresh cement should be used, otherwise the concrete will be weak, and attrition will reduce it easily into dust. The proper proportion of ingredients is one shovelful of cement to two of sand, first well mixed together, and then quickly stirred up with three shovelfuls of screened pebbles or broken stone, and immediately spread upon the floor. The country masons, to whom a cellar without water in it in spring would seem almost abnormal, generally give the concrete an inclination to some point from which it can be drained away to the outside, but we do not intend, and if our directions have been thoroughly enforced we need not fear, that any water will penetrate our walls, and no provision is therefore necessary for carrying it off.

Joinery. After the furnace is set and the concreting finished, and while the plumber is completing his work, the joinery of the house will be going on. The first step is usually the setting of the door-frames, and the superintendent will need to refresh his memory in regard to the specified sizes and heights of the doors, and measure each frame as it is set, or risk finding, too late, that transpositions of the most annoying kind have been made in them. The frames are also very apt to be set out out of square (Fig. 148), so that the door must be subsequently bevelled off to fit them, giving a slovenly appearance to the

Door-Frame.

Fig. 148.

whole work. This should be provided against by rigid testing with the try-square and plumb-rule. Pocket-rules are sold at the hardware stores, containing a level, and a folding steel blade, which can be adjusted so as to form nearly a right angle, and although their usefulness would be much increased if they were more accurately made, the young architect will find them of service. In default of some such tool, the diagonals of the frame may be measured with a string, or a piece of wood: if they agree the frame is rectangular, though not necessarily plumb.

The height, width, and rectangularity of the frames once verified, the position of the rebates should be noted to make sure that the doors will be hung on the side intended. It is usual to mark the swinging of the doors on the plans, but workmen rarely trouble themselves to look at the drawings for information in regard to such matters, and after it is too late to change them the doors are very apt, especially in inferior rooms, to be found opening across stairs or passage-ways, against gas brackets, or in some other inconvenient manner. While considering this point the superintendent may make sure that the doors will be hung on the proper edge, as well as the right side of the partition, by marking the position of the hinges on the frames in accordance with the plans, or perhaps modifying these if circumstances render it advisable.

Swinging of Doors.

Next comes the application of the "standing finish,"—architraves, wainscotings, and bases. Modern moulded work is almost invariably cut with revolving knives, under which it is drawn by fluted cylinders, whose edges, in order to obtain a firm grip of the piece, press so strongly against it as to cause slight transverse indentations on the prominent portions, varying from

Standing Finish.

a quarter to a third of an inch apart, which injure its appearance very seriously unless the marks are subsequently smoothed off with sand-paper. For hard wood even this will not be enough, and the flat surfaces should be dressed with an ordinary plane to prevent the reappearance of the ridges after polishing. Sheathing boards and mill-wrought stock of all kinds for good interior work should be smoothed with the plane in the same way. This adds considerably to the expense, and cheap contractors will shirk it if they can, but it should be insisted upon.

Another way in which the inferior class of builders often try to gain some advantage for themselves is by "splicing" architrave **Splicing of Mouldings.** mouldings (Fig. 149), out of short pieces. As the mouldings come from the mill in lengths of ten to fourteen feet, there is considerable waste in cutting unless some mode is provided for utilizing the short pieces, but the appearance of a spliced architrave is so bad as to make it inadmissible in good work. Horizontal finish, such as bases, wainscots, wooden cornices, and chair-rails, must, however, be spliced, and care will be necessary to see that the adjoining pieces are properly matched, and that the joints do not come in conspicuous situations.

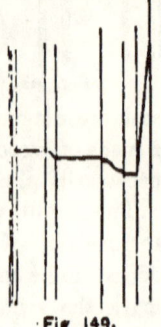

Fig 149.

The stairs will need constant attention to secure a satisfactory result. Before any work is done upon them, the superintendent should examine the plans to make sure that none of the flights **Stairs.** are too narrow or too steep, and that there is ample head-room where any passage is intended beneath them. The draughtsmen who prepare working plans from the architect's sketches sometimes fail to comprehend fully the structure which their drawings indicate, or forget to make the necessary allowance for thickness of floors, stair-timbers, and landings. Where a passage-way or flight of stairs is planned under another flight, the **Head-Room.** clear vertical height beneath the latter may be ascertained by counting the number of risers to the point where the head-room is to be calculated, multiplying this quantity by the height of each riser as found by dividing the total distance from floor to floor by the whole number of risers, and subtracting from the dimension so found at least eighteen inches, which will represent the vertical measurement from the top of the tread, just over the riser, to the under side of the plastering, in ordinary stairs. If the flight is steep, twenty inches, or even more, must be taken, while with long, straight flights it is often necessary to reinforce the stringers with whole timbers, or "carriage timbers," (Fig. 150), set parallel with them, but at a sufficient distance below to clear the inner angle of the steps, which will increase the total vertical depth between top of tread and under side of plaster to thirty inches or more. It should be remembered also that a person in descending a flight of stairs usually leans forward, and that the headway under a trimmer beam which appears ample where the vertical height alone is taken may prove insufficient in execution.

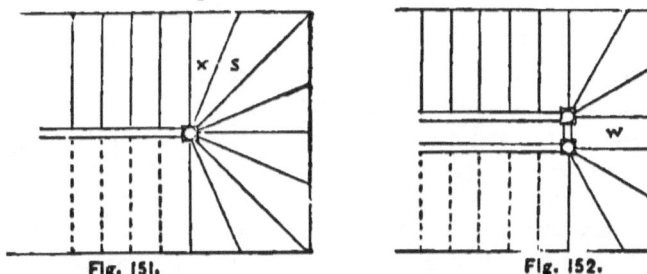

Inferior stairways often show defects of planning, independent of miscalculations in regard to head-room. It is not uncommon to see

Fig. 151.   Fig. 152.

such stairs indicated as are shown in Figure 151, where a person descending in the dark might, with a single step, from S or X, fall three or four feet. The stair-builder would probably take upon himself the responsibility of changing the flight from the "dog-legged" form shown in Figure 151 to an "open newel," (Fig. 152), adding a square "step in the well" W, and putting two instead of three "winders" in each turn, thus making the stairs comparatively safe and convenient, though narrower than the first plan intended; but it is unsafe to depend upon his thoughtfulness to correct errors. Another fault often seen in back stairs is the filling up of the well on one floor by a closet, while the stairway below is left open. (Fig. 153.) The point of the floor, P, is in this case apt to project over the stairs in dangerous proximity to the heads of those passing up or down, and should be protected by sheathing, at least on one side, down to the rail.

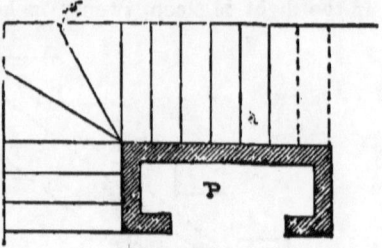

Fig. 153.

The steps should not be too steep. For inferior stairs the risers may be 8 inches and the treads 9 inches, to which the nosing will add 1½ inches more, making the whole width of the step 10½ inches; but this should be regarded as the limit. As the height of the risers is diminished, for superior staircases, the width of tread must be increased; the best rule being that the product obtained by multiplying the measure in inches of rise and tread together should not be less than 70, or more than 75. Seven and a half by ten inches is suitable for ordinary cases; seven by ten and a half is unusually easy, and six by twelve gives an air of old-fashioned luxury to a staircase. Some ancient mansions possess flights which rise only five or five and a half inches at each step, but these are hardly comfortable to our unaccustomed feet.

Height of Risers.

The greatest care should be taken to see that the staircase as executed will correspond with the plans, or, if mistakes in framing or miscalculations in regard to headway should have rendered this impossible, that the difficulty is remedied in the best way. It is a very common experience with young architects to be obliged to modify their designs on this account, and their ingenuity, as well as the patience of the stair-builder, will often be severely taxed to extricate themselves with credit from an unexpected difficulty. The stair-builder himself is liable to errors, and his work should be examined with particular care at the outset, in order, if necessary, to set him right before the progress of the building has made it difficult and expensive to remedy faults which would have been trifling if discovered earlier.

BUILDING SUPERINTENDENCE.   185

Even in putting up the rough "stringers" it is not uncommon to see mistakes made, which if passed over will spoil the effect of the finished structure.

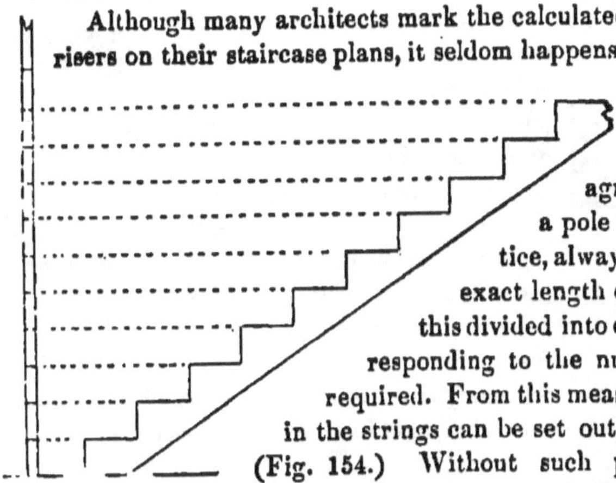

Although many architects mark the calculated height of the risers on their staircase plans, it seldom happens that the actual and theoretical distance from floor to floor agree exactly, and a pole should, in practice, always be cut to the exact length on the spot, and this divided into equal parts, corresponding to the number of risers required. From this measure the notches in the strings can be set out with accuracy. (Fig. 154.) Without such precaution, the strings may, on arrival at the building, be found a fraction of an inch too low, so that the top-most step must be blocked up to a greater height than the rest in order to gain the floor (Fig. 155.) More frequently, the string will be found to have been cut too long, and one of the steps must be made shallower than the rest, or the strings must be allowed to lean backward. (Fig. 156.) Such misfits should be sharply looked out for, and condemned immediately if detected. The cutting of a new set of strings is a small matter, while stairs of varying height, or out of level, are dangerous as well as unsightly.

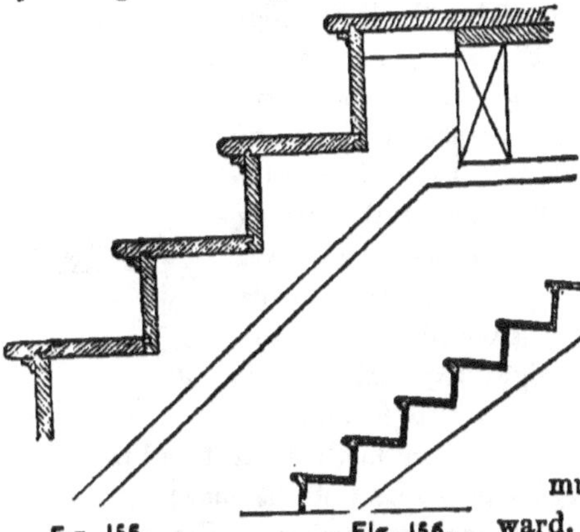

Fig. 154.

Fig. 155. 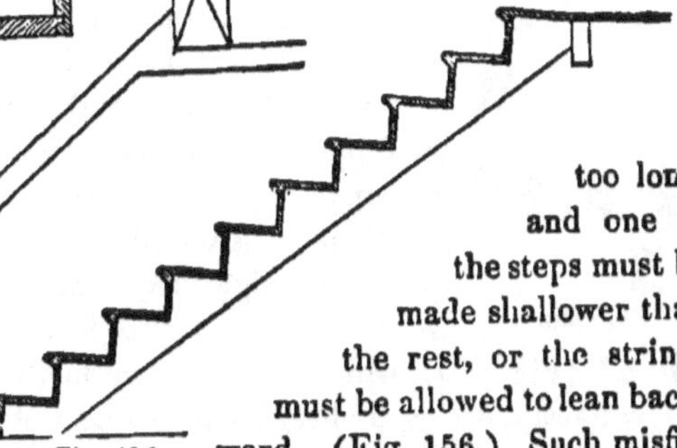   Fig. 156.

Stair-builders and carpenters often content themselves with very frail supports to their work, especially on landings, and this point should receive careful attention. After all is ready for the steps, the risers are sometimes first put on, the back edges of the treads inserted into grooves cut for them in the risers, and the nosings finally put in place (Fig. 157), and glued, securing the whole together. Generally, however, the risers and treads are put together, blocked and glued at the shop, and the steps brought complete to the building, ready for setting in place. In this case the nosings are not ploughed into the under side of the treads, but are simply nailed into the angle formed by the riser and the nosing. (Fig. 158.)

Fig. 157.

Fig. 158.
Fig. 159.

Whether the necessary connection between the steps themselves shall be made by grooving the inner edge of the tread into the face of the riser of the next step above, as shown in Figure 157, or by inverting the process, and ploughing the lower edge of the riser into the top of the tread below (Fig. 159), is a matter about which the practice of stair-builders varies. Those who choose the latter mode justify their preference on the ground that by the shrinkage of the timbers and settlement of the strings, the support may be taken away from the inner edge of the tread, throwing, if this is grooved into the riser, whatever weight may come upon it on the tongue, which is liable to split off (Fig. 160), while by the other method the tread is free to follow its supporting timber, the only result of the movement being the partial drawing out of the tongue in the riser. (Fig. 161.)

Fig. 160.

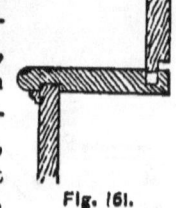
Fig. 161.

In practice, however, the first system is preferable for stairs over which the traffic is as light as is usual in dwelling-houses. When the risers are tongued down into the treads, the tongue necessarily escapes the painting or other finish, so that when it begins to draw out, a streak of a different color is exposed at the bottom of each, forming a very conspicuous defect.

The molded "nosing" of the steps should be formed as indicated on the sectional sketches, the front of the tread being rounded, and ploughed beneath for the insertion of the upper edge of the riser. In "open string" stairs (Fig. 162), where the level top of the tread

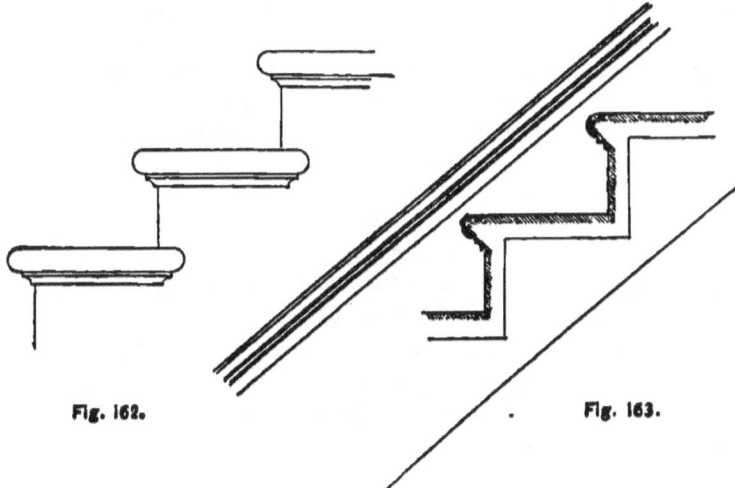

Fig. 162.   Fig. 163.

appears at the end, the nosing is continued across it by means of a piece of wood moulded to the full shape, one end mitering into the angle left for the purpose at the front corner of the tread, while the other "returns on itself" at a point vertically under the edge of the next riser. Before these pieces are finally put on, the dovetails at the edge of the tread, intended to hold the balusters, should be cut out. The balusters may then be fitted **Dovetailing Balusters.** in, and the nosing being nailed firmly in place holds all secure. The best stair-builders tack the pieces temporarily together, so as to insure a perfect fit, before the final nailing. Builders of the poorest class sometimes dispense with the dovetailing of the balusters, and simply fasten them in place by a nail driven diagonally through the foot into the tread after the nosing is finished: this gives a weak as well as uneven balustrade, and should never be permitted. The upper ends of the balusters are almost always secured with nails to the hand-rail, even in the best work.

With regard to the finish around the inner ends of the stairs, the practice of different localities varies somewhat. Whether it consists of a wainscot or a simple base, it is in many places customary to trace upon the lower portion the exact profile of the stairs (Fig. 163), including the nosings, and sink it to a depth of half an inch by means of chisels and gouges. This "wall string" base or wainscot is fixed to the walls before the stairs are put up, and the ends of the steps as fast as put on, are "housed" into the grooves ready to receive them. If nicely done, this is a strong and handsome mode of fitting but the workmanship must obviously be very careful and accurate.

In New England a different mode is adopted, rather easier in execution: for this, the treads and risers of the steps are grooved, before putting together, about ⅜ of an inch from the inner end, and the base or lower member of the wainscot is roughly "scribed" to the profile of the upper surface of the steps, and the lower edge then cut away so as to form a tongue. After the steps are all secured in place, the base is applied and driven home with a mallet.

The proper termination of the rail at the top of a staircase, where, being no longer continued upward, it must be carried across to stop against the wall, is a matter not always considered when the drawings and specifications are made. The best finish is obtained by

Half-Posts. placing a half-post against the wall; but if this is not mentioned in the contract documents, a makeshift, consisting of a round cast-iron plate, with a socket to receive the rail, and screwed to the wall, is likely to be substituted for it.

As soon as the treads and landings are in place, the broken boards of the under floor, and the places cut for the plumbers **Upper Floors.** and gas-fitters should be repaired, and the laying of the upper floors may begin. This is usually commenced in the topmost story, in order that each floor, as completed and planed off,

Spruce. may be swept out and the rooms locked. For floors intended to be carpeted spruce forms the ordinary material north of New York. It is cheap, and has the advantage of being very free from knots and defects, so that a room laid with it looks clean and handsome. The adhesion of the annual rings is, however, very slight in spruce timber, and boards taken off the outside of the log (Fig. 164), which may be recognized by their grain or "figure," like Figure 165, are liable, after being dried by a winter's furnace heat, to splinter up in a most annoying manner; or if the rings do not separate, the boards are likely to curl up (Fig. 166), forming ridges which rapidly cut the carpets laid over them. Both the defects will be avoided by choosing spruce boards in which the figure consists rather of fine parallel lines, indicating that the annual rings are divided in a direction nearly parallel to the radius of the trunk (B Fig. 164). Very white, clear boards with no apparent figure are often cut from sapling trees, but are soft, and liable to excessive shrinkage. In the Middle States pine is the favorite flooring material. It is softer than spruce, and little liable to curl or splinter. Clear pine, however, is very costly, and the second quality, which is generally used, contains small knots and streaks of "blue sap," so that a floor finished with it is not quite so agreeable to the eye as one of good spruce.

Sapling Spruce.

Pine.

Blue Sap.

Fig. 164.

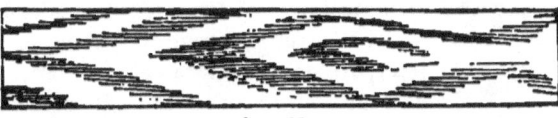

Fig. 165.

Fig. 166.

The mode of laying the boards varies with the locality. The New England carpenters, having a difficult material to deal with, have learned to treat it with great skill. Their upper floors of spruce, in the better rooms, are usually specified to be laid with boards not over four inches wide. Attics may have six-inch boards, but wider ones than these are only permissible in closets and store-rooms. The boards, whatever their width, are "jointed," or planed on the edges until these are made absolutely straight and parallel, then stacked in the kiln or "dry-

Square Joint Floors.

house" until all moisture is evaporated from them, and brought directly from the dry-house to the building. Beginning at one side of the room, they are, or should be, laid in "courses," from end to end of the room, breaking joints as frequently as possible. Where all the boards are of exactly the same width, as should be the case in the best rooms, the joints may, and should be, broken at every course, but as this involves some waste of stock, it is usually necessary to make "straight joints" through three or four courses, before the varying widths of the boards on each side of the joint will add up to an equal sum, so as to admit of its being crossed by the next course. A straight joint of more than four courses should not be allowed in rooms intended to be carpeted, for fear of causing a ridge; and it is hardly necessary to say that all heading joints should come upon a beam. "Flooring clamps" are used to force each board closely up to the side of its neighbor, and it is usual to tack the boards at first with a few nails only, and after all are in place to line them with a chalked string, or straight-edge and pencil, over the centre of the beams below, driving the nails to complete the work on the lines so marked.

Before laying the upper boarding, it is necessary in good houses to spread one, two, or three layers of felt over the under-boarding, in order to prevent air from passing through the joints, and also, by **Felt Deafening.** the interposition of a non-resonant material, to check the transmission of sound. Cane fibre makes a clean, dry material, which will not harbor moths, but a coarse felt of woollen rags is commonly used. Certain varieties are so prepared as to be incombustible, and, especially if used in several thicknesses, may prove valuable in preventing the spread of fire from one story of a house to another. The felt paper made for the purpose from asbestos is the best of these, but is very expensive; the Phœnix paper,

Fig. 167.

made in New York, costs but little more than ordinary felt, and resists fire very effectually, which cannot be said of many so-called "fire-proof" sheathing papers.

In the Middle States, particularly where ordinary floors are laid with a single thickness of boards, it is customary to match the boards (Fig. 167), as otherwise currents of air would come up freely through the joints from the spaces between the beams; and the influence of this habit, more than any real necessity, has made it customary to match also the upper boarding of double floors, even in inferior rooms. **Matched Flooring.** The matching of spruce floors is, however, not to be recommended, as the thin edge of the grooved side of the boards (Fig. 168), is apt to curl up or split off. Pine is better in this respect, but stays in place well enough for an upper flooring without matching.

Fig. 168.

Where hard woods are used for flooring, matching is, on the contrary, essential, since no nails must appear on the surface of such floors, and the only way of securing them to the beams is to drive the nails diagonally through the edges of the boards. **Hard Wood.** This process can, however, be applied only to one edge of each board, since the other is applied firmly against the side of the one which preceded it; and in order to hold the inner edge down, it is necessary to connect it with the outer edge of the preceding board, by means either of tongue and groove, or of some equivalent device. These floors are there- **Blind Nailing.**

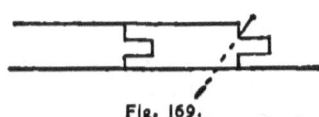

Fig. 169.

fore laid in narrow strips, each with the tongue projecting forward into the room, and the nails are driven diagonally into the upper angle formed by the tongue and the edge of the board (Fig. 169), securing this edge firmly; and the groove of the next strip is forced over the tongue so secured, so as to retain in place its inner edge, while its outer edge, furnished with a tongue, is in its turn nailed.

Parquetry work, such as we are to have in one room, is generally made with much more care than ordinary flooring, and requires special machinery, so that it is best to order it from a regular manufacturer. In some sorts the pieces of hard wood are but half an inch thick, and are dovetailed and glued upon a backing of pine. The patterns are put together in the factory, and sent out in sections some two feet square, which are nailed down like single boards. Simpler hard-wood floors may be made and put down by the carpenter. The oak floor of our hall, in accordance with the wish of the owner, is put down in the French manner with short pieces laid at an

Fig. 170.    Fig. 171.

angle of 45° with the beams, and at right angles with each other. (Figs. 170 and 171.)

Figure 170 shows the most common method, the edges of the pieces being matched, but the heading joints plain. In Figure 171 the heading joints are tongued and grooved by hand as the pieces are laid, so as to fit into the matching upon the edges. We choose the former mode, as the least expensive, and the easiest to execute. All the other floors are simply laid with narrow parallel strips, 2½ inches wide in the best room, 4 inches in the others.

It is of great importance that the under boarding, where a hard-wood upper floor is to be laid over it, should be also of narrow strips, not exceeding four inches at most; if wider boards are used, each one of them will in shrinking gather up, so to speak, a cluster of the narrow hard-wood pieces above it, and draw them tightly together, and although the shrinkage of each hard-wood strip, if well seasoned, is very slight, the movement of the wider board compresses all the joints over it, so as to transfer the total shrinkage to the joints immediately over its own edge. The adjoining wide board of the under flooring acts in the same way, but in the opposite direction, so that in a few months every board below will be exhibited by an inordinately wide separation between the hard-wood strips above, the other joints remaining perfectly close.

The stock for upper flooring will need close examination. Even with spruce it is necessary to see that waney pieces are not smuggled in, and to look out for knots and sap, while hard wood is liable to other defects. Oak for flooring, unless under severe wear, and, indeed, for all kinds of finish, should always be quartered, or "rift," as some say—that is, sawed with two cuts at right-angles with each other, and through the centre of the log, all subsequent cuts being as nearly as possible on radial lines. **Flooring Stock.**

As every one knows, oak is distinguished from all other woods by the "silver grain," or medullary rays, consisting of small bundles of fibres, which shoot out laterally from the centre of the trunk, passing through the annual rings toward the bark. By quartering the .og, these fibres are divided nearly or quite in the direction of their course, and show on the surface of the boards as flecks or irregular silvery streaks, upon a ground of fine parallel lines, formed by the section of the annual rings. If, on the contrary, the log is sawed into parallel slices in the ordinary manner, the middle slice will exhibit the silver-grain, as will also one or two on each side of it. Further from the centre the medullary rays will be divided almost transversely, appearing on the cut surface as nearly imperceptible lines or dashes, while the section of the annual rings will grow broader and broader, showing itself, since the sap-tubes of oak are quite large, as a coarse, rough figure, completely different in appearance from the delicate and silky silver-grain, and liable to a dingy discoloration from the entrance of dust and dirt into the exposed pores. With some varieties, oak sawed in the ordinary way often appears "brashy," or of a very coarse texture, with short fibres which break away easily. **Quartered Oak.**

The manner in which the log is sawn effects also its disposition to warp and curl, which in badly cut oak is very strong. The inner portions of the tree are compressed and hardened by age, so that there is a gradual diminution of density toward the circumference, which is occupied by the soft and spongy sap-wood. The less compact substance naturally shrinks more in drying than that which is nearer the interior of the log, but with boards whose surfaces follow the radial lines the movements caused by dryness or damp are all in the plane of these surfaces, and although the board varies in width, it has no tendency to warp. Those boards, on the contrary, which are cut in lines parallel with the diameter of the log, have one surface which looks toward the back of the tree, and the other toward the heart, and the fibres on one side are therefore slightly softer than on the other, and will shrink more, curling the piece outward with a force proportioned to its thickness.

By keeping constantly in mind these properties of oak, which belong in some degree to all kinds of timber, many annoying defects in hard-wood finish may be avoided. Following the same principle, the Georgia-pine floors for the inferior rooms should be specified of **Rift Hard-Pine.** rift stock; that is, of boards cut like quartered oak, on radial lines. These may be recognized by the figure, consisting of fine parallel lines, in place of the broad mottlings produced by a cut tangent to the annual rings. Hard-pine boards of the latter kind are very liable to splinter, like spruce cut in a similar way, and must be rejected. Hard-pine boards containing large streaks of dark turpentine are also unfit for floors, the turpentine soon crumbling away.

There are one or two points about the hard-wood finish other than the floors, which may be noted. Whitewood, or poplar, is not usually ranked among the hard woods, although it is little inferior in this respect to black walnut or butternut. It has the advantage over them of being clear, dry, and very uniform in texture. The annual rings are almost imperceptible, and the wood is little subject to any warping or checking. For large, solid piazza posts and other heavy out-door work, it is superior to any other material, and inside finish made of it is usually durable and satisfactory. It has the peculiar property of swelling considerably in damp weather, even when perfectly seasoned, to retreat again to its original dimensions under the influence of furnace-heated air; so that doors made of it should not be too tightly fitted. In selecting the pieces care should be taken to exclude those streaked with white sap. The rest of the wood darkens very much after finishing, while the sappy streaks remain white, and soon, by contrast, appear as disfigurements. Black sap, which also occurs, is not generally looked upon as objectionable, but it forms too strong a figure to be admitted in delicately-moulded work.

Hard-wood doors, except those of white wood, are usually veneered upon a core of well-seasoned pine, to prevent warping, and it is necessary to examine them upon delivery, to see that the veneers are of the proper thickness. Those which cover the panels may be $\frac{1}{8}$ inch; over the framing they should be specified $\frac{1}{4}$ inch, although the final planing which such doors undergo generally reduces this thickness somewhat. **Veneered Doors.**

There are innumerable points about the finishing of a dwelling-house which, though trifling in themselves, count for a great deal in the impression which the completed structure will produce upon the owner and his friends, and the architect or superintendent will do well to go thoroughly and repeatedly over the building during the finishing, and make sure that every visible detail is satisfactory before the contractor leaves it. **Detailed Inspections.**

First among the points to be examined is the hanging of the windows. The sashes vary considerably in weight, and unless each one is accurately balanced, which takes both time and care, the sash will not stay in place. **Balancing of Sashes.**
The sash-fasts may also be badly set, so that they will not lock, and nothing short of an actual trial of each sash of every window will serve to make sure that all are as they should be. **Sash-Fasts.**
Door locks and knobs are also very carelessly applied, and there are few houses where all of them work perfectly. The striking-plate, particularly, is apt to be set too high, or too low, or too far into the rebate, so that either the latch or the bolt will not enter the mortise intended for it; while the "roses," or round plates, screwed upon the opposite sides of the door, in which the stems of the knobs move, are rarely placed exactly opposite each other, so that the spindle, instead of being perpendicular to the door, is forced into an oblique direction, causing the knobs to bind and stick in turning (Fig. 172). The knobs, again, are generally put on without inserting the proper number of the thin washers which slip over the spindle for the purpose of filling out the space between the lock and the knobs on each side, and the latter are left loose in consequence, sliding in and out with the touch of the hand in an annoying way. **Door Furniture.**

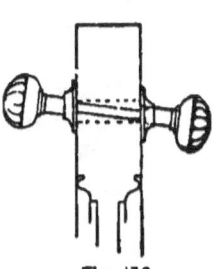

Fig. 172.

Many architects call in their specifications for "swivel spindles," with which the turning of the knob on one side of the door does not affect that on the other side; but except for those front doors which have an arrangement for locking the outside knob separately, the swivel is hardly necessary.

<small>Swivel Spindles.</small>

Chair-rails, picture-mouldings, wooden cornices, and other finish of the same kind will often be applied with the utmost carelessness. Nothing is more common than to see such horizontal mouldings varying very much from their correct position, the workman having put them on by what he would have called "his eye," instead of measuring at short intervals the proper distance from the floor or ceiling.

<small>Chair-Rails, etc.</small>

The young architect or superintendent should train himself to quick observation of all these points. Any defects are sure to be discovered sooner or later by the owner, to the discredit of the one whose business it was to look out for and correct them at the proper season.

<small>An Eye for Defects.</small>

The character of the hardware about a building is also of much importance. The variety of patterns and qualities of locks, knobs, pulls, bolts, hooks, hinges, sash-fasts, and so on, is so great that nothing but a thorough familiarity with the different kinds, and a minutely detailed specification, will protect the architect or superintendent from being occasionally compelled to accept fittings which he does not like, but which the indefinite character of the specification precludes him from rejecting as not in accordance with the contract.

<small>Hardware.</small>

The locks form a very important part of a building. Those used in good houses are generally mortise locks, inserted into a mortise cut in the edge of the door. The centre of the knob should be exactly three feet above the finished floor, and the mortise for the lock, in inside doors, will extend one inch above and three inches below this point. For outside doors, the mortise is generally six inches high — two inches above and four inches below the centre of the knob. In designing the doors, the panelling should be so laid out that the lock-mortise will come beside a *panel*, and *not* opposite a rail of the framing; as, in the latter case, the mortise will cut

<small>Locks.</small>

<small>Proper Position of Knobs.</small>

off the tenon of the rail, weakening the door very badly. For want of attention to this point, young architects often find that doors, in whose elegantly-proportioned panels they take the greatest pride, have to be fitted with handles set either at an immoderate height from the floor, or ridiculously low, to avoid making them altogether unserviceable. This is particularly likely to be the case in copying doors of the last century, either out of books or from actual examples. These show invariably at the level of the knob a wide *rail* instead of a panel, but it must be remembered that mortise locks were not in use then, their place being supplied by rim locks, in which the working parts were enclosed in an iron or brass box, screwed to the outside of the door, a small hole only being bored through the door for the spindle of the knobs. Of course, the tenon not being in this case interfered with, there was no reason why the lock should not be screwed on next the rail, and as many of the ancient rim locks, or the latches which were substituted for them in inferior rooms, were longer than the width of the "style" at the edge of the door, it was an advantage to place it where it could extend back upon the framing. Occasionally, a modern version of the Colonial doors is seen with a mortise lock set opposite a very wide lock-rail; but this must be done by framing the latter with two tenons, far enough apart to give room for the mortise between them, — an arrangement not to be recommended.

Arrangement of Door Panels.

Colonial Doors.

For closet doors, a mortise latch is sometimes used, with either one or two knobs, but no lock or key. The case for this is only about 2¼ inches high, but such furnishings are only suitable for inferior houses, a closet which cannot be locked being as inconvenient as a door with a knob on only one side is mean in appearance. The hand-made locks are far superior to those made by machinery, and also much more costly; but some of the machine-made kinds serve well enough for ordinary purposes. Any good contractor or hardware dealer can furnish the names of the most reliable manufacturers, and the safest course for the young architect is to require a first-class make by name in his specifications. The manufacturers' catalogues will furnish him with all necessary information as to styles, and he should call for exactly what he wants so clearly that there may be no mistake as to his

Closet-Door Latch.

intention, finally assuring himself by inspection that the contract has been carried out.

Reversible Locks. Reversible locks should be chosen, unless the architect is willing to see a door here and there hung on the wrong side to accommodate some carpenter who has selected his locks at random and finds himself short of the proper kind; and the character of the keys should be specified, or he may find a set of locks of tolerably good appearance accompanied by cast-iron keys, tinned or galvanized. For the inside doors of ordinary houses a mortise lock of P. & F. Corbin's make, for instance, with brass face and striking-plate, brass bolts and German silver or plated keys, does well enough. For the best rooms a "fancy" face, formed by grinding the brass in curling forms, may be used; or some expense may be saved by allowing the bolts to be of iron instead of brass. A cheaper lock still has an iron face, lacquered to imitate brass, but there is no real economy in using it.

Keys.

It is hardly necessary to say that locks for interior doors are usually of the simplest construction, the wards of the key merely fitting stationary projections inside the box, which give no security against opening by a skeleton key or a piece of stout wire; but it is not required in such cases to provide against the operations of professional burglars. These locks are commonly used in sets of twelve, the twelve keys differing from each other, but the complete sets being exactly alike, so that in a house with twenty-four doors there will be two keys of each pattern, which may, however, be distinguished by a difference in the finish, one set being bronzed and the other plated, for instance. For outside doors, which must occasionally be left without the security of bolts, "lever" or "tumbler" locks are needed, in which the interior construction is far more complex, and the security, as well as the cost, correspondingly greater. Many varieties of these locks are made, with and without night-latches, and inspection will furnish the best guide as to the arrangement desired. Some front-door locks are so arranged that the outside knob is permanently fixed, but the better ones are furnished with a movement by which it can be held firm, or released if it is desired to allow the door to be opened directly from the outside, without a key. Locks of either kind, of the common

Lever Locks.

construction, are somewhat liable to have the latch become slow in working, so that a sharp bang is necessary to close the door. Age increases this fault, which is only partially cured by oiling, and a more satisfactory service can be had from the patent locks, in which the latch is in two or three parts, one, which projects in front, turning on a pivot as it is drawn against the striking plate, and by the same motion drawing back the others, so that when the whole reaches its place, it slips out into its mortise without any friction of importance. Robinson's patent is the lock of this kind most used, but similar ones are perhaps made by other parties. **Patent Latches.**

Care should be taken, if locks and knobs are procured from different makers, as will generally be the case, to have the hole for the spindle correspond with the actual size. Most locks can be had fitted to either $\frac{3}{8}$ or $\frac{1}{4}$ inch spindles; and although $\frac{1}{4}$-inch is the size commonly used, there are many advantages in having the knobs mounted on $\frac{3}{8}$ spindles, with or without swivels. What shall be the material of the knobs must depend upon circumstances. Brass, bronze, cast-iron, hard rubber, glass, porcelain, celluloid, wood, and various compositions of sawdust and glue, dried blood, glazed earthenware and other substances are used. Among these, dark bronze metal of good quality is the most satisfactory. The light bronze, even when good, is apt to tarnish in rooms not much used, and the soft, inferior bronze wears to a dirty yellow color which is very unpleasant. Moreover, there is in bronze hardware a much greater variety of patterns than in any other kind, and knobs, hinges, bolts, chain-bolts, sliding-door pulls, sash-fasts and other furnishings can be so selected as to match in color and general appearance throughout the building. There is however much difference in the execution and finish of the castings, and it is unsafe to trust to the drawings in the catalogues of unknown makers without seeing samples of the work. **Spindles.** **Knobs.**

Polished brass furniture, where it is fashionable, is very costly, and requires continual attention to keep it bright. Silver-plated brass knobs soon lose their coating, and are becoming obsolete. Cast-iron is used for door-knobs only in a miserable imitation of bronze or brass. Hard rubber and celluloid make durable and pretty furniture for in-doors, but do not bear weathering well.

Glass is a good material, and can be had in great variety: the old-fashioned cut octagonal knobs are the handsomest, but those pressed in various forms are serviceable. The blown-glass knobs, silvered inside, are fragile unless of very good make. Porcelain and the vitrified materials known as "mineral" and "lava" with "hemacite" and some similar substances, are used for inferior rooms. Knobs of wood or its imitations are somewhat liable to become sticky from the softening of the varnish upon them. Whatever the kind used, the specification should not omit to mention a bell-pull for the front door to match the design of the other furniture; and should keep in mind also the sunk pulls for sliding doors, if there are any such.

Bell-Pull.

Bolts are necessary for all doors which need to be rendered tolerably secure against intrusion, and afford more protection than any lock. The neatest and most convenient are the Ives patent mortise bolts, which are set into a small auger-hole bored in the edge of the door, and show only a small key outside. Outside doors are frequently guarded by a chain-bolt, consisting of a strong ornamental chain, attached to the frame, or one leaf of a double door, which can be hooked into a slotted plate on the movable part of the door, and will allow the door to be unlocked and opened three or four inches, for conversation with a person outside, but prevents it from opening further until unhooked. This effectually resists the attempts of tramps to force their way past a servant into the house as soon as the door is unlocked, but as it can easily be dislodged from the outside by a wire, it should be used in addition to, and not as a substitute for, the ordinary bolt.

Bolts.

Chain-Bolts.

Sash-fasts of the ordinary kind are the least effectual of all domestic defences against the operations of burglars. By introducing the blade of a knife between the upper and lower sashes from the outside the lever can be easily pushed back and the window opened, and this is in fact the common mode of entrance for thieves. Of late years the necessity for preventing the movement of the lever from the outside has become so obvious that several devices, more or less perfect, are now in use for the purpose. The earliest form, still much used, has a spring catch with a thumb-piece attached to the inner plate, which, as the lever is

Sash-Fasts.

swung around, is first pushed back and then springs out, holding it in place until it is again pushed back by the thumb. This is convenient, and reasonably secure, but after some years' wear the catch becomes rounded by friction, and the lever may sometimes be forced back over it from the outside by a strong pressure. An improvement on this is a self-locking sash-fast introduced by Hopkins & Dickinson of New York, in which the lever itself is hollow, and contains a spiral spring, acting upon a pin which moves within the lever and has a knob-shaped head projecting at the end. In locking the window, as soon as the lever is turned to its place, the interior pin springs into a hole made for it in the pivot, preventing any back movement until the pin is withdrawn by pulling on the knob. This sash-fast is still very popular, but is closely followed by a simpler and stronger device, the Morris sash-lock, in which the thumb-piece of the lever is movable, and on being turned to its place drops down into a notch made to receive it in the circumference of the plate which carries it. To open the lock from the inside nothing is necessary but to lift the thumb-piece from its notch and turn back the lever by a single motion; but to open it from the outside is impossible. Various styles of all sash-fasts are made, to correspond with bronze, silvered, or the cheaper kind of metal furniture. *Hopkins & Dickinson's Sash-Fast.* *Morris Sash-Fast.*

Hinges form the only other article of importance in the hardware dealer's order. Solid bronze metal, polished brass, silver plate, "Boston finish,"— a brown lacquer over iron, resembling bronze, — black japanned iron, either plain, or with silver or bronze tips, and plain iron, are at the command of the architect. If solid bronze, brass, or silver plate are used, only the best quality, with steel bushings and steel washers should be used; as the softer metal wears out rapidly from the movement of the door. Iron, either japanned or Boston finish, for hard-wood doors, with tips either of the same or of solid bronze or plated, and plain for doors intended to be painted, forms on the whole the best material for ordinary dwelling-houses. Most houses are now fitted with " loose-joint " butts, which allow the door, after opening, to be lifted off and replaced without unscrewing the hinge. With heavy doors, however, there is danger of bending the projecting pin of the *Hinges.* *Boston Finish.* *Loose-Joint Butts.*

hinge during this operation, and it is better to require "loose pin"
butts, in which the pin itself can be drawn out from
Loose-Pin Butts. the top and the door removed and replaced, with as
much ease as in the other case, and greater safety.
Young architects occasionally forget to proportion the size of the
butts to the circumstances of their door-frames and architraves, and
find, too late, that the doors of their best rooms can-
Size of Butts required. not be swung back to the wall. Where the openings
are finished with unusually thick mouldings, Gothic
beads, or pilasters, the proper way is to make a horizontal section
of the door, with its frame and finish, *including* bases or plinth
blocks, and capitals, if there are any; then add the extreme projection of the trim from the plane of the door, to twice the thickness of
the door, and deduct half an inch from the sum; the remainder will
be the minimum width of butt which will hang the door securely
and throw it clear of the mouldings. If the result does not correspond with a regular size of hinge, the nearest size larger should be
specified. Butts are made of several widths to the same height, as
$4'' \times 4''$, $4'' \times 4\frac{1}{2}''$, $4'' \times 5''$, $4'' \times 6''$, and so on; the dimensions being
those of the hinge when opened flat.

For the remaining small items, as coat-hooks, drawer-pulls, and the
like, all that is necessary is to describe clearly in the specification
what is wanted. In certain cases drawer-pulls must
Small Items. be of fancy styles, but for closets japanned iron is
much the best material, and the simpler and smoother the pattern
the better. Architects and builders often go to a small unnecessary
expense in putting fancy cast-bronze or Boston finished pulls on
their cases of drawers, which serve only to bruise and excoriate
the fingers of those who handle them; and fit up rows of roughly-
finished bronze metal hooks, whose edges quickly cut the material
of clothes suspended from them, while the artistic knobs and curves
with which they are adorned always prevent them from being as serviceable as the plain, strong triple hook of japanned cast-iron.

It is important that the young architect should inform himself as
to the character and comparative cost of the various kinds of metallic house-furnishings, and describe distinctly what
Fixing Prices. he requires in the specification, without resorting to
the slovenly practice of specifying that the different articles shall

cost a certain sum per dozen, or per set, or per gross. The actual expense of such goods to the contractor is a very different thing from the cost as set down in the price-lists, and to specify articles of a given price, instead of a given kind, is usually to oblige the owner to pay a large profit on goods which he might have obtained for the net value if they had been distinctly described.

If the architect has been wise enough to demand specific articles of hardware from manufacturers of good reputation, the duty of the superintendent will require little more of him than to see that the order is correctly filled, and that the fittings are properly put on. **Inspection of Hardware.** If, however, the specification is one of the kind that vaguely stipulates that such materials shall be "good," or "neat," or "worth two dollars per dozen," he must prepare himself for a rigid inspection of the goods furnished in accordance with it. The bronze hardware of all kinds may prove to be of soft yellow metal, with a thin bronze finish over it, or even of iron skilfully lacquered or bronzed; the brass faces of locks and bolts may be fictitious, consisting of iron, varnished with yellow lacquer, or brass-plated; the silvered-glass knobs may be of a substance so thin, or so carelessly blown, as to crush in an incautious hand, inflicting frightful wounds; or any kind may be so feebly secured to the metal shank as to come off altogether upon occasion; or the hinges may be destitute of washers, and will soon creak painfully. The cases of cheap mortise locks are often made, to economize material, so short as to bring the knob within an inch of the edge of the door, so that the hand is scraped against the rebate of the frame whenever the door is shut; or sometimes the width as well as the length of the case is reduced, and the knuckles come into painful contact with the key on turning the knob; while occasionally a lock is seen which allows the door to be opened by turning the knob one way only, instead of both ways. The screws furnished for putting on cheap hardware are also generally too small, so that the fixtures are insecurely fastened; and the worst workmen will increase this fault by their fashion of applying them, which consists in driving the screw nearly home with a heavy blow of the hammer, finishing with a turn or two of the screw-driver. Such men also generally show an exasperating indifference to the appearance of their own or other's work, putting on bronze metal or japanned fittings with plain iron screws, instead of blued iron or bronze, and using them of different sizes, or several sizes too large, if necessary to save themselves the trouble of going after suitable ones; screwing hinges, bolts, or plates at random on veneered doors, and if they fail to fit, removing them and screwing them on again somewhere else, leaving two or three sets of screw-holes yawning in the polished surface of the wood, or in a hundred other stupid and blundering ways defacing the building which they help, after their fashion, to complete. Continual vigilance is needed to discover and correct such faults, and the superintendent of a house which on delivery to its owner proves to have all its hardware perfect, well put on, and in good working order, has at least some qualities which particularly fit him for his profession.

While the joiners' work is going on inside the house, the operations of drainage, grading and sodding outside should be com-

Drainage and Grading. pleted, so that the dust incident to them may be laid before the final painting.

Unless, as will rarely be the case in the country, drainage by regular sewers is provided, the first of the outside operations should be the selection of a site for, and the construction of, a

Cesspool. cesspool of some kind. Usually the position of this is marked approximately on the plans, or indicated in the specification, which is necessary in order to lay out the plumbing intelligently; but circumstances will often modify greatly the character of the construction as executed. For most houses, the ancient leaching cesspool or "dry well" is still adopted, as the cheapest means for disposing of house-wastes, but the architect should examine all the conditions with great care before lending his authority to this expedient. If the house is supplied with water from a town or city service, or from springs higher than the building and at a considerable distance, and if the lot on which it stands is so large that the inevitable poisoning of the ground by the soakage of putrefying filth will not affect its inmates or their neighbors,

Contamination of Wells. the leaching cesspool may be regarded, in view of the greater cost and trouble of other devices, as an evil to be tolerated so long as the favorable circumstances continue. If, however, water is to be drawn for use, either in the house or stable of the proprietor or his neighbors, from any well within three hundred feet of the proposed cesspool, and on the same or a lower level, the architect should refuse his sanction to any plan whatever for discharging sewage into the subsoil.

In rocky districts, and in places where deep wells are necessary, a much greater distance should intervene between them and any porous cesspools.

It is positively proved that the typhoid poison contained in refuse thrown upon the ground on a rocky hillside, and washed by the rain into some hidden seam or depression in the ledge beneath the surface, has been carried down with its qualities unchanged, to in fect with very fatal effect a spring, apparently of the purest water, a mile beyond; and it may generally be assumed that with such a subsoil the crevices through which the liquid escapes from a leaching cesspool are, if not the same, at least in communication more or less direct with the seams which, ramifying in all directions, serve to convey water to the wells of the neighborhood. If, again, a well in a porous, gravelly soil is very deep, the extent of the area from which it draws its supply is enormously increased. The pumping out of the excavation in gravelly and clayey soil for a dry-dock near London drained wells at a distance of much more than a mile. After the pumping was discontinued, the wells gradually filled to their normal level, showing that the water was drawn from them to the excavation during the pumping, and that if they had contained foul liquids instead of clear water, and the excavation had been a well instead of a dry-dock basin, the sewage would, even from that distance, have ultimately reached it.

Supposing that all danger of contamination to the drinking water of the house is averted, by the introduction of water either from a public service, or from a spring or well whose *bottom* is considerably higher than the proposed cesspool; this **Leaching Cesspools.** may be excavated of a circular form, in diameter from eight to twelve feet, and of the depth requisite to reach an absorbent stratum, the sides lined with a dry wall of stone or brick, and the top drawn over in the form of a rude dome, leaving a man-hole about twenty inches in diameter at the top, which should be covered with a flat stone. Wooden covers soon rot, forming a dangerous trap. The usual way is to fix the height of the masonry so that the top of the cover shall come about four inches beneath the sod, which may be either carried over it, concealing it entirely, or turned down neatly around the edges of the stone.

In sandy or gravelly soils such a cesspool will dispose of the waste liquids of a house for a long time, but in the course of years the earth around it becomes coated with the fatty deposit from the sewage, and a new cesspool must be dug. Where the sand or gravel is very fine, or mixed with clay, the stoppage of its pores takes place quickly, and as years pass by a continually increasing chain of cesspools, each connected with the previous one by an overflow-pipe, serves to saturate the ground around the house with putrefaction, and the air with malaria. In very clayey soils no leaching whatever takes place, and the cesspool fills up like a tight cistern, a few days' use, with one of ordinary size, causing it to overflow. Such ground often, however, contains strata of porous gravel or sand, and if the excavation can be carried to one of these seams, the cesspool may answer well enough; but if not, and no more favorable spot can be found, a different mode of disposal must be adopted.

The simplest, though not the least troublesome, way of surmounting the difficulty is to drill a hole through the cover of the cesspool, and set over it an ordinary pump, by which the liquid may be pumped out at intervals of days or weeks, according to the capacity of the reservoir, and spread upon the grass, or utilized in the garden. A small tank on wheels, which can be filled at the pump and conveyed quickly to the point where the fertilizing fluid is to be applied, is much used for this purpose. The operation is not offensive, or only very slightly so, since the sewage does not remain in the cesspool long enough for putrefaction to take place, and the results are excellent, but the necessary attention cannot always be given to it, and a neglect which might cause the backing up of the sewage in the drains, to overflow into the cellar of the house, would be a serious matter. To avoid this, it is customary to provide an overflow, through which the liquid can upon occasion escape over the surface of the ground, where although a continued flow would be offensive, its presence is less objectionable than in the house.

Tight Cesspool.

A more automatic arrangement, which has been gaining rapidly in favor of late years, consists in substituting for the surface overflow, to be used only in case of temporary need, a permanent outlet, formed by a series of open-jointed pipes, laid a few inches beneath the surface of the ground, where the liquid exuding from between them will be absorbed, partly by the porous loam which always forms the upper stratum, and partly by the roots of the grass or other vegetation growing upon the surface. If properly arranged, this system is very satisfactory, not only disposing of the house-waste as completely, and with as little attention, as the ordinary leaching cesspool, but accomplishing this in soils where a leaching cesspool could not be made, and what is of even more importance, performing its work for an indefinite period without causing any contamination of the ground, the actively oxidizing property of the air contained in the pores of the top soil serving to destroy with certainty the last traces of organic impurity in the liquid, which if discharged into the ground a few feet below, beyond the reach of atmospheric changes, would retain its foulness for months, if not for years.

Subsoil Irrigation.

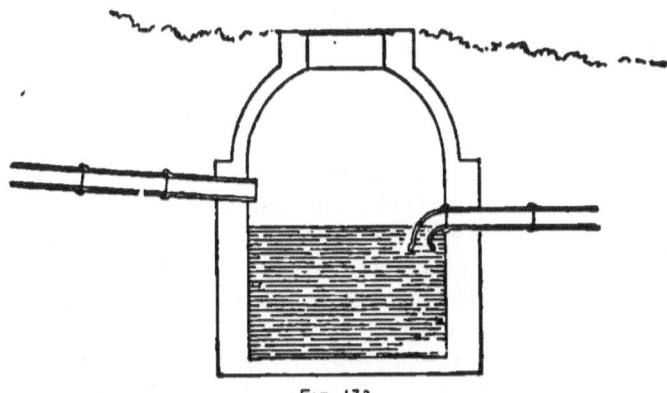

Fig. 173.

An efficient system of subsoil irrigation must consist of two parts: the tight cesspool or tank where the waste matters from the house are retained until they dissolve into a thin, milky liquid, and the network of pipes which receives the sewage from the reservoir, and dissipates it into the ground. The tight cesspool should be constructed as shown in section in Figure 173, of hard brick, laid in cement, circular in plan, about 5 feet in diameter, and 5 feet deep from the mouth of the outlet. The walls should be 8 inches thick in most soils. The bottom may be 4 inches thick, and the top should be covered with a 4-inch dome, with a man-hole 18 or 20 inches in diam-

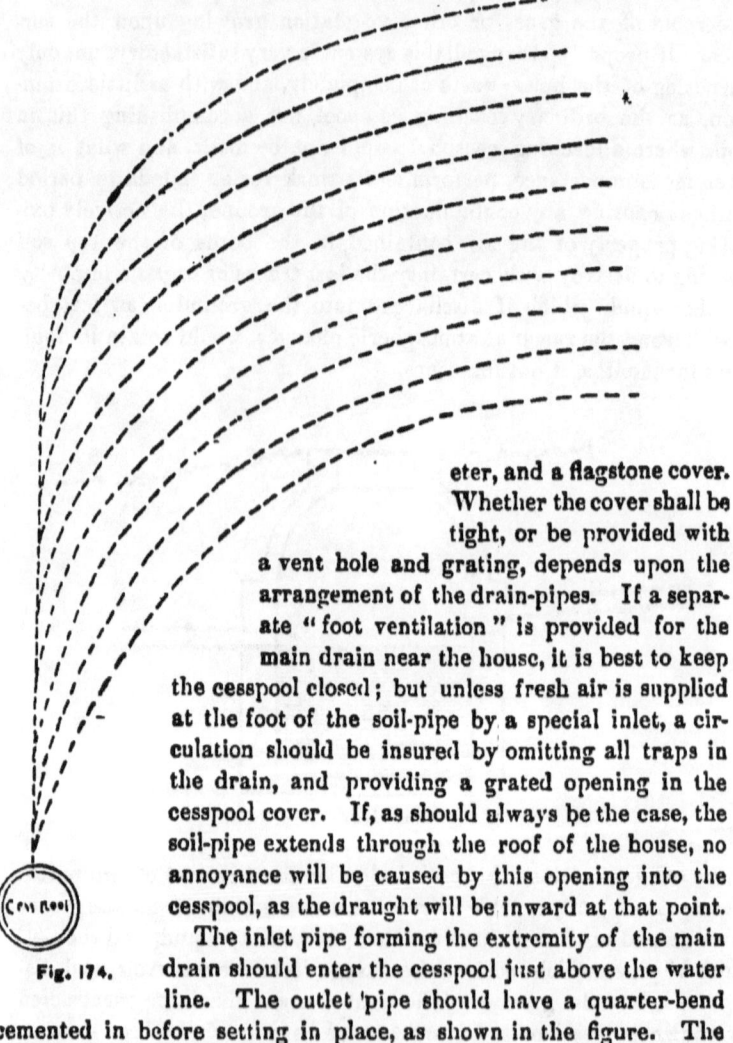

Fig. 174.

eter, and a flagstone cover. Whether the cover shall be tight, or be provided with a vent hole and grating, depends upon the arrangement of the drain-pipes. If a separate "foot ventilation" is provided for the main drain near the house, it is best to keep the cesspool closed; but unless fresh air is supplied at the foot of the soil-pipe by a special inlet, a circulation should be insured by omitting all traps in the drain, and providing a grated opening in the cesspool cover. If, as should always be the case, the soil-pipe extends through the roof of the house, no annoyance will be caused by this opening into the cesspool, as the draught will be inward at that point.

The inlet pipe forming the extremity of the main drain should enter the cesspool just above the water line. The outlet pipe should have a quarter-bend cemented in before setting in place, as shown in the figure. The mouth of the bend, turned downward, then protects the outlet from the scum which floats at the surface of the liquid, and would soon choke the irrigation pipes.

The outlet pipe, after leaving the cesspool, should be laid with a very gentle but perfectly uniform pitch, with branches as indicated in Figure 174.

The best mode is to make it of vitrified pipe, laid with the "hubs" pointing *downward*, instead of upward in the usual way, and having a Y inserted between every two lengths of straight pipe. All the joints should be made tight and smooth with cement, and into the lateral branch of every Y should be cemented a piece of agricultural tile drain. The vitrified pipe forming the main carrier may be 4 inches in diameter, and the agricultural tile should be 2-inch; the Y's forming the connection being of the kind called 2 x 4-inch. Each lateral branch should then be continued by a line of agricultural tiles of the requisite length. If the slope of the lot permits, the outlet pipe should leave the cesspool at least two feet below the surface of the ground, running however in such directions as to bring it, before it begins to ramify, about twelve inches below the sod; and the same or a less distance below the sod should be maintained throughout the system of branches. All the pipes must also be graded to a uniform fall of not more than one inch in twenty-five feet, and to fulfil these two requirements it will generally be necessary to lay all the lines in curves, determined by the irregularities of the surface, and following very nearly a series of contour lines of the ground.

<small>Distributing Pipes.</small>

The tiles forming the lateral branches should be laid $\frac{1}{4}$ inch apart, a bit of paper put over the joint, to prevent earth from sifting in, and the trenches then filled. Either round or sole tile may be used, but sole tile are preferable. If round tiles are employed, they should be laid *without* collars, and pebbles put on each side to keep them from rolling out of place.

The end of the main line of vitrified pipe should not be closed, but a reducer should be inserted, and the line continued with open-jointed tiles, in the same way as the lateral branches. The principal object to be kept in view is the avoidance of any check to the continuous flow of the sewage until it issues from the interstices of the drain tiles, to be absorbed and oxidized in the soil. If any such check is offered by the displacement of a pipe, a sudden dip in a ine, or an abrupt change in direction, the liquid will throw down at that point a copious black sediment, choking the pipe in a few weeks, or, if the main line is closed at the end, filling it like a pocket, and successively cutting off the lateral branches. With the greatest care some stoppages will take place, and it is wise to provide enough lateral pipe to dispose of any possible flow, with a large margin for contingencies. For an ordinary dwelling-house, inhabited by a family of five or six persons, five hundred feet of lateral outlet pipe is the best rule, and the quantity must be increased for a larger household. The character of the subsoil, whether clayey or sandy, makes with this system very little difference, the exudation being absorbed almost wholly by the surface loam, which is always porous enough to take up a certain quantity of liquid.

On account, probably, of the chemical action of the air by which the upper soil is permeated, the sewage distributed through it leaves no trace of itself. Unlike the earth around a deep cesspool, which when uncovered discloses a foul saturation, the soil surrounding irrigation pipes shows after years of service no trace whatever of the organic matter which has passed through it, and if the pipes are carefully laid, either on hard ground, on strips of board, or in the earthenware channel pieces made for the purpose, so as to avoid depressions which may collect sediment, they will work perfectly for an indefinite period in any kind of soil.

The disposition of the surface-water about the house is a matter only second in importance to that of drainage. The main point to be kept in mind is that the ground should everywhere slope at a very sensible pitch away from the building, so as to throw rain-water far enough away from the walls to insure its absorption by the soil, or its harmless removal by surface channels. If at any point the ground should be allowed to pitch toward the house, the water of spring rains will be directed against the walls, and sinking through the soft earth which fills the trenches outside of the masonry will make holes or "gullies" close to the building. If the walls are well drained in the manner previously described the water may pass away without finding an entrance into the cellar, but the stone-work will be soaked, and the loose soil carried down by the torrent is liable to be washed into the drain below, so as to choke it and render it useless. This point is generally understood among country

Surface Water.

contractors, but it is likely to be neglected in practice when the proper grading happens to be inconvenient, particularly under piazzas and porches, where deep holes are often left, to collect the streams of water which soak through from the outside, and direct them against the building.

The lines indicating the surface of the soil as it is intended to be when the work is done should be marked upon the underpinning, bearing in mind that where the ground is to be grassed over two lines will be needed, one showing the grade of the gravel or ordinary earth, and the other that of the loam which must be placed above it to support the proposed vegetation. Where loam enough is at hand it should be spread two feet deep. This will afford a strong, thick growth of grass from seed in a single season. In ordinary cases, however, it is necessary to be contented with a foot or so: less than this cannot be depended upon to produce or sustain a uniform sod. With the usual contracts, where the subsoil is sandy or gravelly, the spreading of the excavated material at a proper grade around the walls, placing the loam on top, with the formation of a gravelled pathway to the doors out of the subsoil material, will constitute all the work to be expected of the contractor; and it will remain for the owner or his gardener to smooth the surface of the loam with a rake, lay a line of sods at the edges of the paths, and sow the rest evenly with "lawn seed," and then drag a bunch of twigs or a tree-branch lightly over it, and leave the rest to nature. Where a bank or terrace forms part of the plan it should be sodded all over; otherwise it will need to be repaired after every heavy rain. Sods should be laid in a bed well soaked with water and kept moist for some days.

It will often be found difficult on sloping ground to keep paths and drive-ways from washing away in severe storms, and even edging them with paved gutters does not always keep their surface in place. Such effects are usually due much more to the working in of water at the sides of the path, under the surface, than to the direct action of the rain, which would have little power to disturb the gravel unless previously loosened by the lateral infiltration of water; and the best remedy is to intercept such infiltrations from the sides, and at the same time drain the subsoil of the path, by means of trenches filled with stones on each side. The trenches are best made narrow, but deep: sixteen or eighteen inches of breadth by two feet of depth.

Where pebbles of suitable size are plenty a narrow path may be completely filled in with stone to within six inches of the surface, which gives good results at considerable expense. Driveways on very costly estates are occasionally built with a layer of broken stone one or two feet deep; or sometimes a single "French drain" is run through the middle. The latter plan costs nearly as much as that of two lateral drains, and is far less effective. It is impossible to take too much pains in the laying out of avenues and paths, and they should be marked through their whole extent with small stakes on each side, three or four feet apart, so as to judge of the effect from all points before work upon them is begun.

When the dust incidental to these operations has subsided, the painting of the exterior may safely be completed. The first coat has, or should have been, put on as soon as possible after the setting of the wood-work in place, to prevent it from warping or "checking" by exposure to the sun, and will now be perfectly dry. The next step will be to fill up all nail-holes and crevices with putty, and for this purpose the heads of all the nails used for securing the exterior finish, including clapboards, must previously have been "set in" to a depth of $\frac{1}{8}$ of an inch or more. Then, beginning with the roof, the house is to be painted downward, so that the portions already finished may not be disfigured by spatterings of a different color from above. With the mineral reds generally used for roofs some of the color is apt to wash down with every rain for two or three weeks after it is put on, so that it is advisable to have this portion completed and dried as long as possible before painting the walls. The colors will generally be chosen by the architect, whose experience in such matters will save him at least from the glaring blunders which amateurs are liable to make; and the general rule is that the smaller the building the lighter its color should be. Tints of green, yellowish, brownish or grayish are by far the most popular, and with reason, since the blending of such colors with those of the surrounding vegetation serves to connect the building with the ground and take away the fragile, portable look which all buildings exhibit whose color contrasts sharply with that of neighboring objects. In rocky districts the gray of the ledges might also be suggested in some portions of the building with excellent results.

Outside Painting.

Choice of Colors.

Where the structure is disproportionately high it is advantageous to make a decided difference in the color of the paint between the first and second stories, bearing in mind that when two different colors are placed side by side they must be of very different shades, one being light and one dark: never both of the same or anything approaching the same depth. The first story may be a dark bronze green and the second a raw-sienna yellow, with very good effect, and a house so painted will seem much lower and more home-like than if only a slight variation or none at all were made in the tint of different portions. Where the proportions of the building are such that any apparent lowering would injure it, two slightly different shades of the *same* color may be used, or the whole painted of one uniform tint. The casings, corner-boards and blinds are usually painted of a darker shade than the rest of the house, the blinds being sometimes the darkest of all, and sometimes of an intermediate tint between that of the "trimmings" and the plain wall; but the architect will be able to judge in each particular case what treatment will most enhance the effect of the design.

In the durability of the various pigments used in house-painting there is less difference than is generally supposed. All colors fade somewhat, and as the darker colors, such as olive and sage green, show the effect of fading more plainly than the old-fashioned paints, buffs and light grays, they are generally thought to change more. The best method of preserving the fresh look of an olive or bronze green is to put it on over a first coat of red: white lead strongly colored with Venetian red, or with Indian red and yellow ochre, will do very well. The green covers it perfectly, taking only a transparent, mellow tone, which is very pleasant, and remains long after two coats of green would have faded to a brownish shade. The most fugitive of ordinary colors for exteriors is vermillion, which soon changes, unless under very favorable conditions, either to a black or white. Yellows bleach out by exposure, and browns acquire an ashy shade, although they are perhaps the most permanent of all ordinary pigments. Mineral reds for roofs soon blacken.

The supervision of the painting work is not so difficult as the examination of the materials used. It is hardly necessary to say that whatever may be the composition of the second coat, the first coat should always contain a large proportion

Supervision.

of white lead; not that it is desirable to modify the color by the ad mixture, but because white lead clings to the wood with far more tenacity than any other pigment, and retains with it not only the other colors which are added to it, but also a second coat of less adherent materials. The best lead will keep its hold long after the oil with which it was mixed has been washed away, forming the "chalky" surface so familiar to us on old buildings, while zinc and other inferior paints blister and leave the wood bare. Something may be learned of the character of the paint by observing its behavior in the pot and under the brush, and still more by seeing it mixed, and observing the names on the cans or kegs from which it is taken. The oil is very likely to be of inferior quality, immense amounts of fish oil being employed to adulterate the linseed oil sold for painters' use. Fish oil dries slowly, never acquiring the hardness and resistance to adverse influences of pure linseed oil, so that the adulterated oil should be avoided, except for painting tin roofs, where its softness has some advantage in enabling the paint to yield to the expansion and contraction of the metal without cracking. The patent mixed or "chemical" paints are convenient, and of carefully selected shades, and are said to last well away from the sea-coast; but for buildings exposed to salt breezes from the ocean most architects prefer colors prepared in the usual way. Oil paints of any kind, after long standing, become "fat" and work less evenly under the brush, so that a fresh mixture of good materials is perhaps always to be preferred.

The paint should be put on by strokes parallel with the grain of the wood, and long, smooth pieces, such as window and door casings, should be finished by drawing the brush carefully along the whole length, so that there may be no break in the lines. No work should be started in the morning which cannot be finished before night; for instance, if one side of the house is begun, it should be completed, if not to the bottom, at least down to some important belt or other division line: otherwise the junction of the portions executed at different times will show as an ugly streak.

Where a building is much exposed to the weather, three coats of paint will not be too much to preserve it; but it is usually better to include only two coats — the priming coat and one other, in the contract. In this way the work will be better done, as there is less opportunity to cover up deficiencies, and the paint is less liable to blister than where three coats are applied at once. After two or three seasons' wear, a third coat may be put on and will stay in place.

It is becoming common to stain or varnish the exterior of wooden buildings instead of painting them. In the shingled Swiss houses the stone basement is frequently painted in two colors, while the wood-work above is always left in its natural condition, to turn gray by the action of the atmosphere, **Exterior Staining** and a similar treatment for picturesque buildings, particularly in rocky situations, often has a charming effect. Under other circumstances, where a greater variety of color is desirable, stains serve to change the tints of the different portions without destroying the transparent richness due to the varied grain of the wood. Linseed oil alone is occasionally used as a dressing, but although the effect is good for a time, the wood soon mildews and becomes black. Oil of creosote, which can be obtained at a low price from any gas-works, gives a blackish stain at once. Linseed oil mixed with umber, raw or burnt, Vandyke brown, sienna, or other colors, is sometimes used for a stain, and the surface then protected with wax or varnish. By applying these pigments, selected as pure in quality as possible, until the desired shade is obtained, and finishing with two coats of wax dissolved in hot linseed oil, a lasting surface of a pleasant texture is produced. Pellucidite, or other good water-proof varnish, answers the same purpose as the wax, and the varnish alone is sometimes used.

The last part of the outside painting is the "drawing" of the window-sashes. It is essential that these should be securely protected against the absorption of moisture, and at least three coats must be applied, of red, yellow, black, bronze green, white or other color, as may be preferred.

For inside work the same materials are used as for the outside, but it is even more necessary that they should be of the best quality. Fish oil, for instance, which will soften on damp days, is very unsuitable for inside use. The puttying **Inside Work.** should be done with great care to avoid unsightly spots, and one coat of oil or paint must be put on before puttying, to prevent the putty from shrinking and falling out through the absorption of the oil from it by the dry wood. In very cheap houses, finished in hard wood or pine of the natural color, two coats of oil often complete the work, and this application answers well enough if a polish is not desired. Where, however, a shining surface is intended, one coat of oil only should be put on to bring out the grain of the wood, followed by a "filling" of patent paste composition, white wax, chalk, oil mixed with pigments of some kind, or with wood-dust, and finished with shellac, pellucidite, or some other varnish, or a wax polish. Of these, shellac, if simply put on in one or two coats, without rubbing down, forms the cheapest and poorest dressing. If, on the contrary, it is put on in three or four coats, rubbed down with emery-cloths dipped in oil after each coat, it forms the most expensive, and the best of all applications. The patent varnishes, of which there are many kinds, are easily applied, and give, with two coats, a good finish. The hard wax polish gives a beautiful surface, which may be renewed, wholly or in part, at any time.

Where a painted finish is intended, the principal point to be observed is the thorough covering with shellac, before priming of knots and pitchy places, which will otherwise discolor the paint over them.

Bad knots cannot be "killed" even by this application, and snould be cut out, and a piece of sound wood set in their place. The most annoying stains come from the minute dots of pitch which often speckle the entire surface of a pine board, but escape notice until after the painting is completed, when each one manifests itself by a yellow stain. Some architects, to make sure of complete protection, specify that the whole of the pine finish shall receive one or two coats of shellac before priming.

Zinc is much used for interior work instead of white lead, and is preferred by some on account of its freedom from the tendency of lead to turn yellow in rooms which are not well lighted. The character to be given to the surface varies with circumstances and fashion. Usually, a dead or "flattened" finish is preferred, and is obtained by mixing the final coat of paint with pure spirits of turpentine instead of oil. Where the paint is exposed to the contact of clothes or fingers, an "oil finish," containing little or no turpentine, is employed, and gives a somewhat glossy surface which can be washed readily. For hotel rooms and other places liable to much wear, a "china gloss," made by mixing the paint with varnish, is often specified for the last coat, or the whole is varnished after completion. Whatever the style of finish, the manipulation should be careful and neat. Every coat of paint except the last should be sand-papered to a smooth surface, and in *each* coat the brush-marks should be so drawn as to follow the lines of the wood-work, without joinings.

If fresco color is used, the most experienced workmen only should be employed to apply it. So much depends upon the consistency of the size, the mode of putting on the color, and other circumstances, that an unaccustomed hand is almost sure to fail.

Fresco.

Hard-wood floors are not easily finished in a perfectly durable and satisfactory manner. The soft, elastic varnishes which answer for the doors and standing finish are, although sometimes employed, unfit for floors, while the hard spirit varnishes, though more durable, will ultimately wear away in certain portions of the floor, leaving the pores of the wood exposed to dirt, which quickly fills them, and after this nothing but replaning and revarnishing will restore them. For this reason, the ancient wax polish, although more troublesome, is perhaps to be preferred. The wax fills the pores of the wood so that, although the surface may be worn away, dust will not enter so long as any wax remains, and by periodical waxing and repolishing the floor may be kept clean and shining until absolutely worn out. An advantageous substitute for the ordinary wax, which is so sticky as to need frequent polishing to keep it bright, is made by mixing it with more or less hard paraffine. Such a compound is sometimes sold for use in dancing-halls in blocks, which are scraped, and the resulting powder scattered over the floor and rubbed into the wood by the feet of the dancers; but the best varieties are softened with turpentine to a paste, which is sold in cans, and needs only to be applied evenly over the floor, and after a few hours' drying polished with cloths, or with weighted brushes, made for the purpose and dragged to and fro over the room. The same brushes are used to brighten the surface when it becomes dull, and any worn spots can be brought back to an equal polish with the rest by a new application of the wax.

Floors.

Glazing. The glazing is usually done by the painter, who sends the sashes to the building with the glass all set, and the superintendent will have little to do except to see that the glass is of the specified quality, and that all the work is left whole and clean. The difference between first and second quality sheet-glass must be learned by observation, and in judging of the glass in a building it must be remembered that it is much easier to obtain small lights free from defects or uneven places than large ones; and that double-thick glass, such as should be specified for all lights larger than about 16″ x 30″ unless plate-glass is used, shows any unevenness of surface more plainly than the thinner sheets.

This last item having been examined, the young architect's duties of supervision will be ended; and it will only remain to review the notes which he ought to have made during the progress of the building, in order to fix in his mind more clearly the observations contained in them, and thereby prepare himself to carry out his next commission with still greater satisfaction to himself and his client. More particularly for the information of persons intending to build, an actual set of specifications for a country house of moderate cost, together with contracts for the same, will follow.

BUILDING SUPERINTENDENCE.

CHAPTER III.

SPECIFICATIONS OF LABOR AND MATERIALS FOR DWELLING-HOUSE TO BE BUILT ON FAIRFIELD STREET, MELROSE, N. Y., FOR JAMES JOHNSON, ESQ., FROM THE PLANS AND UNDER THE SUPERINTENDENCE OF MR. EDWARD TYRO, ARCHITECT, 13 RIALTO STREET, ALBANY, N. Y.

GENERAL CONDITIONS.

EACH contractor is to provide all materials and labor necessary for the complete and substantial execution of everything described, shown, or reasonably implied in the drawings and specifications for his part of the work, including all transportation, scaffolding, apparatus and utensils requisite for the same; all materials to be the best of their respective kinds, and all workmanship to be of the best quality.

Each contractor is to set out his own work correctly and is to give it his personal superintendence, keeping also a competent foreman constantly on the ground, and no contractor is to sublet the whole or any part of his work without the written consent of the owner. The architect or his authorized representative is to have at all times access to the work, which is to be entirely under his control, and may by written notice require any contractor to dismiss forthwith such workmen as he deems incompetent or careless, and may also require any contractor to remove from the premises such of his materials or work as in his opinion are not in accordance with the specification, and to substitute without delay satisfactory work and materials, the expense of doing so and of making good other work disturbed by the change to be borne by the said contractor; and each contractor is also at his own cost to amend and make good any defects, settlements, shrinkage or other faults in his work arising from defective or improper materials or workmanship which may appear within twelve months after the completion of the building, and is to clear away from time to time the dirt and rubbish resulting from his operations, and cover and protect his work and materials from all damage during the progress of the building, and deliver the whole clean and in perfect condition. All work and materials are to comply in every respect with the building laws, city or town regulations and the directions of the Inspector of Buildings, and such building laws, regulations and directions are to be considered as a part of this specification and the contract to which it relates. Each contractor is to give to the proper authorities all requisite notices relating to work in his charge, obtain official permits and licenses for temporary obstructions, and pay all proper fees for the same and for use of water for building, and entrance into sewers or drains, and is to be solely answerable for all damage, injury or delay caused to other contractors, to neighboring premises or to the persons or property of the public, by himself or his men, or through any operations under his charge, whether in contract or extra work.

The contractor for the mason-work is to have charge of the premises subject only to the right of other contractors, the owner and the architect or his representative to have free access thereto, until the [*sill*] is on, and is to provide and maintain all requisite guards, lights, temporary sidewalks and fences during that time; afterwards the contractor for the carpenter-work is to take charge in the same way until the whole is completed.

Each contractor is to carry on his work at all times with the greatest reasonable rapidity, under the direction and to the satisfaction of the architect. The several portions are to be completed on or before the following dates:

    Foundation to be ready for sill................ *November* 1, 1883.
    House to be entirely enclosed.................. *December* 1, 1883.
    Chimneys and piers to be finished..............   "   7, 1883.
    Back plastering................................   "   10, 1883.
    Outside finish completed and interior ready for
        plastering................................. *January* 20, 1884.
    Plastering completed........................... *February* 10, 1884.
    Interior wood-work done........................ *April* 1, 1884.
    Painters' work completed......................   " 15, 1884.

MASON.

EXCAVATION, ETC.

Set proper batter-boards and mark out the building accurately under the direction of the architect. **Batter-boards.**

Take off the sod and loam from site of house and for eight feet additional in width all around. Excavate the cellar to a depth of five feet below the highest part of the ground covered by the building, making the excavation eight inches wider all around than the outside of foundation-walls; **Excavation.** excavate trenches for all walls two feet below cellar bottom, and for footings of piers and chimneys eight inches below cellar bottom; excavate for posts and piers of porches and piazzas four feet below present surface; excavate trench four feet deep and [*one hundred*] feet long for drain-pipe, and excavate cesspool eight feet in diameter and twelve feet deep. Excavate for dry well to each rain-water leader where directed, eight feet from the house and five feet deep, and for trench four feet deep from each dry well to house.

[AREAS, CISTERNS, ETC.]

Separate the loam and stack by itself where directed, and dump the other earth from the excavations wherever directed within two hundred feet of the building. Clear away and remove all rubbish entirely from the premises at the completion of the building. Refill dry wells and around cellar-walls with **Other Excavation.** small stones or gravel. Refill with ordinary earth around cesspools, posts, piers, and pipes. Ram thoroughly or puddle with water all filling material every foot in height; spread and grade neatly the remainder of the material from the excavation as directed, forming gravel-walks and drive-ways neatly, and elsewhere spreading the loam evenly on top, sowing in the best manner with blue-grass seed and rolling, and finishing with two feet in width of the best sods on each side of gravel-walks and drive-ways, and three feet in width around house and piazzas, all to be done in the best manner, properly cared for, watered and kept in order until the house is delivered.

If any blasting should be necessary for making the excavations

above specified [*seven*] cents per cubic foot will be paid by the
owner for blasting, breaking up, and removing the
stone; but all stone so removed which may be suitable
shall be used in building the cellar walls or piers, and for all stone
so taken from the excavation and used in the building, the contractor
shall pay the owner at the rate of [*seven*] cents per cubic foot.

Blasting.

Furnish and lay in the best manner from outside of cellar-wall to
cesspool [*one hundred*] feet of first quality [*Portland, Akron, Scotch*]
five-inch glazed earthenware drain-pipe, all uni-
formly graded, the bed hollowed for the hubs, and
all jointed with clear, fresh Portland cement, and the joints scraped
smooth inside as laid. Leave the line of pipes open until inspected
and approved before refilling the trench. Include in
the line of pipe a five-inch running trap of the same
make, with hand-hole, to be placed not less than six feet from the
house, and the hand-hole closed by tight cover; and include also a
5″ x 5″ T-branch where directed, between the house
and the trap, the branch of the T to be turned upward, and a vertical five-inch pipe of the same make
to be brought to the surface of the ground and covered with a proper
earthenware ventilating cap, all jointed with cement.

Drain-Pipe, etc.

Trap.

Foot Ventilation.

Furnish and lay four-inch glazed earthenware pipes
of the same make, all jointed in cement, from each
dry well to the rain-water leaders. Each pipe to
turn with a quarter-bend at the cellar wall and to be brought upward to the surface of the ground to receive the foot of the leader.

Underground Pipes for Rain-Water.

FOUNDATIONS.

All the lime used in the mason-work throughout to be Extra No. 1
[*Rockland, Canaan, Glen's Falls,*] and all cement except that used for
jointing drain-pipes to be best fresh [*Rosendale, Akron,
Louisville,*] of the [*F. O. Norton*] brand. All sand to
be clean and sharp, and used in proper proportions.
Furnish all materials and build the cellar-walls
18 inches thick to the underside of sills of good ledge or other
approved stone; the first 18 inches to be laid dry in the trenches,
and the remainder to be laid in mortar made with lime and cement

Lime and Cement.

Sand.

In equal parts and clean, sharp sand in proper proportion; the whole to be laid to a line on each face, well bonded, the joints filled with mortar and all to be thoroughly **Walls.** trowel-pointed inside and outside the whole height, holding the trowel obliquely so as to weather the pointing on the outside. Set the best face of the stones outside, both above and below ground. Set stone footings for piers and chimneys and foundations for range and boiler;. Level up carefully and bed the sill in cement-mortar and point up around it inside and outside, and bed and point up around frames of basement windows. Build piers of dry stone for front granite step. Leave openings for drain, gas and water pipes, and fill up around them afterwards.

Build the cesspool with circular wall of dry stone eighteen inches thick. Draw in the top and cover with three-inch planed blue-stone, two feet square, with man-hole and grated cover, set in cement four inches below finished grade, and the **Cesspool.** sod neatly turned down upon it. Build in the drain-pipe properly.

Furnish and set one step of best clear [*Connecticut*] granite at front porch, to be eight feet long, sixteen inches wide on top, and twelve inches high, seven inches to be above **Stone Steps.** ground; the part above ground, top and ends, to be pene hammered.

All the bricks used in the building except for fireplaces, hearths, and setting of range, furnace and boiler to be the best hard common brick, to be carefully culled for facing of chimneys above roof, and all to be new, well-shaped, and **Brickwork.** of uniform size. All to be laid wet except in freezing weather, with joints thoroughly flushed up with mortar, and all well bonded. All brickwork to be afterwards plastered is to have rough joints, other work to have the joints neatly struck; and all work visible outside the house to be washed down after completion with muriatic acid.

Build piers in cellar and for outside work as shown on plans, all to be 12" x 12", laid in mortar made with equal parts of lime and cement and wedged tightly up to underside of timbers with slate chips in mortar. **Piers.**

Build the chimneys as shown on drawings, with flues 8" x 12" or 8" x 8" as shown, of hard brick in mortar made with one part cement to two parts lime to underside of **Chimneys.** roof boarding; above roof to be of selected brick, formed accord ing to drawings and details and laid in mortar made with equal parts of lime and cement, colored with Venetian red to a light red color, and the upper four courses to be laid in clear cement. The brickwork of chimneys to be kept in all cases at least one inch clear of any wood-work. All withs to be four inches thick, well bonded into the walls, and all flues to be carried up separately to the top. Plaster every flue smoothly inside to the top, and clean out at completion, and plaster the outside of each chimney from basement floor to underside of roof boarding. Build in lead flashings, to be provided by the carpenter, and provide and build in eyes and set strong wrought-iron stays, as directed, to all chimneys rising more than fifteen feet above the roof. Provide and set eight-inch iron thimble in furnace-flue, sixteen inches clear below underside of beams, and five-inch thimbles and covers with ventilating arrangement in laundry, two feet clear below ceiling, and in two attics, to be three feet clear above floor unless otherwise directed. Provide and set also 8" x 8" iron cleaning-out door and frame in furnace-flue, two feet above basement floor; and a 12" x 12" door and frame in each ash-pit, close to basement floor; and 8" x 12" black japanned ventilating register in kitchen.

Turn 4-inch trimmer-arches on centres to all fireplaces, to be two feet wide by the length of the breast; turn also trimmer-arches in front of range and wash-boiler to support hearths not less than 20" wide in front of each. Level up with cement-concrete or brickwork, to receive hearths.

Trimmer Arches.

Build the fireplaces with the rough brickwork only at first, making the opening three feet high above top of beams, and putting in two $\frac{1}{4}$" x 2" wrought-iron chimney-bars to each opening, each bar to be eight inches longer than the opening. After the house is plastered provide all materials and build fireplaces and hearths according to detail drawings, and cover securely with boards for protection until the building is delivered. The fireplace in parlor is to be lined with ornamental cast-iron plates, of pattern to be selected by the owner, and to cost ten dollars per set, exclusive of putting up; and to have facings of French majolica tiles, to be selected by the owner and to cost twenty-five dollars per set exclusive of putting up, and hearth of royal blue glazed American tiles in three-inch squares with border of two rows of one-

Fireplaces.

inch black glazed tiles, with one row between of Low's three-inch Chelsea tiles, of pattern to be selected by the owner; the hearth to be 20" wide and 5' 6" long, inclusive of border. All to be executed in the best manner by skilled workmen, the tile facings to be secured in place with polished brass angle-bars, and the hearth to be laid in Portland cement, and all to be thoroughly backed up with brick and mortar. Make the hearth within the fireplace of good face-brick. The dining-room fireplace is to be lined with Philadelphia glazed brick to be selected by the owner, and is to have facing of Italian griotte marble, 4" wide and $\frac{7}{8}$" thick, with cavetto moulding around the opening, and hearth 20" by 5', of American unglazed red tiles without border, all executed in the best manner and thoroughly backed up with brick and mortar, and to have face-brick hearth within the fireplace.

All the other fireplaces in the building are to be of selected pressed brick with borders of moulded brick, as per detail drawings, and hearths of 16" width by length as directed, of pressed brick laid flat, with border of brick moulded with half-round on the edge, mitered at the angles and set with the half-round projecting above the floor, all laid in red mortar and neatly pointed. All fireplaces to be built in the best manner, with $\frac{1}{2}$" x 2" chimney-bars, 8" longer than the opening, to support those shown with square openings, all well backed up, and the joint between old and new work thoroughly broken to prevent the escape of sparks. Provide and set neat ash-grates to all first-story fireplaces, without dampers.

Range. Set the range, to be provided by another contractor, in pressed brick in the best manner, to show a 12-inch pier on each side, and carrying up the face-brick setting to the ceiling, with lintel of rubbed blue-stone, five courses high by the whole length of the breast, to hold the brickwork above. Make hearth to the same 20" wide by the full length of range and piers, of pressed brick laid flat in cement.

Furnace. Do all excavation and other work necessary and furnish all materials, and set in the best manner in pressed brick the furnace to be provided by the contractor for the heating. Make cold-air chamber under furnace not less than eighteen inches deep., all of hard brick in clear cement, with bottom of the same, and make cold-air box ten feet long under cellar floor for

Cold-Air Box. supplying the same, to be 18" deep by 3' wide, with bottom and sides of hard brick in clear cement, and all plastered with cement, and covered with 3-inch flagstones with close axed joints. Connect the cold-air box complete and make tight all around it, and leave the whole in perfect working order.

Wash-Boiler. Provide and set in laundry where directed a 35-gallon [*Steeger's*] best heavy tinned copper wash-boiler with dished soapstone top, grate, ash-pit, and doors complete, and with steam-pipe connected properly into flue. All to be set in pressed brick in the best manner, and to have hearth 20 inches wide by the length of the boiler, of pressed brick laid flat in cement.

Brick Filling. Lay two courses of rough brick in mortar on top of foundation walls behind sill all around the building, and fill up with four courses of the same between beams on top of sill. Lay four courses of the same on top of all dropped girts and caps of partitions which carry beams in every story between the beams and studs, and build a vertical 4-inch wall of the same from the plate all around to underside of roof-boarding. After the partitions are bridged lay one course of brick in mortar between studs on top of all the bridging throughout. Lay one course of brick in mortar on top of under floor in each story around all chimneys and between the furring studs of the breasts, filling the whole space from brickwork of chimneys to outside of studs.

Concreting. Level off the cellar floor, roll or settle thoroughly, and concrete the whole three inches thick in the best manner, the concrete to be made with one part fresh [*F. O. Norton Rosendale*] cement to two parts clean sharp sand, and three parts washed pebbles or broken stone, and the portion not covered by wooden floor to be smoothed off neatly, and all left perfect at the completion of the building.

Miscellaneous. The mason is to assist the other mechanics employed in the building wherever his help is necessary, and is to do all cutting and jobbing required without extra charge and leave all perfect.

PLASTERER.

[*If this is made a separate contract, the full title and the General Conditions should precede the Specification.*]

THE plasterer is to examine and try all ceilings, partitions, and furrings, and is to notify the carpenter of all that are not square, true, plumb and level, and see that they are corrected before lathing, and that all are firm and secure. **Verifying Furrings.**

Back-plaster the whole of exterior walls from sill to plate between the studs, on laths nailed horizontally ⅜" apart to other laths or vertical strips put on the inside of the boarding, all well trowelled and brought well out on the studs, girts, and plate, making all air-tight. **Back-Plastering.**

Lath and plaster basement ceiling **one heavy coat,** well trowelled and smoothed. **One-Coat Work.**

Lath and plaster two coats in the best manner all other studdings, underside of stairs, partitions, furrings and ceilings throughout the building, except in rooms marked "Unfinished" on plans, carrying the plaster to the floor everywhere. **Two-Coat Work.** Laths to be best seasoned pine, free from knots, bark or stains, all laid ⅜" apart, and breaking joint every six courses and over all door and window heads. The first coat of plaster to be of No. 1 Extra [*Rockland*] lime, and clean, sharp sand, well mixed with a half bushel of best long cattle or goat's hair to each cask of lime, thoroughly worked and stacked at least one week before using, in some sheltered place, but not in the cellar of the house; all to be well trowelled, straightened with a straight-edge and made perfectly true, and brought well up to the grounds. The skim coat to be of No. 1 Extra [*Rockland*] lime, slaked at least seven days before using, and washed [*beach*] sand, and well floated. **Laths.**

Run moulded cornice, not over 30" girt, with one enriched member, in Parlor, and plain moulded cornices, not over 24" girt, in Dining-room, Library, and four chambers in second story, and in first-story Hall, carrying two members **Cornices.**

of the hall cornice up the soffit of stairs to second story and around second-story hall; all to be in accordance with detail drawings.

Beams. Form beams where shown, according to detail drawings.

Centres. Plant plaster centres in Parlor, Dining-room and Library, to be 3' in diameter in Parlor, and 2' 6" in Dining-room and Library, all to be made in accordance with detail drawings, and two wax models to be made and approved before casting.

Point up with lime and hair mortar around outside door and window frames; clear away and remove all rubbish from the premises after the second coat of plaster is on; clean the mortar off the floors and sweep out the house and leave all ready for the wood finish; patch up and repair all the plastering at the completion of the building, and leave all perfect.

Whitewash. Whitewash cellar ceiling, walls and piers two coats in the best manner.

CARPENTER.

[*If this is made a separate contract, the full title and the General Conditions should precede the Specification.*]

Scantlings. Sill, 6" x 6", halved and pinned at angles.
Plates, 4" x 6".
Posts at angles and opposite partitions, 4" x 8".
Girts, 4" x 8".
Braces, 4" x 4".
Window studs, 3" x 4".
Door studs, 4" x 4".
All other studding, 2" x 4", 16" on centres.
Partition caps, 3" x 4".
Soles, 2" x 5½".
Girders, 8" x 10".
Sleepers, 6" x 6", 8' apart.

Floor beams, 2" x 10", 16" on centres.
Headers, 4" x 10", and 6" x 10", according to framing plans.
Trimmers, 4" x 10", and 6" x 10", according to framing plans.
Rafters, 2" x 6", or 2" x 8", as marked on framing plans, 20" on centres.
Deck rafters, 2" x 8", 20" on centres.
Hip and valley rafters, 3" x 9", or 3" x 12", as marked on framing plans.
Trimmer and header rafters doubled and spiked together.
Ridges, 1" x 10".
Piazza and porch girders, 4" x 10".
Piazza and porch floor beams, 2" x 6", 20" on centres.
Piazza and porch rafters, 2" x 6", 20" on centres.
Hips and valleys, 3" x 9".
Piazza and porch plate, 6" x 10".
Piazza and porch posts, 8" x 8".

Stock.

Sleepers in basement to be of locust. Piazza and porch posts to be of best well-seasoned dry white pine or whitewood. All other framing timber to be good sound spruce, free from large knots, waney pieces, and shakes.

Framing.

The house is to be full frame, all framed, braced, and pinned in the best and strongest manner, perfectly true and plumb, and in accordance with the framing drawings. No woodwork is to be placed within one inch of the outside of any chimney, and no nails to be driven into any chimney. The underside of sill and ends of girders are to be painted two heavy coats of oil paint before setting in place. The basement beams to be sized upon the sleepers, which are to rest on the concrete. The beams of first-story floor to be notched down four inches on the sill and mortised two inches more into it, bringing the bottom of the beams flush with the bottom of the sill, and to be framed with tenon and tusk into the girders, flush at top and bottom: all to be well spiked to the sill, and the tenons secured to girders with oak pins. Beams of second and third story floors to be notched down four inches on the girts, and spiked to girts and studs, and sized one inch on partition caps, spiking the beams strongly together wherever possible to form a tie across the building. Headers to be framed into trimmers with double tenon and pinned, and all headers which carry more than three tail-beams to have ⅞" joint-bolts at each end in addition. Tail-beams to be framed into headers with tenon and tusk and pinned. All floors to be bridged once in every eight feet

Bridging. with a straight, continuous row of double herring-bone cross-bridging of 1" x 4" pieces, cut in and nailed with two nails at each end of each piece. Beams under unsupported partitions which run parallel with them to be in pairs, set 7½" apart on centres.

Piazza and porch floor to be framed with a 4" x 10" girder from each post or pier to the house, set so that the top of the girder is 1"

Piazza and Porch Floor. below the top of the sill, and pitching away from the house one inch in every five feet. Each girder to be gained one inch into the whole depth of the sill, and to be secured to the sill with ⅞" joint-bolt. Into these girders are to be framed the 2" x 6" piazza beams, all flush on top.

The finished posts of piazza and porch will stand upon the floor,

Piazza and Porch Roof. with tenon 4" long into the girder, and the plates are to be framed into them and pinned. The rafters are to be notched upon the plate and spiked, and strongly secured to house.

Form the cornice of porch and piazza as shown on detail draw-

Piazza and Porch Cornice. ings, all of pine, with planceer, facia, bed-mould, and gutter all around as shown, with leaded joints, two-inch lead goose-necks and [*four*] three-inch [*tin*] conductors where directed, properly supported and entered into the drain-pipes prepared to receive them.

The main roof of the house is to be framed as shown on framing

Main Roof. plans. Rafters to be notched on the plate and spiked. Partitions to be carried up to support roof wherever practicable, and all to be thoroughly tied and made perfectly secure and strong.

Form the main cornice as shown in detail drawings, with 3" x 4"

Main Cornice. gutter all around, facia, and planceer, rebated 1¼" bolt at top of wall, and raking moulding in the angle: all of pine. Gutter to have leaded joints and three-inch lead goose-necks. Put on [*six*] four-inch tin conductors where directed, all strongly secured and properly entered into drain-pipes prepared for them.

Gable finish to be as shown on detail drawings, all of pine, with $1\frac{1}{4}''$ rebated piece at top of wall, to show 8" wide, $1\frac{1}{4}''$ plain planceer put on underside of projecting roof-boards with $\frac{7}{8}''$ furring between and to show 10" wide, and corona and cymatium moulding $3\frac{3}{4}''$ wide in all, to cover ends of roof-boards and edge of planceer, and 3" x 3" bed-mould in the angle. **Gable Finish.**

Cover all the roofs, including those of porch and piazza, with good hemlock boards, planed one side to an even thickness, and well nailed to every rafter, two plies of pine-tarred felt paper, breaking joint, and shingle with first quality 16" sawed cedar shingles, laid $4\frac{1}{2}''$ to the weather, and nailed with two galvanized nails to each shingle. Form dormers, etc., as shown on drawings. Furnish wide counter-flashings of 4-lb. lead for the mason to build into joints of chimneys, shingle in wide zinc flashings to turn up against the brickwork as high as the counter-flashing will allow; then turn down the counter-flashing, dress close, and cement perfectly tight against the brickwork. Shingle in wide zinc flashings in valleys and around dormers, and put on wide zinc apron to protect junction of piazza and porch roofs and house wall, and warrant all tight for two years from the completion of the building. Make all tight under dormer sills and around dormers. **Roofing.**

Make scuttle 2' x 2' in roof where directed with rebated frame 4" high, and cover hung with strong strap hinges, and to have iron bar fastening and fixtures to keep it open at any desired angle. Frame and cover to be tinned and all warranted tight. **Scuttle.**

Enclose the walls with good hemlock boards planed one side to an even thickness, and two thicknesses of good felt-paper breaking joint. Shingle the whole of the walls above belt at second-story floor level with first quality 16", sawed cedar shingles, laid 6" to the weather and nailed with two common nails to each shingle. Where the shingles come against door or window casings nail only at the side next the casing, with two nails. **Walls.**

Under window sills and elsewhere where exposed the nails to be galvanized. Shingles in gables to be laid alternately long and short, without selecting for uniform width, the difference in length to be 1½".

Cut Shingles. Cut shingles, of uniform width and to pattern as per drawings, to be used in panels where shown on elevations.

Clapboards. All other portions of the wall to be covered with sap-extra pine clapboards, all laid to a perfectly even gauge of not over 4½", and all nailed to every stud with galvanized nails, set in for puttying.

Zinc. Put strips of zinc 3" wide around all window and door casings throughout, running under casings and clapboards or shingles.

Form the belt at second-story floor level according to detail drawings by putting on furrings over each stud on top of the under boarding and putting a second boarding outside, brought **Belts.** to a feather edge at the top. The lowest of these boards to be planed on the edges and the shingles to be brought down over. Finish under the edge with a 2⅜" bead. Under the moulding will be a belt 1¼" thick and to show 6", with the bottom edge rebated.

Make also belt at level of window-sills where shown, of a 2½" cove and fillet moulding, bevelled on top to correspond with the pitch of the window-sills, and put on over the shingles, which are to be gauged to correspond. Form the front edge of these window-sills so as to continue the moulding without any break.

Form the base as shown on detail drawings, on sides next to piazza and porch to be 1¼" thick, rebated on top, and to show 6" high **Base.** above piazza floor, with small bevel, stopping against half-posts of the balustrade, scribed against the wall. In other places the base will be formed by 1¼" feather-edged piece standing out from the wall, the clapboards to be brought down over it, and a 1¾" bead to run underneath. All of good, seasoned white pine.

There will be corner-boards only in first story where clapboarded shingles being in all cases brought out to the angles. All corner-boards, door and window casings to be $1\frac{1}{4}''$ thick, and of widths as marked on drawings, all of good, seasoned pine, and the top edge rebated. **Casings and Corner-boards**

The porch will stand on $12'' \times 12''$ brick piers to be built by the mason. Piazza to stand on red cedar posts, $4'$ in the ground, furnished and set, exclusive of excavation and refilling, by the carpenter. Floors to be framed as described above, and to be covered with $\frac{7}{8}''$ rift hard-pine boards, not matched, laid close joint and well nailed and the outer edges rounded. Finish under edge of floor with a planed board $10''$ wide, case the posts in front with planed boards, and fill in between them, and between piers under porch, with jig-sawed work of $\frac{7}{8}''$ pine boards down to the ground, finishing with board, and with mouldings broken around to form panels, all as shown on drawings. **Piazza and Porch.**

Complete the porch and piazza as per detail drawings, the balustrades, braces, turned, carved and ornamental work to be all of good, seasoned white pine; and ceil the underside of roofs, and of all projections of oriels, balconies, etc., with $\frac{7}{8}''$ matched and beaded spruce sheathing not over $4''$ wide, on furring strips $12''$ on centres, formed into panels by $\frac{7}{8}'' \times 4''$ bevelled strips, planted on. Case over the plate with $\frac{7}{8}''$ pine with $\frac{3}{8}''$ bead on each edge, and finish the junction of piazza ceiling and main wall with $2''$ cove moulding neatly stopped.

All outside steps except the granite lower step at front entrance to have $\frac{7}{8}''$ pine risers and $1\frac{1}{4}''$ rift hard-pine treads with rounded nosings returned at the ends; all to be supported on $2'' \times 12''$ spruce strings $12''$ on centres, the outer strings to be planed or cased, and the foot of the strings to be notched upon the granite step or upon a $4'' \times 4''$ piece supported by two red cedar posts $4''$ in the ground. Enclose under ends with vertical $\frac{7}{8}''$ pine strips, jig-sawed as shown on drawings, with bevelled base. **Outside Steps.**

Make bulkhead entrance to cellar, with plank steps on plank strings. all planed, and cover of $4''$ matched pine boards, battened, hung with heavy strap hinges bolted **Bulkhead.**

on, in strong plank frame, and all made tight, and furnished with strong bar fastening.

Other Outside Finish. All other outside finish is to be of good, thoroughly seasoned white pine in strict accordance with elevations and detail drawings, and all leaded where necessary to make it weather-tight. Project beams for bays, oriels, etc., and do all furring and other work for completing the whole in the best and strongest manner.

Priming. The carpenter must call upon the painter to prime all exterior finish before putting up or immediately afterwards, and is to replace all work warped or cracked.

The Hall, Laundry and Water-Closet in basement will have single floor of first quality $\frac{7}{8}''$ rift hard pine, square joint, not over 6'' wide.

Inside Flooring. All other floors throughout the building are to be double, the under floor to be of good hemlock boards, planed one side to an even thickness, well nailed, and two plies of good felt paper to be laid between all upper and under floors. Under flooring in Hall, Parlor and Bath-room to be not over 3'' wide; in Kitchen, not over 6'' wide; elsewhere to be any width.

The Kitchen, Back Vestibule, Store-room and Butler's Pantry on first story, and Back Hall in first, second and third stories to have **Upper Floors.** upper floor of first quality rift hard pine, not over 6'' wide, matched and blind-nailed. The main Hall on first story to have upper floor of first quality selected quartered $\frac{7}{8}''$ oak, $2\frac{1}{2}''$ wide, matched, laid in herring-bone pattern, as per detail drawing, and blind-nailed. The Parlor is to have upper floor of $\frac{7}{8}''$ second quality white pine, not over 6'' wide, matched, but not blind-nailed, with border of [*Dill's*] parquetry, 20'' wide, in maple and cherry, of pattern No. [*twenty-six*] all laid in the best manner and warranted not to shrink. The Bath-room in second story is to have upper floor of alternate strips of maple and black walnut, not over 4'' wide, matched and blind-nailed.

All other rooms and closets throughout the building to have upper floor of $\frac{7}{8}''$ second quality pine, matched but not blind-nailed, not over 6'' wide in first story and second story chambers; not over 9' wide elsewhere. Make mitred borders to all hearths, registers, and staircase openings.

All under floors to be thoroughly repaired and cleaned before upper floors are laid.

All hard-wood flooring to be of the very best selected perfectly seasoned stock, and all upper floors to be kiln-dried, laid breaking joint in every course, well strained and nailed to every beam with twelve-pennies, and all neatly smoothed off by hand and scrubbed out at the completion of the building.

Put on grounds for $\frac{7}{8}''$ plastering in first story and $\frac{3}{4}''$ elsewhere, and put on all angle-beads. Cross-fur all ceilings, including basement ceiling, with $1'' \times 2''$ strips, $12''$ on centres. Cross-fur rafters in finished attic rooms diagonally, with strips $12''$ on centres; fur out attic outside walls with studding to give $4'$ vertical height, and fur down attic ceiling to $9'$ clear height. Set the grounds for vertical sheathing $4'$ high in Laundry, Kitchen and Bath-room; for panelled wainscot $3'$ high in first and second story Hall and Vestibule, and stairs from first to second story, and for bases elsewhere of heights as specified. **Grounds and Furring.**

Fur chimney-breasts with $2'' \times 4''$ studding set flatways, to be everywhere $1''$ clear away from the brickwork; and fur outside stone walls of Laundry in basement with $2'' \times 4''$ studs set flatways. Fur for beams, arches, etc., as required. All furrings, grounds and angle-beads to be perfectly strong, true and plumb.

Set all partitions with $2'' \times 4''$ studs $16''$ on centres, try with a straight-edge, and straighten and bridge before plastering. All partitions except those that stand over each other to stand on a $2'' \times 5\frac{1}{2}''$ piece, and all partitions to have $3'' \times 4''$ cap. Where a partition stands over another, the studs of the upper partition must stand on the cap of the partition below — not on the floor nor on the beams. Truss over all openings in partitions which extend through more than one story, or carry beams; and strongly truss all partitions not supported from below, to take the weight off the middle of the beams. **Partitions.**

Bridge all partitions in first and second stories with two rows of angular bridging of $2'' \times 4''$ pieces cut in **Bridging.**

and nailed with two nails at each end of each piece. Reverse the direction of the bridging pieces in each row. Line the pockets for sliding doors with planed boards.

Saw-dust Filling. Fill in with saw-dust or planing-mill chips between beams around all water supply pipes which run in the floors, making all tight.

INTERIOR FINISH.

All the stock for interior finish of every kind is to be of the very best quality, thoroughly seasoned and of selected grain, and well **Stock.** smoothed, sand-papered and kiln-dried before putting up. Ash to be best Indiana calico figure, and whitewood to be free from white sap. The Laundry, Kitchen, and Back Halls throughout, including back stairs, are to be finished in hard pine, excepting doors, which are to be of whitewood; the Front Hall and Vestibule in first story to be finished in quartered oak, except stairs to second story, which are to be cherry. The Parlor is to be finished in maple and the Dining-room in ash. The Bath-room in second story is to-be finished in black walnut and maple. All other rooms and closets in first and second stories to be finished in whitewood. The hall and large front room in attic are to be finished in white pine for varnishing. All other rooms and closets in attic are to be finished in pine to paint.

Framed Partitions. Make panelled partitions under front stairs as per detail drawings, the framing to be $1\frac{1}{8}''$ thick, moulded, with raised panel $\frac{7}{8}''$ thick, all bead and flush on back, all of quartered oak.

Front Hall and Vestibule in first story, stairs from first to second story, and second-story hall, to have panelled wainscot 3' high, ac-**Wainscot.** cording to detail drawings, with framing $\frac{7}{8}''$ thick, moulded, and raised panels $\frac{7}{8}''$ thick, and moulded cap, but no base. To be of quartered oak in first-story Hall, cherry on stairs, and whitewood in second story.

The Laundry in basement and Kitchen in first story are to be sheathed 4' high with $\frac{7}{8}''$ matched and beaded vertical strips of hard **Sheathing.** pine, not over 4" wide, without base, but with neat bevelled cap. The Bath-room in second story is to have $\frac{7}{8}''$ matched and beaded vertical sheathing, 4' high, of alternate strips of black walnut and maple, 4" wide, with neat bevelled cap of black walnut and $\frac{5}{8}$" bevelled black-walnut base, 4" high. Plough the sheathing into cap of bath-tub.

The Parlor and Dining-room are to have moulded base, $\frac{7}{8}$" x 10", according to detail drawings, to be of maple in Parlor, and ash in Dining-room. All second-story rooms except Bath-room to have $\frac{7}{8}$" x 10" bevelled base of whitewood. All other rooms and closets to have $\frac{7}{8}$" x 8" plain board base, of hard pine in back halls and back stairs throughout; whitewood in remaining parts of basement, first and second stories; pine to varnish in attic hall and large front room, and pine to paint in other rooms and closets in attic.

Bases.

Plough all bases together at the angles and put them on before the upper floor is laid, and allow $\frac{1}{8}$" extra below top of floor.

All doors and windows in first-story Hall and Vestibule, Parlor and Dining-room, and hall and all chambers in second story, to have 1$\frac{1}{8}$"x 4" architraves, moulded as per detail drawings. To be of quartered oak in Hall, maple in Parlor, ash in Dining-room, and whitewood in second-story hall and chambers. All other doors and windows to have $\frac{7}{8}$" x 4" plain square board architraves, of black walnut in Bath-room, and elsewhere of wood to correspond with finish of room.

Architraves.

The architraves of doors and windows in Parlor, Hall and Dining room will have corner blocks with carved rosettes according to detail drawings; all others to be mitred. All moulded door-architraves throughout will have plain plinth-blocks. Architraves in Parlor, first-story Hall, Vestibule, Kitchen, Laundry and back halls to be carried to floor; elsewhere to be cut $\frac{8}{16}$" short, to allow for carpet.

Plinth-blocks and Corner-blocks.

All stool-caps to be $\frac{7}{8}$" thick, with round edges, of wood to correspond with the finish of the several rooms; and in rooms with moulded finish to have moulding and be according to detail drawings; elsewhere to have bevelled board 4" wide under.

Stool-cups.

Brad all architraves, bases and other moulded work in the quirks of the mouldings, and set in all finish-nails for puttying. No splic ing of any architrave will be allowed, and joints of bases must be carefully matched.

Doors. Front and kitchen outside doors to be made of pine according to detail drawings, the front door to be 2½" thick, moulded both sides, with panels raised and carved outside only as shown; the kitchen door to be 2" thick, moulded both sides, with plain panels raised one side only.

Inside Doors. All doors in first-story Hall, Parlor and Dining-room, including each leaf of double doors, to be 1¾" thick, 8' high, 8-panelled according to detail drawings, with flush mouldings and raised panels both sides. All to have thoroughly seasoned pine cores, veneered with quartered oak in Hall and Vestibule, maple in Parlor and ash in Dining-room. Doors opening between rooms finished in hard wood are to be veneered to correspond with the rooms; but doors opening from rooms finished in hard wood into closets or inferior rooms are to be veneered both sides with the hard wood. Doors from Hall to Vestibule to be similar to other doors in Hall except that the four upper panels are to be filled with stained glass in lead work, to be selected by the owner, but paid for and set by the carpenter, and to cost $2.00 per square foot, exclusive of setting. All other inside doors throughout the house to be 1½" thick, of solid whitewood. To be 7' 6" high throughout second story, 7' high in Attic, Kitchen and Basement. Second-story doors to be 5-panelled according to detail drawings, with flush mouldings and raised bevelled panels both sides. Attic, Kitchen and Basement doors, except sash-doors, to be 4-panelled square with raised panels both sides.

Sash doors. The door from Laundry to Basement hall way will have sash in upper part, in 6 lights, glazed with best ¼" ribbed plate glass; the glass and glazing to be furnished by the carpenter.

Door Frames. All doors to have 1⅜" rebated and beaded frames of wood to correspond with the finish of the room, and all to have ⅞" hard-pine thresholds. Veneer frames where necessary to show different wood on each side. Sliding doors between Parlor and Hall to have astragal and hollow on the meeting-styles. Double front outside doors to meet with the ordinary bevel, but to have half of a turned colonnette planted on one leaf to protect the joint. Vestibule doors and double doors between Dining-room and Hall to meet with ordinary bevel, and moulding over joint.

Windows. The carpenter is to furnish all window frames and sashes, and is to deliver the sashes and sash doors, excepting those which he is himself to furnish ready glazed, to the contractor for painting and glazing, and bring them back to the building when completed; and is also to deliver all frames for cellar windows and doors when required by the mason for building into the walls.

Cellar Windows. All windows in Basement except in Laundry are to have $1\frac{3}{4}''$ rebated pine frames and $2\frac{3}{4}''$ sills, and $1\frac{1}{2}''$ pine sashes hinged at the top, with galvanized hooks and staples to keep them open, and strong japanned iron button fastenings; and to have heavy galvanized wire netting with $\frac{3}{8}''$ mesh nailed securely on outside of frame. Make frame only for cold-air box openings, covered with galvanized netting in the same manner.

Fixed Sashes. The small windows in Laundry Closet in Basement and in Coat-closet, Pantry, and China-closet in first story, are to have rebated plank frames and $1\frac{1}{2}''$ pine sashes, screwed in tight.

Double Hung Windows. All other windows throughout the building are to have boxed frames with pockets, 2″ sills pitching $1\frac{1}{2}''$, and ploughed for shingles or clapboards as required, all of pine except beads and pulley-styles, which are to be of hard pine; and $1\frac{3}{8}''$ clear pine sashes in lights as shown, with moulded sash-bars and counter-checked meeting rails, all to be double hung with best steel-axle capped brass-faced pulleys, and best shoe-thread sash-line and best iron weights, and well balanced.

Inside beads to be of hard pine throughout, and all put on with round-headed blued screws. Frames in rooms finished with hard wood to have a strip of corresponding wood veneered on the inner edge of the frame.

Make and put up dresser in Kitchen as per drawings, all of hard pine. To be 5' wide and 8' high, including neat cornice, with two cupboards under with shelf in each, and four shelves above, enclosed by $\frac{7}{8}''$ sash doors to slide past each other on metal tracks, with sheaves complete; all to be furnished ready glazed with first quality sheet-glass by the carpenter and fitted up in perfect order.

Dresser.

Fit up the China-closet as shown on drawings or as directed by the owner, with stand for sink and cupboard under, and four drawers each side; large cupboard with two shelves; six glass shelves 6'' wide on one side, with open fronts, but to have brackets and neat fancy turned standards; and five shelves on the other side, 14'' wide and enclosed with $\frac{7}{8}''$ sash doors in front to slide past each other on metal tracks, all to be furnished glazed with first quality glass and fitted up with sheaves complete by the carpenter. All the work in China-closet to be of whitewood. Fit up slide in partition with porcelain pull to run sideways.

China-Closet.

Fit up Pantry with three barrel-cupboards, with neat panelled doors and lifting covers, one case of three drawers, and four shelves running all around. All of whitewood.

Pantry.

Fit up Coat-closet in whitewood, with one case of four drawers, a shoe rack with eight compartments, each 8'' square, and two rows of hooks.

Coat-Closet.

Make stand for wash-bowl in Bath-room, with cupboard under, and panelled door, and four drawers each side, all of black walnut. Case the bath-tub with black walnut in one long panel on each exposed side with cap of the same. Case the cistern in the same way, finishing the angles with a neat quarter-round. Fit up water-closet with seat, 8'' bevelled wall-strips and flap only, strongly supported, both seat and flap to be hung with nickel-plated brass hinges and screws: all to be of black walnut. Put black-walnut strip and hooks on one side of the room only.

Bath-Room.

Basement water-closet to be fitted with whitewood in the same manner above specified for the one in bath-room, and cistern to be cased with whitewood panelling. Put whitewood lining all around above seat, 15'' high.

Basement Water-Closet.

Make stands for other wash-bowls shown on plans as above specified for the one in bath-room, but all to be of whitewood instead of black walnut. **Other Wash-bowls.**

Fit up all closets not specially described above as marked on plans or as directed by the owner. All closets in second story more than 18" deep, and two closets in attic, are to have, unless otherwise indicated, a case of three drawers, with three shelves over, and two rows of hooks. The remaining closets to have two rows of hooks, and one shelf above. All drawers to have neat panelled or moulded fronts, and to run on hard-wood centre strips. **Other Closets.**

Make tank in attic where directed, 4' long, 2' wide, and 4' deep inside, of 1¾" planed pine plank, with splined joints, to be lined by the plumber. Make cover to the same of matched and beaded pine sheathing, battened on the under side, with rounded edges, and hasp and padlock fastening. **Tank.**

Make strong frame to support soapstone sink in Kitchen and wash-trays in Laundry, with 2¾" ornamental turned legs, and put grooved draining-shelf over sink, and neat covers to wash-trays, all of whitewood. Put beaded strips of whitewood on walls and ceilings of Kitchen and Laundry, and black walnut in Bath-room, for pipes to run on, and case over such pipes as may be directed with neat casings of wood suited to the rooms, screwed on with brass screws. **Kitchen Sink, etc.**

Provide and hang first quality 1⅛" outside blinds to all windows, divided and hinged with care so as to fold back neatly and without interfering. All to have rolling slats in the lower half only. Divided blinds, and those of fixed windows, to be hung with wrought-iron L-hinges; all others to be fitted complete with patent Automatic blind-awning fixtures, and all to have Washburn's patent ring fasts, except those for fixed windows, which are to have Shedd's patent wire fasts, so that they can be opened from the outside. **Blinds.**

Furnish mosquito guards to all outside doors and windows. Those for the windows to slide outside the sashes on beads put up for the

Mosquito Nets. purpose. Those for the doors to be hung on the outside, with springs to keep them closed, and brass hook and staple fastening. All to be made in the best manner with frames of clear seasoned pine, $\frac{3}{4}''$ thick, covered with suitable wire netting, and all to be stained, varnished, fitted, and marked complete by the carpenter, and neatly stored in a convenient place in the attic.

Outside Windows. Furnish outside sashes to all exterior windows, to be of clear seasoned pine, $1\frac{1}{2}''$ thick, glazed with first quality double-thick glass, in lights to correspond with inner sash, packed with listing around the edges, and arranged to be secured by round-headed screws from the inside to small permanent brass plates set in flush with the outside of the casing and firmly screwed to it. Six plates and screws are to be provided for each window, and for convenience in fixing projecting screws are to be set in the outside sash, by which it can be hung from the edge of the upper plates while the other screws are adjusted. All to be made in the best manner, glazed, painted three coats, fitted and marked complete by the carpenter, and safely stored in a convenient place in basement.

Setting Mantels. The carpenter is to set all mantels in the best manner, and case over for protection until the house is delivered, removing the casing only for painters' work, and replacing afterwards.

Mantels. The carpenter is to make mantels to all fireplaces, and also shelves with brackets in chambers which have no fireplaces, in the best manner, in strict accordance with detail drawings and the directions accompanying them. To be of first quality thoroughly seasoned stock to match the finish of the rooms, and all well bolted and dowelled together.

Step-Ladder. Make strong step-ladder of planed spruce to scuttle in roof: to be movable.

Shelves. Fit up shelf along one side of Laundry and Kitchen, $4\frac{1}{2}$ feet above floor, strongly supported, and put one row of hooks on beaded strip under the same. Put up also neat roller for towel in Kitchen, and hook strip over sink, and put 20 feet run of shelving in furnace-cellar in the most suitable place, and three shelves for batteries and gas-meter where directed by the electrician, bell-hanger, and gas-fitter, and swing shelf where directed.

Make in a suitable place a strong double box of 1¼" matched and beaded pine, to contain ash and garbage barrels, with division between. Each part to have battened door in front, with good lock and two keys, for removing barrels, and lifting cover on top, hung with brass butts, and with brass hook and staple fastening. **Ash and Garbage Boxes.**

Put base-knobs with inserted rubber to all doors, of wood to match finish of room. **Base Knobs.**

Fit up strong flap-table in Kitchen and one in Laundry, of pine. **Flap Tables.**

Make two good coal-bins in cellar as directed, to hold ten tons each. **Coal-Bins.**

Build temporary privy for the workmen, to be cleared away and the place cleaned out, filled up, and graded over at the completion of the house. **Temporary Privy.**

Cut the floors for registers and hearths as may be requisite, and fit borders neatly around, and cut as required for plumbers, gas-fitters, and other workmen, repairing neatly afterwards. Assist other workmen employed in the building, furnish centres, patterns for bays, lintels, and rough furring as much as may be needed. **Miscellaneous.**

HARDWARE.

The sliding-doors between Parlor and Hall are to be hung in the best manner with Prescott's patent balance hangers complete; to have Russell & Erwin's solid bronze sunk handles, pattern No. 332, dark finish, and bronze astragal-face sliding-door locks and pulls of the same make, pattern No. 333. **Sliding-door Furniture.**

Butts. The front outside double doors are to be hung with 6" x 6" Russell & Erwin's fancy solid bronze dark finish acorn loose-pin butts of pattern No. 15, three to each leaf, with steel bushings and steel washers.

All other doors in first-story Hall, Vestibule, Parlor and Dining-room are to be hung with 5" x 5" bronze acorn steel-washer japanned loose-joint butts, three to each door or leaf of double doors.

All other doors throughout the building are to be hung with 4" x 4" japanned loose-joint acorn butts with steel washers, two butts to each door.

Locks. The front door is to have Enoch Robinson's patent front-door mortise lever-lock, with brass or bronze face and striking-plate and night-latch, with one key to the large lock and four to the night-latch.

All other doors throughout the building to have Russell & Erwin's, Corbin's, or Nashua Lock Co.'s 5-inch mortise-locks, with brass face and striking-plate, brass bolts, and German-silver or plated keys; no two keys in the house to be alike. The outside kitchen door is to have in addition a Yale rim night-latch, with two keys.

Knobs. The front outside door is to have Russell & Erwin's fancy solid bronze dark finish 2½-inch knobs, pattern No. 923, on both sides of one leaf only. Doors in basement, kitchen and attic to have 2¼-inch best lava knobs with bronze roses and escutcheons. All other doors throughout the building to have Russell & Erwin's 2¼-inch fancy solid bronze dark-finish knobs, pattern No. 933. Double doors to have knobs on one leaf only.

Bell-Pulls. The carpenter is to furnish bell-pulls to front and kitchen outside doors, to correspond with the door-knobs.

Bolts. The outside front and vestibule doors, and double doors between first-story Hall and Dining-room, are to have bronze-metal flush-bolt at top and bottom of the leaf which has no lock, and front outside and vestibule doors will have in addition a strong solid bronze chain bolt. The kitchen outside door, door at head of basement stairs, door of basement water-closet, and doors of bath-room and all chambers in second story, to have Ives's patent mortise-bolts with bronze roses and bronze keys.

All cupboard doors to be hung with brass butts and to have brass slip-latches. **Cupboard Doors.**

Drawers of wash-bowl stands to have handsome brass drop-handles. All other drawers to have plain japanned iron pulls. **Drawer-Pulls.**

All double-hung windows to have Morris's patent self-locking sash-fasts, to be of solid bronze in first-story Hall, Vestibule, Parlor, and Dining-room; bronzed iron with plated drops elsewhere, and two pulls on lower sash to correspond. **Sash-Fasts.**

Put heavy triple hooks of japanned cast-iron in closets and other places specified: to be in two rows unless otherwise expressly directed, and to be 8" apart in each row; those in the upper row to be set over the middle of the spaces between those in the lower row. **Hooks.**

All brass hardware to be put on with brass screws, plated with plated screws, bronze or bronzed with bronze screws, and japanned with blued screws. **Screws.**

STAIRS.

[*This is very commonly made a separate contract, and in that case the General Conditions should precede.*]

The front stairs from first to second story are to have open string, moulded nosings returned at the ends and carried around well-room, $\frac{7}{8}$-inch risers and treads, housed into the wainscot on the wall side, the treads ploughed into the risers, and risers ploughed into underside of treads; $1\frac{3}{4}$-inch fancy turned balusters, two to a tread and around well-rooms in the same proportion, all dovetailed at the foot and tenoned into underside of rail; $2\frac{3}{4}$" x $3\frac{1}{2}$" double moulded hand-rail; 4" x 4" solid turned, chamfered and fluted posts at angles, with half-post at upper termination of rail, and 5" x 5" fluted and carved boxed post at foot: all to be strictly **Front Stairs.**

in accordance with detail drawings. The posts and rail are to be of best Spanish mahogany, all the rest to be of cherry. Outside string on stairs, and face-board around well-room, to have moulding to cover joint with plaster.

Back Stairs. The back stairs from first story to attic are to have open string, rounded nosings returned at the ends and carried around well-rooms; $\frac{7}{8}$-inch risers and treads, the treads ploughed into risers, and risers ploughed into underside of treads, and both treads and risers to be ploughed for the base on the wall side; $1\frac{1}{4}$-inch plain round balusters, two to a tread and around well-rooms in the same proportion, and mortised at top and bottom; $2'' \times 2\frac{3}{4}''$ plain moulded hand-rail, and $3\frac{3}{4}'' \times 3\frac{3}{4}''$ solid turned and chamfered post at each angle and at foot, with half post at upper end of rail: all to be according to detail drawings, of hard pine throughout.

Cellar Stairs. The stairs to cellar are to go down under the front stairs; to have $\frac{7}{8}$-inch risers and treads with rounded nosings, ploughed for base; $1\frac{1}{4}$-inch plain round balusters, two to a tread, and mortised into treads and underside of rail, and $2'' \times 2\frac{3}{4}''$ plain round hand-rail: all to be of hard pine.

All stairs are to be framed and supported in the best and strongest manner, on $2'' \times 12''$ spruce strings $12''$ on centres: all to be thoroughly wedged, blocked and glued in the best manner, and left clean and perfect. The stair-builder is to put on all the face-boards and nosings around well-rooms, and is to furnish and put on plain square hard-pine base on wall side of back stairs and cellar stairs, ploughed into treads and risers, and is to do all the work of housing the front stairs into the wainscot.

BELLS.

[*This is often included in the carpenter's contract. If it is separated the General Conditions should precede the Specification. If electric gaslighting is introduced (see specification below) the bells may be included in the contract with it.*]

Put in electric bells as follows, with annunciator in kitchen, battery, insulated copper wires, push-buttons to match finish of rooms, and bells of five different tones, all put up in the best manner and warranted for three years :

Bell from front door to ring in Kitchen and in Attic hall.
Bell from Kitchen outside door to ring in Kitchen.
Foot-bell from Dining-room to ring in Kitchen.
Bell from Parlor to ring in Kitchen.
Bell from second-story Hall to ring in Kitchen.
Bell from second-story Hall to ring in Attic Hall.

The pulls for the outside front and Kitchen doors only will be furnished by the carpenter.

All wires to be run behind the plastering.

ELECTRIC GAS-LIGHTING.

[*The General Conditions should precede the Specification unless this is joined with some other contract.*]

Wire all the gas outlets in the building with the best insulated copper wire, all concealed behind the plastering. The drop-light in first-story Hall is to be wired for automatic burner, to light from wall of Hall near door to Vestibule, and also from east wall of chamber over Dining-room, just under the bracket outlet. The bracket at foot of cellar stairs is to be wired for automatic burner, to light from wall at head of stairs. All other outlets are to be wired for pull-burners.

When directed, after the fixtures are in place, put automatic burners on first-story Hall lantern and on bracket at foot of cellar stairs, with buttons as above specified; and put four pull-burners on Dining-room chandelier, six on Parlor chandelier, four on brackets in Parlor, three in each of three principal chambers in second story, of which two will be on the mirror light, and one on bracket as directed; one on second-story Hall bracket, and one on bracket in Bath-room. All to be of the best pattern, perfectly tight against any escape of gas; all fitted up in the best manner, with battery complete, and to be kept in good order free of expense for two years.

PAINTING AND GLAZING.

[*If this is made a separate contract the General Conditions should precede the Specification.*]

Outside.   Oil hard-pine piazza and porch floors and treads of steps two coats.

Stain.   Stain sheathing under piazza and porch roofs with one coat of light oil of creosote.

Paint.   Paint all roofs one coat of pure Venetian red in oil, finishing with a second coat of pure Indian red.

Paint all other outside wood and metal work two coats, of three tints as directed: the first story to be one shade, the second story another, and the doors and trimmings a third. Paint also the mouldings on belts, etc., where directed, in Venetian red.

Paint the blinds three coats of color as directed.

Paint sashes three coats of color as directed.

Inside.
Floors.   All hard-wood floors and borders, including hard-pine floors in Kitchen and Basement, are to have one light coat of oil, and to be finished with hard wax or Butcher's Boston Polish, put on in the best manner and well rubbed. Other hard-pine wood-work, and all ash and whitewood finish, to be filled with oil filler, and to have two coats of Pellucidite, rubbed down with emery cloth and oil. The maple finish in Parlor is to be filled with oil filler and to have three coats of white shellac, rubbed down with emery cloth and oil. The cherry stairs are to be stained in the best manner to imitate mahogany, filled with oil filler and finished with two coats of Pellucidite, rubbed down with emery cloth and oil. The mahogany posts and rails are to be filled and varnished with two coats of best copal varnish. Oak wood-work in Hall is to be oiled one light coat, and finished with Butcher's Boston Polish, or hard wax, well rubbed. The pine in Attic hall and large room is to be oiled one coat and varnished with two coats of copal varnish. Other woodwork in attic is to be painted three coats of pure zinc white and oil, to finish with a plain oil surface, not flatted.

Sashes. Inside of sashes to be stained cherry color and varnished.

Pipes. Varnish all exposed lead and brass pipes and bands with one coat of white shellac.

All materials are to be of the very best quality. Pure linseed oil only is to be used. The body for inside work is to be pure zinc white, and for outside work to be the best French ochre. No lead to be used for outside work unless expressly directed, and in that case to be pure [*Jewett's*], [*Union*] or [*Salem*] white lead.

Puttying. Putty-stop thoroughly and smoothly all work inside and outside after the first coat and before the last coat, coloring the putty to match the wood after darkening. Use wax suitably colored instead of ordinary putty, wherever wax finish or Butcher's polish is specified. **Wax.**

Knots. Cover all knots, sap and pitchy places with strong shellac, and kill knots or pitch with lime where necessary. Sand-paper all inside work, rubbing with the grain, and clear out all mouldings before the first coat, and sand-paper after each coat of paint, shellac, Pellucidite or varnish except the last.

GLAZING.

[*This is almost invariably included in the contract for painting.*]

Glaze all inside and outside sashes, except those specified to be furnished ready glazed by the carpenter, in lights as shown on drawings or as directed, with first-quality double-thick French or German glass; all well bedded, puttied, back-puttied, and tacked, and all repaired at the completion of the building, thoroughly cleaned and left whole and perfect.

PLUMBER.

[*If this is made a separate contract, as it usually should be, the full title and the General Conditions must precede the Specification.*]

There will be **Fixtures.**

In Basement: Set of four white earthenware Wash-Trays.

Cold Supply to Wash-Boiler.

One [*Hellyer's short Artisan Hopper*], with seat attachment and preliminary flush.
Sill-Cock.
Furnace Supply.
In First Story: One Soapstone Kitchen Sink with draining shelves and wall-plates.
One 50-gallon Bath-Boiler.
One Pantry Sink.
In Second Story: One Bath.
Three Wash-Basins.
One [*H. C. Meyer & Co.'s*] Brighton Water-Closet.
In Third Story: Tank.

Iron Pipes. All iron pipes, including both waste and air pipes, to be of the best quality, [*Mott's*] or [*Brady's*] make, with all proper fittings; the horizontal portion of the main soil-pipe in basement to be double-thick; all other iron pipes to be single-thick; and all to be thoroughly coated inside and outside in the best manner with asphaltum before putting up, and the outside to have a second coat afterwards: all to be put up in the best and strongest manner with iron hooks and stays, and the joints caulked with oakum and melted lead.

Lead Pipes. All cold-water supply, air, and waste pipes under 2" are to be best drawn lead, and to weigh as follows:—
Rising-Main: ⅞ inch to weigh 4 lbs. per foot.
Other Supply-Pipes:
    1-inch to weigh 5 lbs per foot.
    ¾ " " " 4 " " "
    ⅝ " " " 3 " " "
    ½ " " " 2 " " "
Waste, Air and Overflow Pipes:
    2-inch to weigh 5 lbs per foot.
    1½ " " " 3 " " "

Brass Pipes. Hot-water pipes throughout to be of National Tube Works' best seamless drawn plumbers' brass tubing, with brass fittings, all put together with red lead in the best manner and made perfectly tight.

BUILDING SUPERINTENDENCE.

All lead and brass pipes are to be put up in the best manner on boards set in place by the carpenter. The lead pipes are to be secured with hard metal tacks or brass bands and screws, and brass pipes with brass bands. No hooks are to be used. The hot and cold water pipes are to be kept at least ½" apart everywhere. All joints in lead pipes throughout are to be wiped joints, no cup-joints to be permitted anywhere; and all brass pipes are to be put up with right-angled turns so arranged as to allow free expansion and contraction. All connections between lead and iron pipes and traps are to be made with cast-brass ferrules, of the same size as the lead pipes, soldered to the lead pipes with wiped joints, and caulked with oakum and melted lead into the iron pipes.

The waste-pipes are to run as follows: —

Waste-Pipes. The main 4-inch soil-pipe is to be extended through the north side of cellar-wall, and jointed air-tight with clear Portland cement into the drain-pipe outside. From this point it is to run with uniform pitch along the cellar-wall, and under floor of basement water-closet, to a point in Laundry under second-story water-closet, with 4-inch Y-branch to receive trap of basement water-closet, and 3-inch Y-branch for waste-pipe from Laundry wash-trays, then turning up on Laundry wall with 4-inch Y branch and 4-inch brass trap-screw caulked in at the turn for cleaning out the pipe, and running straight up on Laundry wall through the Kitchen floor and Bath-room with 2-inch Y-branch below Kitchen floor for waste-pipe from Kitchen sink, 4-inch Y-branch below Bath-room floor for pipe from trap of water-closet, and 2-inch T-branch above all other connections for air-pipe from traps; and thence straight up through and two feet above the roof; the top to be protected with brass wire netting, and a flange of 16-oz. copper, 18" square, to be soldered on to protect roof, shingled in and warranted tight.

Three-Inch Waste. Carry a 3-inch iron pipe from the connection provided for it in Laundry with uniform pitch on north wall of Laundry to the corner, there turning, with Y-branch and 3-inch brass trap-screw for cleaning out, accessible from the vegetable-cellar, and continuing with the same pitch on east wall of Laundry behind wash-trays, with Y-branch for pipe from trap of wash-trays, to a point nearly under pantry sink in first story,

and there turning, with Y-branch and 3-inch brass trap-screw at the turn for cleaning out, and continuing up on Laundry wall to China-closet, with 1½-inch Y-branch under floor for waste from pantry sink, and up through China-closet and closets over it, with double 1½-inch Y-branch under second-story floor for wastes from wash-basins, and 2-inch T-branch for air-pipe from traps above all other connections and thence straight up through the roof and 2 feet above, with brass wire netting on top, and 18" x 18" flange of 16-oz. copper soldered on and all warranted tight.

Iron Air-Pipes. Carry a 2-inch iron air-pipe, all caulked air-tight in the same way as other iron pipes, from trap of basement water-closet beside the main soil-pipe to the connection with soil-pipe provided for it in Bath-room, with 2-inch T-branch above Bathroom floor for connection with lead air-pipe from traps of sink in Kitchen, and wash-basin, bath and water-closet in Bath-room. Carry a separate 2-inch iron air-pipe from trap of Laundry wash-trays beside the 3-inch waste-pipe to the connection provided for it in second story, with 1½-inch T-branches to receive air-pipes from trap of pantry sink and second-story wash-basins.

Supply-Pipes. The plumber is to apply for ⅝-inch service-pipe from street main to house, and pay all charges, if any, for it, and is, when called upon, to furnish and fit up temporary cock in cellar to supply water for building. As soon as the house is plastered he is to complete the permanent supply-pipes as follows:

Rising-Main. Carry ⅝-inch four-pound rising-main up on cellar-wall and walls of Laundry, Kitchen and Bath-room to tank in attic, with stop-and-waste cock at cellar-wall, and branches as follows:

In Basement: — ⅝-inch branch to Sill-cock.
 ½-inch branch to Furnace Supply.
 ½- " " " Water-closet cistern.
 ⅝- " " " Laundry wash-trays.
In Kitchen: — ⅝-inch branch to Kitchen sink.
 ⅝-inch branch to China-closet, running on Kitchen ceiling, and dividing in China-closet into three ½-inch branches, one of which is to supply the pantry sink, and the other two the wash-basins in second-story chambers.

In Bath-room: — ⅝-inch branch to bath; the same to supply wash-bowl by a ½-inch branch.

½-inch branch to Water-closet cistern.

Every branch from the rising-main is to have separate stop-and-waste cock in a convenient place, and so arranged that the pipes can be completely drained of water; and each of the three branches in China-closet is also to have separate stop-and-waste cock, to shut off the pantry sinks or either basin and empty the pipes at pleasure. All stop-and-waste cocks to be finished brass ground cocks of the best quality. **Stop-and-Waste Cocks.**

Line under and behind all pipes and traps above the first floor with eleven-ounce zinc, the horizontal joints to be well soldered, and the upright joints to be lapped. This lining is to enclose completely all pipes, to protect ceilings and walls from any defect or leak, present or future. The lining to be graded to certain points, and to be connected with ⅜-inch drip pipes to run to basement, and empty over cistern of basement water-closet; the end of the pipe to dip below the water-line of cistern. **Pipe Lining.**

Line with 16-oz. tinned copper in the best manner the tank in attic, 4' x 2' x 4', to be furnished ready for lining by the carpenter. Supply from the rising-main, with ⅝-inch finished brass compression ball-cock and 6-inch copper float, and put in 1¼-inch boiler valve, with ⅞-inch pipe to boiler, and 1½-inch lead overflow pipe, to be carried down beside rising-main and to empty over kitchen sink, with the end over the sink turned up to form a trap. **Tank.**

Hot water will be supplied to Laundry wash-trays, Kitchen sink, pantry sink, bath, and three wash-basins. **Hot Water.**

Furnish and fit up in Kitchen, beside range, a 50-gallon [*S. D. Hicks & Son*] first-quality, warranted copper bath-boiler, on Lockwood pattern cast-iron stand, to be connected to water-back of range with 1" brass pipe and ground-plug sediment-cock union coupling, and to have waste-pipe carried to the waste-pipe from kitchen sink. The boiler is to have three brass pipe-couplings on top, one for the supply from tank, and the **Boiler.**

others for two separate ¾-inch hot-water supplies. One of these is to be carried on Kitchen ceiling, with ⅝-inch branch to Kitchen sink, thence to Bath-room, with ⅝-inch branch to bath and ½-inch branch to wash-basin, and to continue on Bath-room wall and ceiling to connect with the coupling provided on the boiler for circulation, and a ½-inch expansion pipe to be carried from the highest point of the circulation pipe two feet above the tank, and turned down over the mouth of the overflow pipe. The other ¾-inch hot-water supply-pipe is to be carried on Kitchen ceiling to China-closet, with ⅝-inch branch down on Kitchen wall to Laundry wash-trays, and ½-inch branch to pantry sink. Beyond the China-closet it is to divide into two ½-inch branches, extending to wash-basins in second-story chambers, without any circulation or expansion pipe.

Hot-Water Stop-and-Waste Cocks. Put three finished brass stop-and-waste cocks on hot-water supply-pipes in China-closet, to shut off pantry sink and wash-basins in second-story chambers separately, at pleasure, and drain the pipes. Put two cocks of the same kind over Kitchen sink, to shut off the sink or the Bath-room fixtures.

Wash-Basins. Furnish and fit up in Bath-room and two chambers in second story, where shown or directed, three best 16-inch white earthenware overflow ground wash-basins. Each to be supplied with hot and cold water through ½-inch pipe, and No. 4 extra silver-plated cast-tube compression basin-cocks; and to waste through plated socket and strainer, with plated plug and chain-stay, bolted to the marble, and plated safety-chain No. 1, 1¼-inch lead pipe to main wastes with 1½-inch lead S-trap close to the outlet, and brass trap-screw, and 1½-inch vent-pipe to main air-pipe. Connect the overflow with the traps by 1½-inch lead pipe, entering below the water-line, to prevent circulation of air.

Cover each basin with 1¼-inch, best quality, blue-veined Italian marble slab, dished, with all free edges ogee moulded, and with ¾-inch wall-plates of the same marble, 15" high, with ogee-moulded edges. Basins to be secured to marble with three brass basin clamps and bolts to each, and the joint to be made tight with plaster-of-Paris.

Furnish and fit up in Bath-room, where shown or directed, one 6-foot 20-oz. [*Steeger's*] tinned and planished copper overflow bath, to be supplied with hot and cold water through ⅜-inch pipe and ⅜-inch extra silver-plated compression bath bibb-cocks; and to waste through plated socket and strainer, with plated plug and chain-stay, and plated safety-chain No. 2, with 2-inch lead pipe to main waste, and 2-inch lead S-trap, with brass trap-screw and 1½-inch lead vent-pipe to main air-pipe. Connect the overflow to the trap by 1½-inch lead pipe, entering below the water-line. **Bath.**

Furnish and fit up in China-closet, where shown or directed, one [*Steeger's*] 24-oz. tinned and planished copper 14" x 20" flat-bottomed overflow pantry sink; to be supplied with hot and cold water through tall extra silver-plated upright core compression pantry-cocks, with screw nozzle for cold water; and to waste through plated socket and strainer, and 2-inch brass Boston waste-cock, with silver-plated plate and lever, and 1½-inch lead pipe to main waste, with 4-inch round trap and 4-inch brass trap-screw, and 1½-inch lead vent-pipe to main air-pipe. Connect the overflow to the trap by 1½-inch lead pipe, entering below the water-line. **Pantry Sink.**

Cover with 1¼-inch best quality, blue-veined Italian marble slab, dished, with all free edges ogee-moulded, and with ⅞-inch wall-plates of the same marble, 15" high, with ogee-moulded edges.

Furnish and fit up in second-story Bath-room, where shown, one [*H. C. Meyer & Co.'s*] white earthenware Brighton water-closet, and wooden cistern, lined with 16-oz. tinned copper, 16" x 16" x 24", inside measurement, with [*Meyer's*] 4-inch brass cistern-valve, brass compression or Fuller ball-cock and 4-inch tinned copper float, cranks, lever, and plated safety-chain No. 2, polished black handle complete, and 1¼-inch 2½-lb. lead pipe to basin, and 1¼-inch overflow carried into basin supply. To be connected to branch of soil-pipe by 4-inch lead pipe and brass ferrules, and a 2-inch lead vent-pipe to be carried from this pipe to main air-pipe. Connect the ventilation-pipe from back of closet by 2-inch pipe to range flue in chimney, making all tight. **Water-Closet, Second Story.**

Furnish and fit up in Basement, where shown, one [*Hellyer's*] white earthenware short Artisan Hopper water-closet, with flexible metallic connection, and wooden cistern lined with 16-oz. tinned copper, 16" x 16" x 24", inside measurement, with 24-oz. two-gallon service-box, **Water-Closet, Basement.**

brass compression ball-cock and 4-inch tinned copper float, and [*Meyer's*] 4-inch brass cistern valve, cranks and all attachments complete for automatic seat-supply with preliminary and after flush, and 1¼-inch 2½ lb. lead pipe to basin with 1¼-inch overflow connected to basin supply. To waste through 4-inch lead pipe to branch of soil-pipe, and a 2-inch lead vent-pipe to be carried from trap to main air-pipe.

Slop-Safes. Put Mott's enamelled iron slop-safe, of suitable shape, over bowl of each water-closet.

Furnish and fit up in Kitchen, where shown or directed, one best-quality soapstone kitchen sink, 24″ x 48″ x 8″, with grooved soapstone draining shelf and soap-dish, and soapstone back 16″ high; to be supplied with hot and cold water through ⅝-inch pipe and ⅝-inch finished brass compression-cocks, the cold water cock to have screw nozzle; and to waste through 6-inch brass cesspool and strainer, and 2-inch lead pipe to main waste, with 6-inch round lead trap and 4-inch brass trap-screw, and 2-inch lead vent-pipe to main air-pipe. Sinks to stand on wooden frame furnished by the carpenter.

Kitchen Sink.

Furnish and put up in Laundry, where shown or directed, a set of four white earthenware wash-trays, of the Morahan Ceramic Co., one to have rubbing board formed in the porcelain. All to be set complete by the plumber, and each to be supplied with hot and cold water through ⅝-inch pipe and ⅝-inch finished brass compression wash-tray cocks. To waste through best 1¼-inch silver-plated wash-tray strainers and couplings, with plated plugs, chain-stays, and plated safety chain No. 2, and to have one 6-inch round lead trap for the set of four trays, with 4-inch brass trap-screw, and separate 1¼-inch lead waste-pipe from each tray, all entered into the trap below the water-line, with 2-inch lead outlet pipe from trap to main waste, and 2-inch lead vent-pipe from trap to main air-pipe.

Wash-Trays.

Supply the wash-boiler, to be furnished and set by the mason, with cold water only, through ⅝-inch lead pipe, carried through top of boiler and neatly finished, with ⅝-inch finished brass ground stop-cock on the pipe, in a con-

Wash-Boiler Supply.

venient position over the boiler. There will be no waste to this boiler.

Bore through the sill of house under middle of front bay, or elsewhere where directed, and put on outside a ⅜-inch plated compression sill-cock, with screw for hose. **Sill-Cock.**

Furnish and fit up on pier next to furnace a ½-inch finished brass compression bibb-cock for drawing water. **Furnace Supply.**

Line under all fixtures of every kind about the basement with 4-lb sheet lead turned up 2 inches all around, and to have convex brass strainer and 1-inch drip-pipes carried to basement and emptying over cistern of basement water-closet; the end of pipe to dip below water-line of cistern. **Safes.**

All the work is to be done in the very best, neatest and most thorough manner, tested by turning the water on to each part. All defects to be made good at the completion of the building and all left perfect and warranted for two years.

As soon as the pipes and traps are ready the traps are to be filled with water, and the whole system tested by closing all air-pipes and other outlets, and pouring 5 ounces of oil of peppermint into the top of the main soil-pipe, followed by two or three gallons of hot water, and immediately closing the top of the soil-pipe. **Peppermint Test.** This is to be done in the presence of the architect or of some person appointed by him, and if any odor of the peppermint is detected in any part of the house the plumber is at his own expense to search for and find the defect or defects which may have allowed the vapor of peppermint to escape, and make them all good. After repairing the defects so discovered, the test is to be repeated in the same way, and any further defects so detected are also to be made good at the plumber's expense, until the whole is satisfactorily proved to be tight and perfect, and all work is then to be restored and replaced in good order; the whole cost of making such repairs, and of removing and replacing other work disturbed to obtain access to the pipes, with all expense of making good any damage so occasioned, and indemnity for delays to other contractors, to be paid by the plumber.

GAS-FITTING.

[*The gas-fitting is often included in the plumber's contract.*]

Piping. PIPE the house for gas in the best manner in accordance with the regulations of the Gas Company, with outlets as marked on plans, [*seventy-two*] outlets in all. All pipe to be best wrought-iron, and all fittings under 2-inch to be of malleable iron. All to be put together with red lead, capped, tested and proved perfectly tight before any plastering is done, and the caps left on. All pipes are to be laid with a fall towards the meter, which is to be placed [*in furnace-cellar near west window*] and all are to be well secured with hooks and bands.

The gas-fitter is to call upon the carpenter to do such cutting of timbers as he needs, but no beams are to be cut, notched or bored at a greater distance than 2 feet from the bearing; drop-lights where requisite to be supplied by special branches. All nipples to be of the exact length for putting on fixtures without alteration, and all to be exactly perpendicular to the wall or ceiling from which they project. Bracket outlets in all Halls and passages, Parlor, Dining-room, Kitchen, Bath-room, Furnace-cellar and Basement Water-closet, to be exactly 5' 6" above finished floor; elsewhere to be exactly 4' 9" above finished floor. Mirror light outlets to be 8' above finished floor.

Connecting Meter. Apply and pay all necessary charges for service-pipe from street main to inside of cellar-wall, and connect the service-pipe and house-pipes with the meter in the best manner, with stop-cock on the street side of the meter, and leave all perfect.

HEATING.

Furnish and set where shown or directed in Basement one best quality [*No.* 24 *Peerless*] brick furnace, including galvanized-iron inverted cone top and covering bars, 8-inch galvanized-iron smoke-pipe, tin hot-air pipes, register-boxes, registers, soapstone borders, plastering rings, wire nettings and dampers complete, and including also all cartage, transportation and labor of every kind except only mason and carpenter work.

In First Story: The Front Hall is to have 12-inch pipe and 12" x 15" register.

Size of Pipes and Registers

The Back Hall is to have 12-inch pipe and 12" x 15" register.

The Parlor is to have 12-inch pipe and 12" x 15" register.

The Dining-room is to have 12-inch pipe and 12" x 15" register.

In Second Story: The Chamber over Dining-room is to have 10-inch pipe and 10" x 14" register.

The Chamber over Parlor is to have 10-inch pipe and 10" x 14" register.

The Chamber over Kitchen is to have 9-inch pipe and 9" x 12" register.

The Chamber over Hall is to have 9-inch pipe and 9" x 12" register.

The Dressing-room is to have 8-inch pipe and 8" x 10" register.

The Bath-room is to have 8-inch pipe and 8" x 10" register.

Each room is to have an independent hot-air pipe and all hot-air pipes to be double where they pass through floors or partitions, or behind furrings. The pipes are to be carefully arranged so that all may draw equally, with easy turns at every change of direction. All register-boxes are to be double; all tin-work to be of XX bright tin, and all wood-work within 1" of any hot-air pipe, or within 16" of the smoke-pipe, to be protected with pieces of bright tin, securely nailed on. All registers are to be of [*Creamer's*] make, all placed where shown on plans or as directed; the one in Parlor to be nickel-plated; all the others to be black japanned.

Make and put up a cold-air box of galvanized-iron as shown by blue lines on plan, to be 22" x 31" in section, all riveted together in the best manner, and strongly secured to cellar ceiling. **Cold-Air Box.** The mouth at each end to be flanged out to protect the joint with the frame, and to have strong galvanized-wire netting, ¼" mesh, over the opening. Put a sliding damper of galvanized-iron in each end, and make door 24" x 30" for cleaning out, with button fastening.

Regulator. Furnish and put on in the best manner [*White's*] automatic regulator, to act both upon the check-draught and the damper in the smoke-pipe, and leave all in perfect working order.

Clean up the iron-work and leave it neat at the completion of the building, and furnish a shaking handle, poker and iron shovel.

Range. Furnish and fit up where shown in Kitchen one No. 5 [*Carpenter's*] Range with water-back and couplings, and plate-warmer, all to be set complete, including all transportation and labor except mason-work. Clean up the range and leave it neat at the completion of the building, and furnish at the same time the usual list of tin and iron ware.

Laundry Stove. Furnish and set up where directed in Laundry a [*Walker's*] No. 1 stationary laundry-stove, with galvanized-iron smoke-pipe and zinc-covered stove-board complete: all to be left neat and perfect.

CONTRACTS.

CHAPTER IV.

Not the least important of the young architect's duties is that of guarding the interests of his employer by means of a clear and explicit contract with each of the mechanics employed on his building. In general the architect himself will have to draw up these instruments, the ordinary printed forms being quite inadequate for the purpose, and he will find the task not always an easy one. While it is incumbent on him to secure the best and safest terms for his employer that the contractors will agree to, he has no right to use unfair means to induce them to sign their names to stipulations of which they do not fully understand the meaning; and if he wishes to be able to enforce the contract in case of need he cannot be too careful to express in the plainest language every point upon which it may subsequently be necessary to insist.

In order to provide for all possible contingencies it is necessary to define the rights and duties of the parties under a great variety of circumstances, and a good contract will for this reason be somewhat long; but this is better than a condensation which leaves loopholes for evasion or dispute.

It must not be forgotten that contracts are often entered into unthinkingly, which may prove very disadvantageous for one of the parties, and the architect must be careful to protect his principal from such mishaps. Thus the making of a bid for work by a mechanic, and its unconditional acceptance by the owner, or the architect for him, constitute a valid agreement, by which the mechanic is bound to do the work, and the owner to pay for it, perhaps without any stipulations as to the time of completion, the terms of payment, or other very important matters, which may have to be decided afterwards by costly litigation. For this reason many architects accept a tender only upon the condition that a satisfactory contract shall be signed by the party offering it; but such a conditional acceptance does not bind the latter, who is then at liberty to withdraw his bid, if he chooses, at any time before the final agreement is signed. If, therefore, any tender should be particularly advantageous, it may be desirable to prevent its withdrawal by a prompt and definite acceptance, and in order that this may carry with it by implication the consent of the parties to at least the more important clauses of an ordinary contract, it is usual to employ the device of General Conditions, which are prefixed to every specification, and constitute a part of it; and since the tender is always for doing work or furnishing material according to the specification, the general conditions will be included in the terms to which the bidder offers to conform, and will be binding upon him if his bid is accepted. As the general conditions relate to the duties of the mechanic, not to those of the owner, the former will usually be very willing to exchange the one-sided agreement constituted by them, under which, for instance, the owner would not be required to make any payments before the completion of the work, for a new one, defining the rights of both parties; and the new contract may, and should, comprise the substance of the general conditions of the specifications, expressed in nearly the same words.

A building contract is usually divided into two portions, the first of which is the simple agreement of one party to do a certain work for the other within a certain time, in consideration of a certain payment, which the latter promises to make; while the second comprises the conditions which explain, or modify the principal agreement. The first portion, comprising the essence of the contract, should be drawn up with special care, and although expressed in the briefest possible terms, it should be made to include the points regarded as most important. In English contracts the principal agreement is generally written without any punctuation except periods at the ends of the sentences, the object of this being as much to enforce clearness of expression as to prevent the possibility of fraudulent or careless change in the sense by the alteration of points, and the practice seems to be a useful one. The conditions which follow may have ordinary punctuation or not, but it is not easy to make them intelligible without it.

Whether contracts should be sealed by the signers or not is perhaps doubtful. A simple agreement, in which a consideration is expressed for each promise, would not require seals, but among the many and various stipulations of a building contract some might possibly be regarded as promises to which the consideration did not apply, and a seal would be required to make these valid: so that it seems to be a reasonable precaution to affix them.

Seals.

In some states building contracts must be recorded by the proper public officer, and the law in this respect should be ascertained by architects practising beyond the boundaries of their own state. The statutes in regard to the liens of mechanics and material-men also vary in different states, and are frequently changed, so that some care is necessary to secure owners absolutely against loss by the dishonesty or bankruptcy of the builder. In most states the right of mechanics to file liens against a building or estate to recover the amount of their wages for work upon it is barred at the expiration of thirty days from the time that they cease their labor, so that if the record after this time shows no lien to have been filed, the contractor may be paid the balance due him in full, without fear that the workmen can demand an additional sum. Where sixty days or more are allowed by law for filing liens, the last payment to the contractor must be deferred until after the lapse of this term, allowing a few days more to give opportunity for examining the record. Liens for materials furnished to a contractor can in most states only be claimed by giving notice to the owner of the building in which they are used, before they are delivered, that he will be held responsible for the price of them, and he can then retain the amount out of the payments to the contractor, or he can notify the dealer that he will not receive them, and the latter can then only look to the builder for his pay.

With these explanations, the clauses in the following form of contract will be sufficiently clear. The insurance clause may need modification according to circumstances, but the important points are to make sure that the builder's interest will be covered, so that in case of loss he may not be thrown into bankruptcy, to the injury of the owner; and to define clearly the mode in which the policy shall be taken out and paid for, so that the building may not be left unprotected through any misunderstanding between the owner and contractor as to each other's duty.

CONTRACT FOR BUILDING

Made this *fifteenth* day of *September* in the year 1883 by and between *James Johnson* of *Albany* in the County of *Albany* and State of *New York* party of the first part and *Thomas Smith Richard Smith* and *Henry Smith* of *Melrose* in the County of *Rensselaer* and State aforesaid *copartners doing business under the firm name and style of Smith Brothers* builders party of the second part.

The said *Smith Brothers* party of the second part for themselves and each of them and each of their heirs executors administrators and assigns hereby covenant and agree to and with the said party of the first part his heirs and legal representatives in consideration that the party of the first part agrees to perform the consideration hereinafter mentioned to make erect build and finish for the said party of the first part his heirs and assigns a dwelling-house on land of said party of the first part on *Fairfield Street* in said *Melrose* including all the *excavation and grading mason-work plastering carpenter-work roofing painting and glazing but exclusive of plumbing gas-fitting and heating* and to furnish all the materials of every kind labor scaffolding and cartage for the full completion of the said *excavation and grading mason-work plastering carpenter-work roofing painting and glazing* and they agree to supply said work and materials in strict accordance with the drawings and specifications made by *Edward Tyro* of said *Albany* architect which said drawings and specifications are and are to be taken and deemed to be a part of this contract by both the parties thereto and all things which in the opinion of the said architect may fairly be inferred from such drawings and specifications to be intended without being specially stipulated are to be taken as expressly specified and all the materials are to be supplied in sufficient quantity and where the quality is not otherwise described in the specifications the best quality of materials is specifically implied throughout and the said *Smith Brothers* party of the second part covenant to perform the whole of their agreement in the best most substantial and most workmanlike manner subject to the directions from time to time and to the satisfaction of the said architect or his successor appointed by the party of the first part and to deliver their work completely finished on or before the *fifteenth* day of *April* next.

And the said party of the first part hereby promises and agrees in consideration of the promise of the party of the second part to perform the foregoing covenants to pay to the said party of the second part the sum of *six thousand four hundred and ninety-seven dollars* ($6,497) in four several payments as follows:—

The first payment to be *Fifteen Hundred* Dollars ($1,500) when the roof is on and boarded.

The second payment to be *Fifteen Hundred* Dollars ($1,500) when all the outside work including piazza and porch is done and painted one coat and the mason-work and plastering finished and the sashes in.

The third payment to be *Fifteen Hundred* Dollars ($1,500) when the standing finish is done and the upper floors and stairs completed and the second coat of paint on the outside.

And the balance *thirty-three* days after the said work shall have been completely finished and delivered and accepted by the said party of the first part unless some defect shall meanwhile have been discovered in the said work.

Provided however that no payment shall be made except upon the certificate of the said architect or his successor that the work for which said payment is to be made is properly done and that the payment is due such certificate however not exempting the party of the second part from liability to make good any work so certified if it be afterwards discovered to have been improperly done or not in accordance with the plans or specifications and provided further that prior to each payment by the party of the first part a satisfactory certificate shall have been obtained to the effect that the estate or building upon or for which the work is done is at the time when the payment is made free from all mechanic's liens and other claims chargeable upon said building or estate and incurred by said party of the second part.

And it is hereby further agreed by and between the said parties hereto that the drawings and specifications are intended to coöperate so that any works shown on the drawings and not mentioned in the specifications or *vice versa* are to be executed by the party of the second part without extra charge the same as if they were both mentioned in the specifications and shown on the drawings. The said party of the first part or the said architect with the consent of the party of the first part shall be at liberty to order any variations from the drawings or specifications either in adding thereto or diminishing therefrom or otherwise however and such variations shall not vitiate this contract but the difference in cost shall be added to or deducted from the consideration of this contract as the case may be by a fair and reasonable valuation and the architect shall have power to extend the time of completion on account of alterations or additions so ordered such extension to be certified by him to the party of the first part at the time when such order for alterations or additions is given. Orders for changes which do not affect the cost of the work may be given by word of mouth but no order for any change which increases or diminishes the cost of the work or affects the time of completion shall be valid unless given in writing.

Neither the whole nor any portion of this contract shall be assigned or sub-let by the party of the second part without the written consent of the party of the first part.

If the said party of the second part shall fail to complete the said works including all variations should such be made at or before the time agreed upon with such extension if any in the case of extra work as may have been made and certified by the architect then and in that case the said party of the second part shall forfeit and pay to the said party of the first part the sum of *Ten* dollars ($10) for each and every day that the said works shall remain unfinished after the time agreed upon for their completion unless in the opinion of said architect such delay could not with reasonable diligence and prudence have been avoided or foreseen by the said party of the second part the sums so forfeited to be retained as liquidated and ascertained damages out of any money that may then be due or owing or may thereafter become due or owing to the said party of the second part on account of *their* work and materials under this contract.

If the said party of the second part shall become bankrupt or insolvent or assign *their* property for the benefit of creditors or become otherwise unable *themselves* to carry on the work or shall neglect or refuse to do so at any time for six days in the manner required by the architect or shall refuse to follow his direction as to the mode of doing the work or shall neglect or refuse to comply with any of the articles of this agreement then the said party of the first part or his agent shall have the right and is hereby empowered to enter upon and take possession of the premises with the materials and apparatus thereon after giving two days' notice in writing and thereupon all claim of the said party of the second part *their* executors administrators and assigns shall cease and the said party of the first part or his agent may after using such of the materials already on the ground as shall be suitable provide other materials and workmen sufficient to finish the said works and the cost of labor and materials so provided shall be deducted from the amount to be paid under this contract.

All materials shall be the property of the party of the first part as soon as they are delivered on the ground subject only to the right of the party of the second part to remove surplus materials at the completion of the building but no materials are to be paid for before they are set in place in the work.

The said party of the first part shall keep the said building at all times fully insured against loss by fire for the benefit of whom it may concern and in case of loss the indemnity shall be divided between the parties hereto according to their respective interests in the property destroyed. The said party of the second part shall be solely responsible for all loss failure or damage from whatever cause to the said works loss by fire alone excepted until the whole is delivered and accepted by the party of the first part and shall give all necessary assistance to the other mechanics employed in the building and shall be solely responsible for any delay to their operations or damage to their work or materials or to neighboring property or to the persons or property of the public by the workmen or through the operations of the said party of the second part.

And for the faithful performance of each and every the articles and agreements hereinbefore contained the said parties hereto do hereby bind themselves their heirs executors administrators and assigns each to the other in the penal sum of *Three Thousand* Dollars ($3,000) firmly by these presents.

In Witness whereof the said parties hereto have hereunto set their hands and seals the day and year first above written.
In presence of

EDWARD TYRO { JAMES JOHNSON, [Seal.]
to both.   { SMITH BROTHERS. [Seal.]

CONSTRUCTION OF A TOWN-HALL.

CHAPTER V.

THE third division of our subject will treat of a more complicated piece of construction than either of the preceding, and will lead us into the consideration of stresses and the resistances of materials in a way which should be familiar to every young architect. The habit of analyzing designs with regard to their merit as structures is one of the most valuable that such persons can form. To say nothing of the importance of acquiring that constructor's instinct which is the best safeguard against the slips that often bring discredit upon the most highly trained men, this same constructor's instinct, built up, so to speak, by the constant consideration and appreciation of weights, thrusts and resistances, is the main element of power in architectural design. A certain amount of prettiness, and even of theatrical picturesqueness, is within the reach of the architect who cares and knows nothing about the concealed construction which is to hold up his pretended arches and lintels, but the mind soon tires of detected sham, while it finds ever increasing pleasure in the masterly ease and originality which come from the skilful handling of brick, stone and wood as materials to be frankly used in satisfying the various conditions of modern structures. With us, unfortunately, certain important classes of building, such, for instance, as city dwelling-houses, have fallen into a routine of construction, which, if not positively vicious, offers at least few opportunities for the solution of novel problems; so that it is all the more important for the younger members of the profession to seek out and study special constructions, in order to keep their minds prepared for the difficult programmes which will sooner or later be set before them.

How this may be done can best be shown by means of an example which is to be considered in the way in which an intelligent architect would either study the details of his own design, or would criticise the work of another. The problem chosen is that of a town-hall, measuring seventy-five by one hundred and fifty feet, containing municipal offices in the basement and first story, and a single large room above for public entertainments. This room is divided into auditorium and stage by a brick wall, the proscenium-arch being a real construction of stone. The thrust of the arch is resisted on one side by the weight of a tower, which rises to a considerable height above the roof, and contains a fire-proof staircase, as well as a ventilating shaft, and on the other by the wall of a small dressing-room. Opposite the proscenium-arch is the gallery, under which is the main staircase, with ante-rooms on each side. The auditorium is covered by an open-timber roof, with hammer-beams and curved braces, while the stage and gallery have roofs of simpler construction. The exterior walls are of brick, with strings, lintels and other work of stone, and all interior walls are of brick, except a few light framed partitions.

In order to shorten the calculations as much as possible, the floor-plan is made extremely simple; the object being merely to illustrate principles. The structure is to stand on piles, driven through made ground to a stratum which, although not possessing great resistance, forms the best foundation within reach; and as it is essential to cut off the piles under the water-line, which is here fifteen feet below the curb, this distance will be the height of the foundation-walls, including the footings. As usual in all buildings, the plan of the upper story determines that of the sub-structure, and the distribution of the weights and thrusts of the proscenium-arch and roof must be the first step toward the laying out of the foundation-plan.

We will make our provisional plan without windows or partitions, since the latter will depend somewhat upon the distribution of the former, and these again upon the spacing of the roof-trusses, which is yet to be decided. Figure 175 shows therefore the outline of the main walls of the first story, and Figure 176 those of the second story; the tower and the other cross-wall which serves as abutment to the great arch being indicated, as well as the out line of the roof.

As we wish, for the sake of greater security against fire, to cut off the upper portion of the stage entirely from the auditorium, by carrying the proscenium-wall through the roof, the profile of the roof will determine that of the wall, which will, however, project a foot or more beyond the slates. This being fixed, together with the heights of the various stories, we must design the proscenium opening, taking into consideration the proportions of height and width which we desire it to have, and the consequent pressure which must be resisted. This will give us the length of the cross-walls which form its abutment on each side.

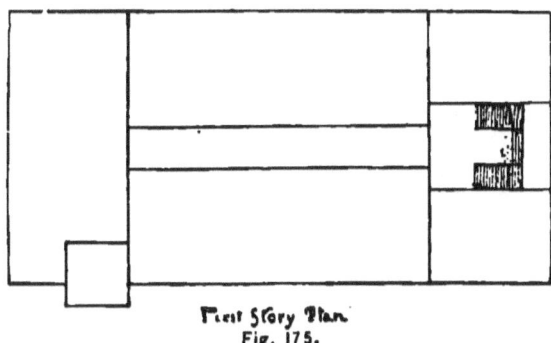
First Story Plan.
Fig. 175.

We may next study the main walls, first determining the thickness required for stability under the given conditions of height and length, and then adding to this whatever additional mass may be necessary, either in the form of extra thickness or of internal or external buttresses, to resist such horizontal thrust as may be exerted by any portion of the roofing. This will involve the consideration of the trusses for various portions of the roof, and the fixing of such forms and spacing as best meet the conditions to be fulfilled. The choice among the different modes of buttressing may depend here on the

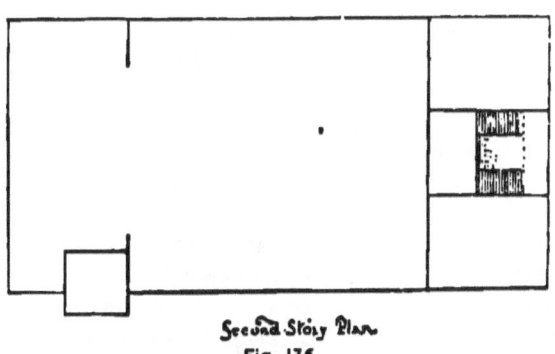
Second Story Plan.
Fig. 176.

effect which we wish to secure in elevation, since so far as interior convenience or acoustic effect is concerned it makes little or no difference whether the pressure of the roof is resisted by the mass of a plain wall, by internal pilasters, external buttresses, or the weight of pinnacles.

As the form of the wall above, and the distribution of the weights upon it, whatever may be its plan, must substantially determine those of the wall below it, we shall now have the main features of the exterior masonry, and can go on confidently with the study of the elevations. The thickness and position of the interior walls and piers will, however, depend somewhat on the mode of supporting the floors, and this point should be generally fixed upon at the outset, more accurate determinations being made subsequently. There is not much danger in such a building of imposing too great a crushing strain upon any of the materials, but it is best to keep the strength of brickwork and stone in mind, and to guard against diminishing unduly the size of piers between openings, either external or internal.

In order, however, that our conception of the problem to be solved may include all the essential conditions, we must, before proceeding beyond the most general outline of our design, know something of

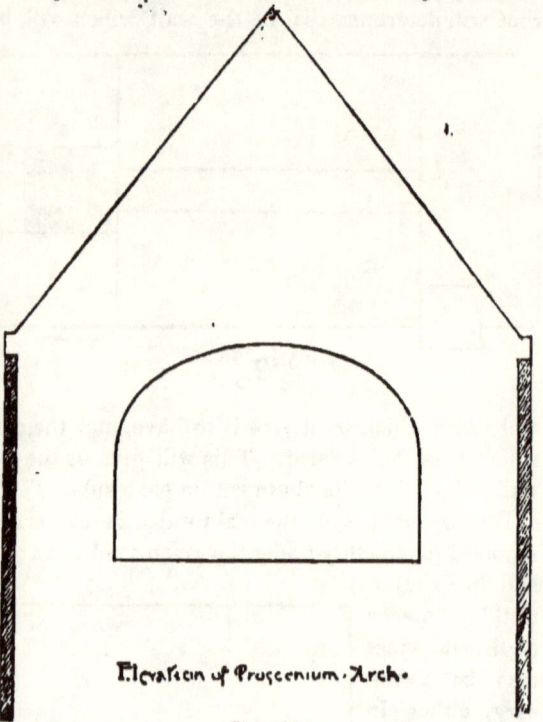

Elevation of Proscenium Arch.

. Fig. 177.

the foundation which is to support our building, and this will be all the more necessary in the present case, since the bearing stratum to which the piles are to be driven is not of very firm consistency.

If we proposed to put an unusual weight on any part of the foundation, by carrying the tower to a considerable height, or in any other way, it would be necessary, besides driving trial piles at different points within the site of the intended structure, to show the hardness of the bearing stratum, to ascertain also its thickness, and the character of the material below it, by boring through it with an auger which will bring up in its hollow a portion of the formations which it traverses, but our building being considerably lighter than others in its neighborhood, which have stood for years safely on foundations about which no special precautions were taken, we need hardly fear the disruption of the bearing stratum itself under the weight of the building, provided the load is so distributed over the piles that none of these will settle after the structure is completed. Just how much weight can safely be placed on each pile under our building cannot be known until they are actually driven, since the consistency of a bearing stratum such as we have to deal with varies greatly in different parts of a given area, but we can form an approximate opinion by observing the driving of those for works in the neighborhood. We will therefore inspect for a few minutes the operations taking place in the cellar of a new block of stores near by. The piling for one of the party-walls is just started, and we observe that the length necessary to reach the bearing stratum is about thirty-five feet. The piles are of spruce, most of them perfectly straight, although we notice some crooked ones being hastily rolled to an inconspicuous position as we approach. We see one swung under the hammer of the machine, and the driving begun. The pile sinks rapidly, and at a nearly uniform rate, through the gravel filling, until its head is within two feet of the surface of the ground, when the blows are seen to meet with greater resistance; the hammer rebounds, and the ground is felt to quiver, while the sinking of the pile under each stroke is but five or six inches, instead of a foot or so. The shaking of the ground indicates that the bearing stratum in that place is thin, and there is danger that it may be broken through, so the foreman orders a man to climb half-way up the machine and detach the hammer by hand, so as to lessen the distance through which it falls, and with it the force of its impact upon the head of the pile. Stepping up to the machine, we now ascertain the rate at which the pile sinks under the blows, by marking on one of the standards with an old nail the position of the top of the hammer after two successive blows, and measuring the distance between them. For two blows the sinking is five inches, then it falls to four, and then to three inches. The next stroke drives it three inches more, and the foreman gives the signal to stop. As the machine is being shifted to a new place we inquire the weight of the hammer, and learn that it is 1650 pounds, and we ascertain, also, that the height from the ground to the point at which it was detached for the last few blows is 15 feet. We have now the data for determining the weight which can be placed upon the head without fear of its sinking under it, by Sanders's formula, $\frac{FH}{8S} = W,$

in which $F =$ Fall of the hammer in inches.

$S =$ Sinking at last blow in inches.

$H =$ Weight of hammer in pounds.

$W =$ Safe weight in pounds.

In our case $F = 180$; $H = 1650$; $S = 3$; and the formula gives $\frac{180 \times 1650}{8 \times 3} = 12375$ pounds, which is the safe weight required. The next pile, with the same fall and the same hammer, sinks only two inches at the last blow, which, if substituted in the formula, would give $\frac{180 \times 1650}{8 \times 2} = 18562.5$ pounds as the safe weight. Most of the piles, however, give a minimum sinking of three inches, further driving increasing rather than diminishing the rate, by the penetration, as the foreman informs us, of the bearing stratum; and the corresponding safe load of 12,375 pounds must be accepted for the present as the standard upon which our calculations should be based.

We notice that the piles for the party-wall are being driven in a double row, spaced two feet apart from centres in the direction transverse to the line of the wall and three feet from centres in the same direction as the wall; and, curious to learn what load the unknown builder proposes to place on them, we ask to see the plans, which are in a tool-chest on another part of the ground. These show a row of three-story stores with finished basements, each 23 feet wide from party-line to party-line. The walls scale 45 feet high from the curb-level, and we know that the depth from this level to the top of the piles must be 15 feet. As the party-walls have no pressure of earth to support, they have but two courses of stone footings; one, immediately above the piles, being 3 feet wide and 18 inches high, while the second is 2 feet wide and also 18 inches high. This is equivalent, in a rough estimation of weights, to an addition of 4 feet 6 inches to the height of the wall, which, above the footings, is of brick, 12 inches thick, making the virtual height of the wall 64 feet 6 inches. At 112 pounds to the cubic foot, which is a fair estimate for brickwork, but low for stone-work, the weight of each foot in length of the wall will be 7,224 pounds. To this must be added the weight of the floors and roof and the probable loads upon them, which in a retail country store may safely be taken at 100 pounds to the square foot. The roof, which in this case is covered with tar and gravel, will require about the same allowance. The clear span of the floors is 22 feet, and the party-wall supports half the floor on each side of it, or 22 feet in all. Counting that of the basement, which is framed just like the others, there are four floors besides the roof, and the total floor and roof load supported by each foot in length of the party-wall will therefore be $5 \times 22 \times 100 = 11000$ pounds. The weight of each foot of the wall itself, as we just ascertained, is 7224 pounds, so that the total pressure at the bottom of the wall is 18224 pounds to each foot in length. The piles being spaced 2 feet apart transversely, and 3 feet longitudinally, there are but two piles under every 3 feet in length of the wall; so that each pair of piles has to support $3 \times 18224 = 54672$ pounds of wall and floor load, making 27336 pounds to each pile.

But we have ascertained by Sanders's formula that the safe weight upon the piles under the conditions which we find to exist is but 12,375 pounds; so that if the plans are carried out they will be loaded to more than double their safe capacity. After satisfying ourselves that our calculations are correct, we call the attention of the foreman of the pile-drivers to them. He shrugs his shoulders at first, and mutters something about "book-larnin," and "practical men," but is finally brought to admit that if the building had been his, he would, considering the softness of the ground, have put a triple instead of a double row of piles under the wall, and made the spacing in each direction 2 feet, which is as near together as long piles can be driven without danger that they may force each other up from their solid bed on the bearing stratum. This would give 3 piles for the support of every 2 feet of wall, and would bring the load on each within the safe limit. "However," the foreman adds, in a burst of confidence, "I aint got nothin' to do with that. They give me the piling plans, and I go by 'em; and anyhow, I guess these are only speculation buildin's." That they are likely in this case to fulfil the condition required of them — to hold themselves up until they are sold—we cannot deny, and we forbear to meddle further with what is clearly not our business.

Nevertheless, we should be sorry to run the risk of seeing in our own building after a few years the evidences of incipient settlement, the joints of the brickwork opening, and the lintels and interior plastering cracking; and we resolve to keep the weights on the piles within the safe limit.

We are now ready to put the drawings into definite shape. Beginning with the proscenium-arch, whose form we wish first to determine, we consider what requirements of width and height it is necessary to conform to, and reflect that as the hall will often be used for balls, fairs and other entertainments, which will fill it to its utmost capacity, it is desirable to make the proscenium opening as wide as possible, in order that the stage may not be too much separated on such occasions from the auditorium. Allowing only a small projection on the inside of the tower wall, with an abutting wall on the other side of the arch of the same width, we shall have an opening of 44 feet in width, which is ample for our purpose. The height, however, is restricted by the consideration that the height of ordinary theatrical scenery is limited, and it is desirable to avoid the necessity for wide "sky borders" to fill the space between the top of the scenes and the soffit of the arch; while it is also of importance in checking the spread of fire from the upper portion of the stage to the auditorium not to make the proscenium opening unnecessarily lofty. A glance is sufficient to show us that the arch, loaded as it is most heavily at the crown, must have considerable rise, unless it is made of segmental form, which would involve a thrust beyond the power of the limited abutments to support with safety. We will for the first trial give the arch an elliptical shape, making the rise $14\frac{1}{2}$ feet, with 44 feet span, and taking the depth of the arch-stones at 3 feet. Laying out the arch and the wall above it in elevation at a large scale we have the elements necessary for determining approximately its stability. A diagram of one-half the arch is sufficient, as it is symmetrically shaped and loaded, so that the line of pressure will be the same on each side of the centre.

This line will show us the direction of all the forces which act upon the arch and its abutments, and if it fulfils two essential conditions the arch and its abutments will be stable; if not they will be dangerously weak or fail altogether.

These requirements are :—

1. The curve of pressures in the arch must lie wholly within the middle third of the voussoirs.

2. The line of pressure prolonged through the abutment must strike well within the base of the abutment.

The reasons for these will be readily seen. If the line of pressure passes through the central line of the voussoirs, the crushing strain due to it will be equally distributed over the surface of each joint, but any deviation from a central position gives an inequality in the distribution of the strain which increases very rapidly with the variation of the pressure-curve from the central line. So long as the curve remains within the middle third of the depth of the voussoirs, the strain upon each joint is one of compression only, although it may be unequally distributed; but when it reaches the limit of the middle third, the crushing strain at the nearest edge of the joint is twice as great as when equally distributed, while that on the more remote edge is reduced to zero; and beyond this, while the compression of the nearer edge is increased to a hazardous extent, the strain on the other edge passes into one of tension, which, if there should be any opportunity for movement, will open the joints, and bend and dislocate the arch until it falls. This is a fatal defect, and the boundaries of the middle third of the voussoir, beyond which the pressure line cannot pass without producing a tensile strain either at the extrados or intrados, must be strictly regarded.

The second requirement, that the resultant of all the forces acting upon the abutment must strike within its base, is obviously a necessary one, for otherwise the effect of the combined pressures would be to overturn the abutment, as often occurs with arches carelessly designed. If the abutment were a solid and unyielding mass, it would be stable if the pressure curve fell anywhere within the base, even at the extreme edge; but in practice the resistance is always given by masonry of some kind made up of small blocks, united by mortar or cement which may be compressed in a greater or less degree; and the effect of a pressure applied too near the edge of such a mass is to crush or distort it, and finally to disintegrate it, so that the usual rule is to require that the pressure-curve in an abutment of stone or brick work, standing on a horizontal base, shall strike the base at a point not nearer to the outside face of the abutment than half the distance between this outside face and the point where a vertical line passing through the centre of gravity of the abutment would intersect the base.

To determine the line of pressure for our arch we will take the following method, which is sufficiently accurate, and is applicable to arches of any form, and loaded in any manner.

Figure 178 shows one-half of the arch and the wall above it in elevation, to scale. We begin by dividing the arch and its load into slices by vertical lines. The slices may

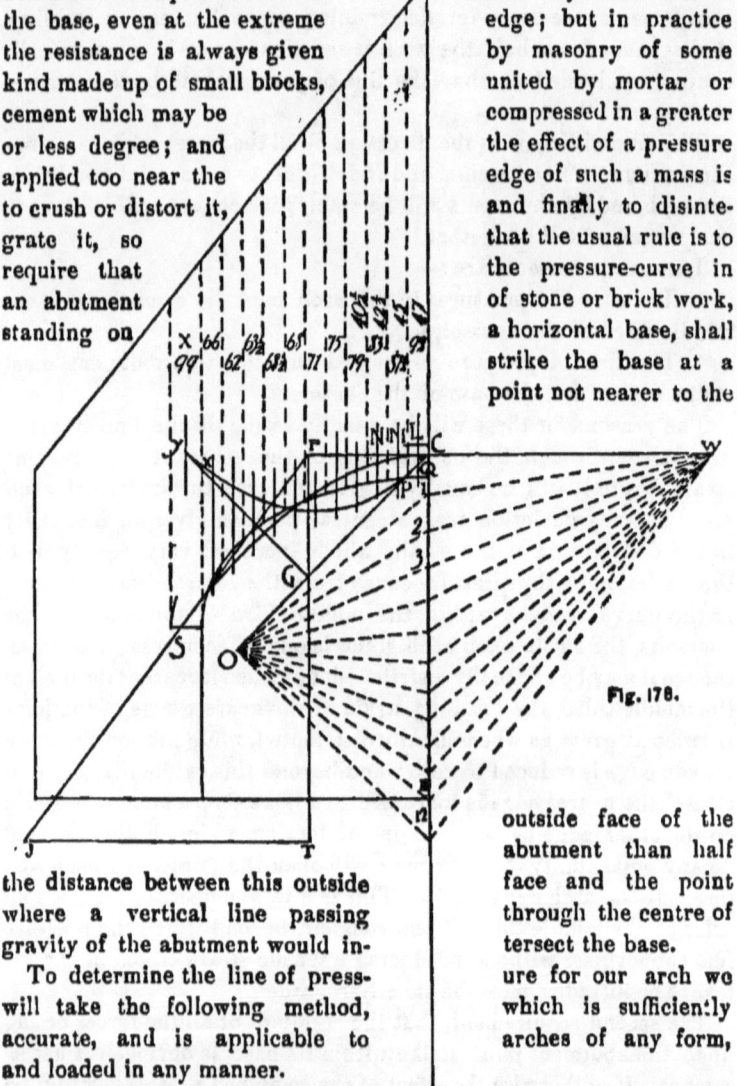

Fig. 178.

be of any width, but it saves trouble in computation for a large arch to make them each two feet wide as far as possible from the centre line. If the span of arch is not a multiple of 2 feet, the width of the last slice but one will be a fraction of two feet. The inner edge of the last slice, marked X, must just touch the springing of the arch, and the outer edge must touch the extrados of the first voussoir. In the present case, the bed of the first voussoir being horizontal, and the depth three feet, the width of the slice X will also be three feet.

We have now to find the vertical line in which lies the centre of gravity of each of these slices, and from that the vertical drawn through the centre of gravity of the entire half-arch and its load. If the slices are taken small enough, the centre of gravity may without appreciable error be assumed to lie in a vertical drawn midway between the lines bounding the slice, and we have only to draw the short lines shown in that position. The relative weight of each slice is next to be obtained, and as in a wall of homogeneous masonry this will be proportional to the areas of the slices, we need only calculate these, leaving the thickness of the wall and the weight per cubic foot as constant factors, to be supplied in case we wish to determine any actual pressure from the relative ones which the diagram will give.

Measuring with the scale the length of the two vertical sides of each slice, dividing their sum by two and multiplying by the width, will give their areas in square feet, which we mark as shown.

We have next, for the sake of simplifying our work, to make two assumptions, which experience shows to be justified, although they have no theoretical foundation. One of these is that the pressure-curve passes through a given point at the crown of the arch; and the other, that it also passes through a given point at the springing. If these two points are fixed, the rest of the corresponding curve is easily found, and for ordinary purposes we can safely suppose them to be so by the adhesion of the mortar and general inertia of the masonry. For a semi-circular or semi-elliptical arch, which naturally tends to rise at the haunches and descend at the crown, it is usual to make these points coincide with the outer limit of the middle third of the voussoir at the crown and springing, as at C and S in the figure. For pointed and segmental arches, which under ordinary circumstances tend to rise at the crown and descend at the haunches, the points fixed should coincide with the *inner* limit of the middle third of these voussoirs. If, however, a pointed or segmental arch is most heavily-loaded over the crown, so that its natural disposition to rise at that place is counteracted, the fixed point may be at the outer third, either at the crown alone, or both at crown and springing, as may seem best suited to the circumstances.

In the case of our elliptical arch, the points C and S being the ones assumed to be fixed, we will prolong the centre line of the arch and load indefinitely downward, and then space off upon it in succession the weights of the slices of the arch and its load, or rather, of the areas which stand as the abbreviated form of those weights. Any scale may be taken, as these dimensions have nothing to do with those of the arch itself. At the scale we adopt, 92, the number representing the area of the first slice, will extend from C to 1; $87\frac{1}{2}$, the second slice, from 1 to 2; and so on, 11–12 representing the last slice.

Next, take a point O, at any distance to the left of the line C–12, and opposite its middle point; draw $O\ C$, $O\ 1$, $O\ 2$, and so on, to $O\ 12$; the simplest way of doing this being to draw $C\ O$ and 12 O at 45° with the vertical, which will give O at their intersection.

Draw now, from the intersection of $C\ O$ with the centre line of the first slice, another short line, $P\ Q$, parallel to $O\ 1$, until it intersects the centre line of the second slice; then from this point, parallel with $O\ 2$, to the centre of the third slice, and so on, the last line, $X\ Y$, being parallel to $O\ 11$. From Y draw a line downward parallel to $O\ 12$, until it intersects $C\ O$ at G. A vertical drawn through G will pass through the centre of gravity of the arch and its load.

If the half-arch and its load were in a single piece, supported at S, they would be acted upon by three forces, namely : —

A horizontal pressure, proceeding from the key of the arch, and caused by the push of the other half-arch; the force of gravitation, acting vertically on the line of the centre of gravity; and the reaction of the abutment, which serves to oppose the other forces and maintain the whole in equilibrium. These three forces must meet at a common point, since otherwise they would not balance each other; and as the horizontal and vertical forces are fixed in direction, this point must be at their intersection, or at F. We have already assumed that the line of pressure in the abutment, which is the same as the line of reaction against the pressure, shall pass through S; hence, as F is already found, $F S$ must show the direction of the thrust at S. We have yet to find the amount of the pressure; but we know that, like every other oblique force, it can be decomposed into a vertical and a horizontal pressure, which will be represented by the adjacent sides of a rectangle of which the original force forms the diagonal; and as the vertical component is obviously equivalent to the sum of the weights of the small slices of the arch and its load, which furnish the only vertical pressures, and are already laid out from C to 12, we have simply to take $F T$, equal to $C\ 12$, and draw the horizontal $T J$, intersecting the prolongation of $F S$ at J. Then $J F$, at the scale to which the other pressures are drawn, shows the amount and direction of the oblique reaction acting through S, and applied at F, and $F T$ and $T J$ show the amount of the vertical weight and horizontal thrust by which it is balanced at that point.

The arch and its load not being, however, a solid mass, but composed of small parts, mutually wedged against and supporting each other, the actual direction of the pressures is not the broken line $C F S$, but a curve, or rather a curved series of short straight lines, coinciding with $C F S$ only at the extremities C and S, and varying in direction with the gradually accumulating weight of the successive slices of the arch and its load from the centre. To facilitate the drawing of the diagram, we prolong $F C$ horizontally, and draw 12 W parallel to $J F$. As $C\ 12$ is equal to $F T$, $C W$ is equal to $J T$, and, like it, represents the horizontal thrust at the key of the arch, while 12 W represents the oblique pressure at the springing S. The pressures at successive points in the arch will then be represented, both in direction and amount, by lines intermediate between these two, and if we draw to W lines from the points 1, 2, 3, and so on, which correspond to the weights on the central lines of the slices of the arch and its load, we shall have the successive directions of the curve of pressures at its intersection with those lines. Nothing then remains but to draw $C L$ horizontally, $L M$ parallel to 1 W, $M N$ parallel to 2 W and so on to S, where the line will coincide with that previously found.

By referring to the diagram, it will be seen that the compressive strain upon the arch-stones grows continually greater from the crown to the springing, the horizontal component remaining always the same, while the vertical component increases.

Having found the line of pressures, it remains to see whether it is contained in the middle third of the voussoirs. A glance shows us that this is not the case, and the arch is impracticable.

If it were built it would bend outward at R, and sink at the crown; the inner edge of the voussoirs would crush at the points where they are crossed by the pressure-curve, and the whole would fall.

There is nothing for it but to design a new arch, and as the curve of pressures varies with every form of arch, a new curve must be constructed for each. After trying an ellipse of greater rise, and then a circle, all in vain, we are led to the pointed arch, as being the only one adapted for a large span, with the maximum load on the crown, and by laying out such a curve, as shown in Figure 179, we succeed in passing the curve entirely through the middle third of the voussoirs, taking these at three feet long.

We have now three more points to determine: 1. Whether the abutment is sufficient to resist the thrust of the arch safely. 2. Whether the pressure upon any arch-stone will be so great as to risk crushing it; and 3. Whether the inclination at which the pressure is applied upon any voussoir, or any course of masonry in the abutment, is so great that the superincumbent mass will be in danger of sliding on it, instead of simply pressing against it. As we have seen, the direction of the thrust of the arch at its springing is shown by the line FS, and if the arch and its load were required to be held in equilibrium by an inclined column, for instance, the line JS would show the position of the axis of the column. We have here, however, to resist the thrust, not a rigid support, but a mass of considerable weight, which will add a vertical pressure due to this weight, to the inclined thrust, modifying its direction as well as its amount, and we must find the modified direction of the pressure before we can determine whether it will fall upon the base of the abutment so far within its outer face as we have found to be required for perfect safety.

Strictly, the modified line of thrust through the abutment would be a curve, since the vertical component accumulates as we follow the pressure line away from the springing of the arch; but for our present purpose we need only ascertain the point and direction of its application at the base of the abutment. To do this, it is sufficient, instead of dividing the abutment into successive portions and calculating the modification in the thrust due to the weight of the arch, to regard the whole abutment as a single mass whose weight will give the vertical component which we need to fix the final direction of the thrust. Neglecting the slice of the abutment between its inner face and a vertical line dropped from the extrados of the arch at the springing, since the weight of this does not affect the thrust, the remaining portion will have a trapezoidal shape, and we proceed to find the position of its centre of gravity by drawing the two diagonals of the trapezoid, and finding the middle

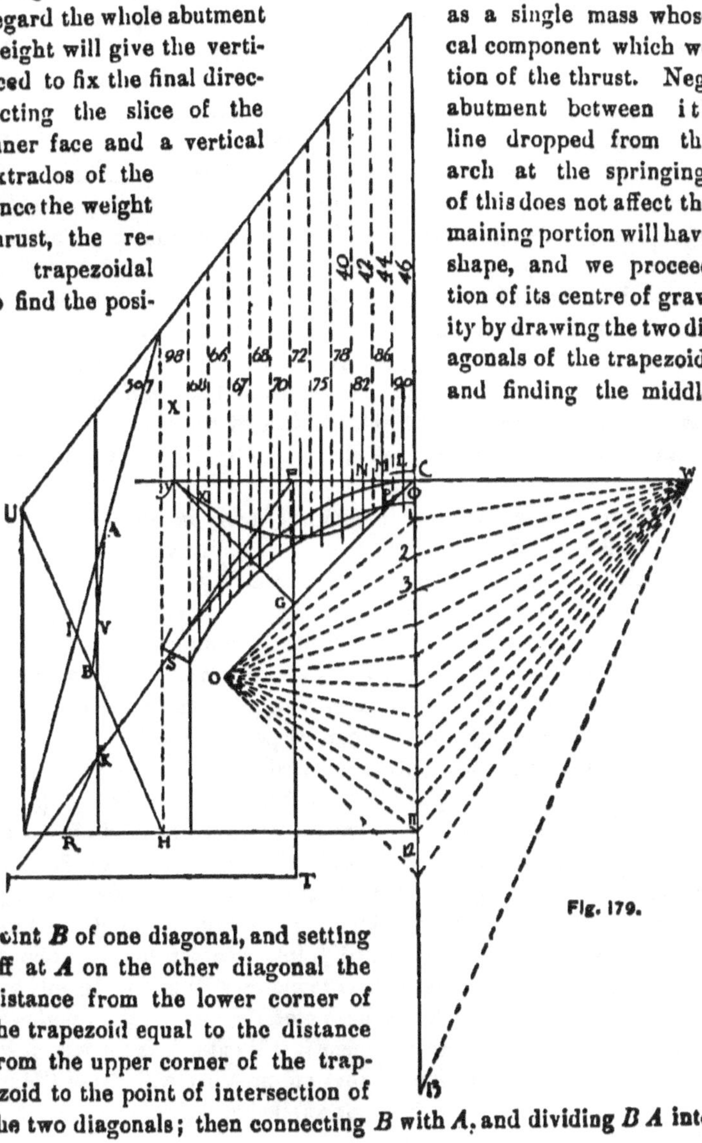

Fig. 179.

point B of one diagonal, and setting off at A on the other diagonal the distance from the lower corner of the trapezoid equal to the distance from the upper corner of the trapezoid to the point of intersection of the two diagonals; then connecting B with A, and dividing BA into three equal parts, the first point of division, V, showing the position of the centre of gravity, as required. From V we now drop a vertical line intersecting the pressure-line SJ at K; this will give the point of application of the vertical component of the new pressure-line. To determine the amount of this component, we measure the area of the abutment trapezoid as has already been done with the slices of the arch and its load, and obtain 507 as the result. We lay this off from 12 to 13, on the same vertical line that measures the vertical pressures of the slices of the arch and its load, and at the same scale, and then draw 13 W, which gives the direction and amount of the total combined pressures of the arch, load, and abutment. Transferring the direction of the final pressure so as to intersect K, the actual point of application, we find that it will strike the base of the abutment at R, which is nearer the vertical dropped from the centre of gravity of the abutment than half the distance between this vertical and the exterior face of the abutment, and the abutment may therefore be relied upon as stable under the given pressure.

We must now test the second point of safety in our arch, and ascertain whether the pressure at any given joint is greater than the stone can be relied upon to resist. The greatest pressure, as we see at once from the diagram, is at the springing line. Scaling the line 12 W, which represents this pressure, we find it to measure 1140, at the scale of this part of the diagram. This, however, being expressed in terms of superficial feet, must be multiplied, to find the pressure which it represents, by the number of pounds which a portion of the wall one superficial foot in area will weigh. The wall is 16 inches thick, and at 112 pounds per cubic foot the weight required will be 149 pounds. The total stress at the springing will be therefore $149 \times 1140 = 169860$ pounds. The area of the joint is $36 \times 16 = 576$ square inches, and the pressure will therefore average $169860 \div 576 = 295$ pounds, which is far within the limit of safety.

The determination of the third point, whether the direction of the pressure at any joint is such as to cause sliding, can be only approximately made, since the adhesion of the mortar, the roughness of the stone, and many other elements, will enter into the actual result, but we may safely assume that no pressure will cause sliding of the stone voussoirs which is not applied at a greater angle than

32° with a normal to the direction of the joint. Our pressure-curve shows that the angles of application of the stress are all well within this limit, and we need feel no uneasiness in regard to the voussoirs. With respect to the joints of the abutment, however, we may feel some anxiety, as the direction of the pressure for the courses nearest to the springing of the arch forms an angle of somewhat more than 32° with the vertical; but the adhesion of the mortar to brickwork is far greater than to stone, and the true angle of safety is correspondingly increased, so that if we take the precaution of delaying the removal of the centering on which the arch is built until the mortar in the abutment is well set, we need have no apprehension as to the result.

Having fixed upon the shape of the proscenium-arch, and made sure of the stability of the abutments, we will next determine the construction of the roof, which is intimately connected with the disposition of the wall supporting it. For the sake of simplicity, we decide to make the ridge of the roof level throughout, varying only the interior construction to suit the means of support at our disposal, or other requirements.

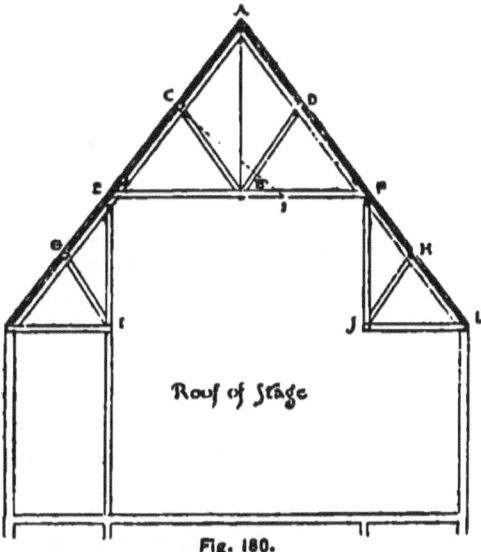

Fig. 180.

Over the stage, where no sacrifice need be made to appearance, we will use the simplest and cheapest devices, covering the central span, of 45 feet, with a truss of the form shown in Figure 180, and the rooms at the side with plain lean-to roofs, with tie-beams, and uprights next the wall, to prevent lateral pressure upon it, and a strut to prevent the sagging of the principal rafter. This rafter will form a continuation of the rafter of main truss, so as to bring the surface of the roof in one plane. The auditorium will be covered by an ornamental roof in one span, and the upper portion of this will be continued over the gallery at the rear, the ante-rooms on each side of the gallery being covered by the gallery floor.

We will first investigate the simplest roofing, that over the stage. As both the large and small trusses are furnished with horizontal ties at the foot, they can have no tendency to spread, and therefore exert no thrust upon the walls; so that we shall only need to ascertain the strains upon the timbers and determine the necessary sizes. This we will do here only for the central truss, A E F, the principle being the same for all.

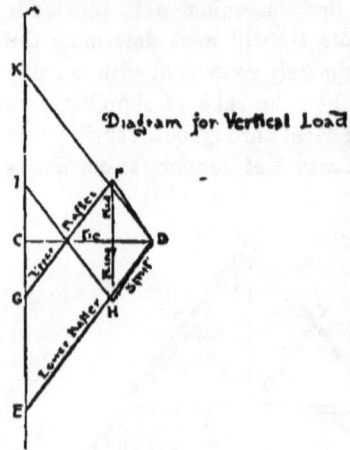

Fig. 181.

We have first to find the total weight which the roof must sustain. The length of the rafter A E is by the scale 30 feet, and as the trusses are spaced 12 feet apart from centres this rafter will have to sustain $12 \times 30 = 360$ square feet of roofing, and whatever extraneous pressure there may be upon this area, such as that of snow and wind. The other rafter, A F, will have the same stress to bear.

It will be best to consider first the vertical stress produced by the weight of the roof, including that of snow upon it, taking afterwards the oblique stress caused by wind. The weight of the roof itself, which consists simply of trusses, purlins, common rafters, boarding and slate, without plastering beneath it, may be taken at 15 pounds to the square foot. If the roof were flat, or nearly so, a load of wet snow might occasionally be added to this, amounting to forty pounds per square foot as a maximum, but our roof being inclined at an angle of about 52° with the horizon, the snow falling upon it would slide off as it accumulated, and a snow load of 15 pounds to the foot may safely be taken as the greatest to which it will ever be subjected. This makes the total weight per square foot of roofing 30 pounds, and the rafter A E must therefore be calculated to sustain a vertical stress of 360 × 30 = 10800 pounds. The load on the other rafter, A F, will be the same, making the whole vertical pressure on the truss 21600 pounds. In Figure 181, express this weight by a vertical line from A to B, at any scale, say 1000 pounds to the inch. Find the centre, C, of this line. As the truss and its load are kept up by the supports on which they rest, the upward force of these supports, or, as we should say, their reaction, is just equal to the weight imposed upon them; or, in the diagram, half the roof A F and its load presses downward with the force $A\ C$, and is held up by the force $C\ A$, while the weight on the other half, A E, is $C\ B$, and is sustained by $B\ C$.

We must now make another division of the vertical line $A\ B$, to indicate the proportionate part of the whole weight borne by each joint in the truss. Looking first at the rafter A F, we see that the joint D must suffer twice as much strain as either F or A, because it sustains an area of roofing extending on each side to a point midway between it and the next joint, while A and F being held, one by the opposing rafter and the other by the support at the foot, each carry only the portion between them and the point half-way to D. Hence in the diagram, if $A\ C$ expresses the whole weight on A F, $A\ K$ will indicate the portion borne at F, $K\ I$ will show that upon D, and $I\ C$ that upon A. Then $C\ G$ will show, in the same way, the strain at A upon the other rafter, A E; $G\ E$ that upon C, and $E\ B$ that on E. We have now all the data from which to determine the stresses on the other pieces of the truss, each of which plays a part in sustaining the total load. Beginning at the foot F, of the rafter A F, we find it to be the point of application of four different forces, the first being the reaction of the support on which the truss rests at that point, indicated by $C\ A$ on the stress-diagram; the second being its own portion of the weight, shown, as we have just seen, by $A\ K$; the third being an oblique pressure passing down the rafter, and the fourth a horizontal pull from the tie-beam. The direction and amount of each of these may be obtained from the diagram as follows: Starting from C, we pass upward to A, over the distance which represents the reaction of the support F, and in the direction of that reaction; then down again to K, over the space, and in the direction, corresponding to the share of the vertical load supported by the joint F : from K we draw a line, KD, parallel with the direction of the rafter A F, of such length that another line, drawn from its further extremity, parallel with the direction of the tie-beam F E, will just meet the point C, from which we started. The length of the line KD, according to the scale to which the diagram is drawn, will then give the number of pounds of longitudinal stress along the rafter from D to F, and DC will be the tensile strain upon the tie-beam between F and B.

To distinguish between the tensile and compressive strains we will indicate the former in the diagram by a light line, and the latter by a heavy line as shown.

In the same way we find the stresses upon the pieces around the joint D, in the middle of the rafter A F. We know already the stress upon the piece D F, which we found just now to be KD on the diagram; but as the compressive strain upon this piece, which was a downward push upon the joint F, is an upward push upon the joint D, we must now trace it in a direction reversed from that previously found, and starting from D on the diagram, follow it upward to K. From K we have another known force, the vertical load upon the joint D, which we ascertained at the beginning to be equal to KI, or twice as great as AK. Following this strain, then, downward to I, we have left two unknown forces, that on the upper portion of the rafter D A, and on the strut D B, both of which are applied at D, the joint whose equilibrium we are tracing. These are found in the same way as before, drawing IH parallel to A D, until HD parallel to B D will close on D. HD is then the compressive strain on the strut, and IH that on the upper portion of the rafter, and both are to be indicated by heavy lines.

The next strain to be determined is that on the king-rod A B. We have assumed that the vertical pressures are the same on each side of the roof, and the stress-diagram will therefore be symmetrical, and GF will represent the stress on one upper rafter, and HI that on the other. At the joint A, these two stresses, together with the vertical load $IC + CG$, or IG and the pull of the king rod, include all the forces applied at that point, and starting with the known stresses HI and IG, GF, symmetrical with HI, will be the strain on the upper rafter C A, and FH, drawn parallel with the king-rod, and connecting F and H, will represent the tensile strain on the king-rod.

Measuring with the scale the forces thus indicated, we shall find as follows: —

STRESSES FOR VERTICAL LOAD.

Tie 6400 lbs. Tension.
King-Rod 6000 " "
Lower Rafter 10300 " Compression.
Upper Rafter 7000 " "
Strut 3600 " "

We have now to consider an additional series of strains, — those due to wind-pressure. Of course the wind may blow upon either side of the roof, but by calculating the stresses due to a pressure on one side, we shall have all the data required for extending it to the other.

We will suppose the wind to blow from the left in Figure 180. As the general direction of wind-movement is nearly horizontal, the maximum pressure in a direction normal to the plane of the rafters occasioned by it increases as the pitch of any given roof rises, and in a certain ratio to the angle which the rafters make with the horizon. The angle of the present roof being 52° it will be safe to assume a maximum wind-pressure in a direction normal to that inclination of 44 pounds to the square foot, which will give as the wind-pressure supported by the rafter $A E$ in Figure 180, which is 30 feet long and spaced 12 feet from the next rafter, $30 \times 12 \times 44 = 15840$ pounds. This, as the wind will only blow on one side at a time, will give the total wind-pressure on the whole roof, although its *direction* may be reversed. We lay this off in Figure 182 from A to B, at the same scale as in the preceding diagram, A B being drawn in a direction perpendicular to the inclination of the left-hand rafter. The next thing is to lay off on A B the points showing the proportionate portions of the pressure borne at the several joints, and also the reactions of the supports at E and F. We can easily see that, as in the case of vertical pressure, the joint C bears half the strain on the rafter, A and E bearing one-quarter each, which will give A C on the diagram as the pressure at A, $C E$ as the pressure at G, and $E B$ at E. To find the reactions of the supports we must consider that the tendency of the oblique force of the wind on the whole truss, A E F in Figure 180, is to turn it about the point E, with a leverage which will be proportional to the distance from E of C, which is the centre of the rafter, and forms the point at which the pressure, uniformly distributed over the roof-surface, may be assumed to be concentrated. We can see that the support F will, with this particular form of roof, be most severely strained, and the exact proportion between the loads borne by each support can be readily found by drawing from C a line perpendicular to the rafter A E, and striking the straight line connecting the points of support at 1; then measuring the distances 1 F and 1 E, which will give, *inversely*, the relative pressures borne by F and E. In this case 1 F is ⅓ of the whole distance E F, and 1 E is ⅔ of the same, from which we infer that F bears ⅔ of the wind-pressure, and E only ⅓. Applying this to Figure 182, we divide A B by the point D, at ⅓ the distance from B to A. Then B D will represent the reaction of the support E, and D A that of the support F. Beginning with the joint E of Figure 180 we trace the stresses in the same way as before with vertical strains. From E on the diagram we pass down to B for the direct pressure, then upward to D for the reaction of the support E, then horizontally to G for the tension on the tie-beam, and down, parallel with the rafter, to E, the point of beginning, for the compressive strain on the rafter at its foot.

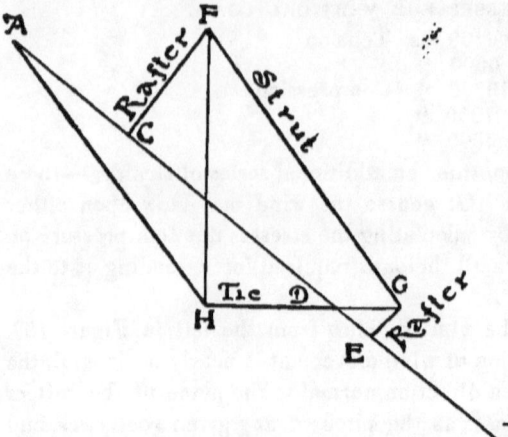

Fig. 182.

To find the strains at the joint C, in the middle of the rafter, we start at C in the diagram and pass downward to E for the proportion of the wind-pressure borne at C, then up to G for the reaction of the lower portion of the rafter, then upward, parallel with the direction of the strut, to F, until a line drawn downward from F, parallel with the upper portion of the rafter, will close on C, the point of beginning. This will give us the stress due to wind upon each portion of the rafter, the strut, and the portion of the tie-beam nearest the wind, as follows, scaling them from the diagram.

STRESSES FOR WIND-PRESSURE. WIND LEFT.

Left-Hand Part of Tie 1800 lbs. Tension.
Left Lower Rafter   900 "   Compression.
" Upper Rafter   3300 "   "
Left-Hand Strut   8300 "   "

All these stresses would be reversed by a change in the direction of the wind.

There are still other strains to be found on the other pieces of the truss, but the reader can easily finish the diagram for himself. It is enough to say that the next point to be investigated is that at A in Figure 180, then D, and then F, and finally, the investigation of B will serve to check the correctness of the others. It will be found that the strut B D receives no stress of any kind from wind-pressure on the opposite side; that the portion B F of the tie-beam suffers a *compressive* strain of about 3300 pounds; that the rafter A F is equally strained throughout its whole length with a compressive force of about 8000 pounds, and the king-rod A B is subjected to a tensile stress of 6,500 pounds.

As the tie-beam E F is strained by tension, to the amount of 1800 pounds, at one end, and by a compression of 3300 pounds at the other, it would seem that the net effect of the pressure would be a compressive stress equivalent to the difference between the two, or 1500 pounds. This, is, however, an unsafe inference, the two kinds of stresses acting to a certain extent separately, instead of neutralizing each other, so that the prudent method is to take the largest amount of stress of the principal kind, without deduction for the neutralizing effect of the opposing, but inferior forces. In the case of the rafters, which are all strained in the same way, but of which we find that the one away from the wind is the most strained, we will provide for resisting the greater stress, which will make us safe against the lesser one. This will give us a corrected table of wind-pressure strains, which we will place side by side with those due to vertical pressure, adding them together to find the sums which will give us the total stress acting *along* each piece which that piece must be calculated to bear safely. For shortness we will mark tension as — and compression as +.

STRESSES ON TRUSS. — *Vertical Load and Wind-Pressure.*

Piece.	Vertical Load. lbs.	Wind-Pressure. lbs.	Total. lbs.
Tie, E F, (Fig. 180.)	− 6,400	− 1,800	− 8,200
King-Rod, A B	− 6,000	− 6,500	−12,500
Each Lower Rafter, C F, D F	+10,300	+ 8,000	+18,300
Each Upper Rafter, A C, A D	+ 7,000	+ 8,000	+15,000
Each Strut, B C, B D	+ 3,600	+ 8,300	+11,900

We can now obtain the sizes of the timbers and rods which will safely sustain these stresses.

The tie-beam, E F, suffers a tensile strain of 8200 pounds. The safe tensile strength of spruce timber, which we suppose to be the material of the truss, should not, for such a roof, be taken at more than 1000 pounds to the square inch, and a timber of 8.2 square inches sectional area, or about 2″ x 4″, would give the resistance required.

There is, however, another consideration which enters into the calculation of the size of the tie-beam. Not only does it keep the feet of the rafters of the truss from spreading, but, in virtue of its horizontal position, it is also a beam, or rather a pair of beams, each having a span equal to the distance from the king-rod to the wall-plate, about 20 feet, and burdened with its own weight, which tends to break it by a transverse strain. This strain is entirely independent of the longitudinal stress along the timber, and must be provided against separately, by increasing the size of the timber, so as to give additional fibres for resisting the bending strain, which those fibres engaged in tensile resistance to the longitudinal stress cannot deal with. We will try, therefore, a 3″ x 4″ timber in place of 2″ x 4″, and see if it gives us strength to meet all the stresses, transverse as well as longitudinal. Supposing 2″ x 4″ of this to be occupied in resisting the direct tensile force, we shall have remaining a beam 1″ x 4″, 20 feet long, which must sustain the weight of the entire timber, 3″ x 4″ x 20″, which, at 45 pounds to the cubic foot, will be 75 pounds, uniformly distributed over the beam.

The simplest formula for transverse strength of rectangular beams is $\frac{b d^2 C}{s L} = W$ in which

b is breadth of beam in inches.
d is depth   "   "   "   "
C is a constant, which for spruce is 450.
s is the factor of safety, which should be 6.
L is the length of the beam in feet.
W is the safe centre load

Applying this, and remembering that the distributed load may be with safety twice as great as the centre load, we shall have $\frac{1 \times 4^2 \times 450 \times 2}{6 \times 20} =$ 120 pounds safe distributed load. As the weight of the timber is only 75 pounds, we have here a surplus of transverse strength of about 60 per cent, but it is hardly worth while to make the stick any smaller. In fact, wooden tie-beams are ordinarily made of far greater dimensions than this, and with reason, for they are very liable to be used for supporting the weight of partitions either above or below, or are subjected to other extraneous transverse strains which they are less able to resist than other beams, which have no special work of their own to do, while their strength is further impaired by the mortises by which the rafters and king-rod are framed into them. The king-rod, AB, is of wrought-iron, and endures a simple tensile stress of 12500 pounds. The usual estimate for the safe tensile strength of wrought-iron is 10000 pounds per square inch, and as the area of a circle is .7854 of that of the square in which it is inscribed, the diameter of a round rod to hold safely 12500 pounds will be $\sqrt{\frac{12500}{10000 \times .7854}} = \sqrt{1.59} = 1.26$, or about $1\frac{1}{4}$ inches.

The upper and lower rafters and the struts are all subjected to compression, and their dimensions will be found by the formulas for wooden columns. Taking the lower rafters first, with a stress of 18300 pounds, we may, as they are only about 15 feet long, take the compressive strain which they will bear safely in the direction of their length at 400 pounds per square inch of sectional area. To sustain 18300 pounds will therefore be required a timber of 46 square inches sectional area, and we may use a stick 4″ x 12″, 5″ x 10″, or 6″ x 8″, as may be most convenient. As in the case of the tie-beam, a small surplus of strength must be reserved to resist the transverse strain due to the weight of the rafter itself, which tends to bend it inward, but the slight excess of these dimensions over those strictly necessary will be sufficient for the purpose. If, however, a purlin were placed, as is often the case, bearing directly on the rafter, midway between its foot and the strut, a very important transverse strain, equal to the wind-pressure on a portion of the roof $7\frac{1}{2}$ feet wide and 12 feet long, together with the component normal to the direction of the rafter of the vertical pressure due to the weight of the same portion of the roof and its load of snow, amounting in all to about 6000 pounds, would have to be resisted by the rafter, acting as a beam loaded at the centre, in addition to the compressive stress acting in the direction of its length, and its dimensions would need to be increased accordingly.

The same observations apply to the upper rafter, and even to the strut, in some instances, but if we are sure that we will not forget this caution in case of need, we may, for the present, continue to study our truss as it is intended to be built, without transverse strains on any of the pieces except those due to their own weight. The upper rafter, having to sustain a compressive strain of 15000 pounds will then need, by the same rule, a sectional area of 37½ square inches, and a 4" x 10", 5" x 8" or 6" x 7" stick would give the required strength with a surplus for resisting the transverse strain of its own weight. The strut, which sustains 11900 pounds, may be 4" x 8" or 6" x 6". This completes the schedule of timbers for the trusses over the stage.

The trusses over the auditorium and gallery require to be submitted to similar processes, which it is unnecessary to describe at length. The former, however, as shown in Figure 183, not being tied at the foot of the rafters, exerts a thrust upon the walls at each side, which would certainly overturn them if not counteracted, and in order to be able to provide the necessary resistance we must know the direction and amount of the thrust.

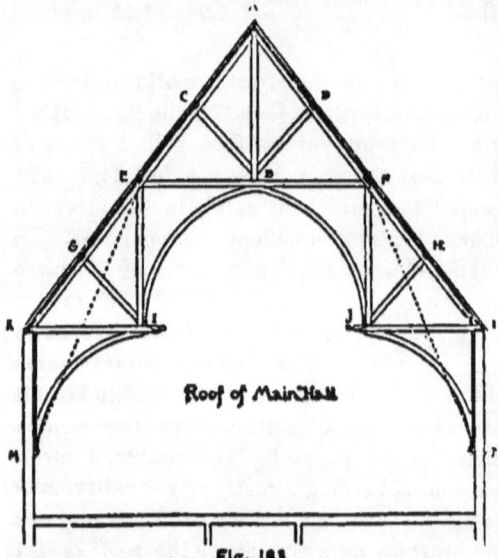

Fig. 183.

A glance at the diagram shows that the truss really consists of three portions, the upper part, A E F, forming a rigid frame,

BUILDING SUPERINTENDENCE. 295

resting upon two other frames, E K M I B and B F L N J, the latter of which are inclined inward from the wall, and tend to push it outward at their foot, and, disregarding the various pieces composing these inclined frames, which serve principally as braces against deformation by wind-pressure, we may consider each as equivalent to a single straight timber, or inclined column, acting as shown by the dotted lines E M and F N. It is easy to ascertain the thrust at the foot of this column, by laying off *A B* in Figure 184, equal, by any required scale, to the vertical pressure on E, that is, to half the weight on the truss E A F; then drawing *A C* on the diagram, parallel to E M, and *B C* horizontally intersecting *A C* at *C*, *B C* will then show by the scale the horizontal outward thrust on the wall at the point M, in Fig. 183. This being ascertained we must now find the shape and size of buttress, if any, which is necessary to keep the walls from being pushed over by the thrust. We can best do this graphically, as follows:—

Beside the diagram for thrust in Figure 184 draw a section of the wall. The scale of this section, which is in a certain number of feet to an inch, has nothing to do with that of the stress-diagram, which is in pounds to an inch, and any scale may be used. Fix the position of the section so that, by its scale, the point *C*, which represents the place of application of the inclined force of the lower part of the truss, supposed to rest upon a corbel projecting from the wall, will come at the proper distance from the wall-surface, and draw a line at *Y*, representing at the same scale the floor-line, which, if the wall is anchored to the timbers of the floor, would be the point about which it would

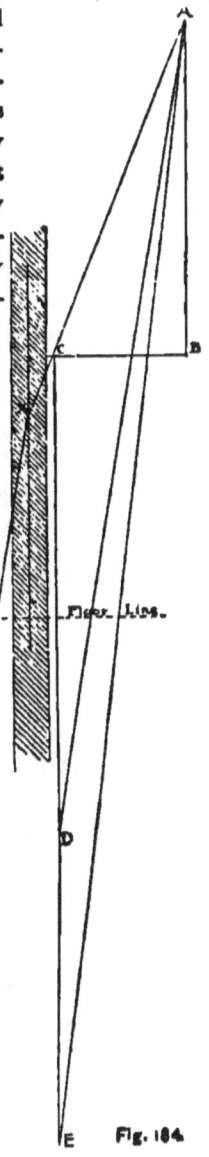

Fig. 184.

revolve in overturning. Next find the weight of a portion of wall extending from the floor-line up to K in Figure 183, and equal in width to the space between the windows, which we suppose to be five feet. Add to this the actual weight of the portion of the root supported by K E, Figure 183, not including any allowance for snow or wind. The sum will give the vertical pressure which combines with the oblique pressure E M to change its direction in its passage through the wall. Laying off now this vertical pressure downward from C, in the diagram of Figure 184, to the same scale of pounds as the other pressures, we find that it extends to D. Draw now $A D$, which will give the direction and amount of the total combined pressures at the floor-line, and $X Y$, drawn parallel to $A D$ from X, the point where $A C$ prolonged strikes the plane of the centre of gravity of the wall, will give at Y the actual position of the intersection of this modified pressure-line with the base-line of the movable portion of the wall. The point Y falls outside of the wall, showing that its unassisted stability is not equal to the oblique pressure upon it, and that it will be overturned.

There are three ways of adding to the wall the requisite support. The most obvious of these is the addition of exterior buttresses, the weight of which will serve to deflect the pressure-line more directly downward, at the same time that their position will improve the stability of the wall by removing the point about which the pier must revolve, in order to overturn, beyond the intersection of the pressure-line with the base. The second resource is the construction of interior buttresses, the weight of which will also serve to deflect the pressure-line to a direction more nearly vertical, at the same time removing the point C, in Figure 184, horizontally away from the wall, until the point Y is brought within the base. The third method consists in piling up masonry in the form of pinnacles above the wall at the proper places, increasing by their weight the vertical component of the total pressure, until the line falls within the base of the wall, with little or no help from buttresses, either exterior or interior. This would be not only a perfectly legitimate and safe construction, but perhaps the most economical of any, since the weight of all the masonry extraneous to the wall itself would be applied in increasing the stability of the pier, while buttresses, either interior or exterior, must be continued to the ground, although the portion below the floor-line, much the largest part of the whole, serves in this case only as a support, without adding anything to the stability of the wall above the floor-line. There would probably be, however, some objection on the part of the building committee to such an unusual construction, and as we have ourselves some fear that masses of snow sliding down the roof might push the pinnacles off, with disastrous results, we will abandon the idea of employing this method. Of the two others, that of inside buttressing seems the less adapted to the circumstances, as the projection of the buttresses would obstruct the side aisles of the hall. If the room were planned with high walls and flat ceiling, the acoustic advantage of these projections would be sufficiently important to outweigh the objection to them as obstructions, but in the present case the shape of the roof, the echoes from which would be broken up and dissipated by the net-work of trusses and the regular succession of braces springing from the corbels, which would intercept the waves of sound conducted along, as well as reflected from the walls, give all the security against unpleasant reverberation which could be obtained by interior projections, and it will be best on other accounts to avoid them by placing the buttresses on the outside.

We will first make trial of a buttress of the shape and size shown in elevation in Figure 185, and in plan in Figure 186. Finding first the weight of the buttress from the floor-line to the top, we add this to the weight of the wall as a part of the ver-

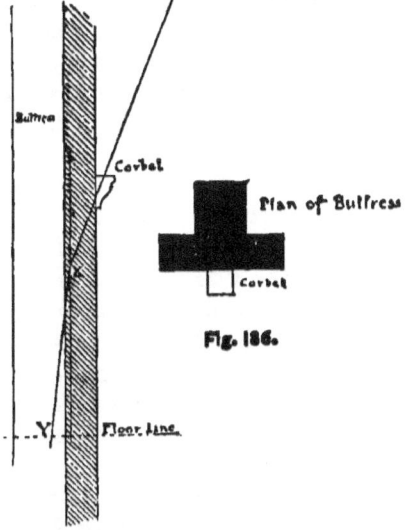

tical force in Figure 184, where it is represented by DE, CD representing the weight of the wall exclusive of the buttress. The whole vertical force will therefore now be CE, and AC, the oblique thrust of the truss, remaining the same, the new resultant force will be represented, in direction and amount, by $A E$. If then, we draw a line, parallel to this new resultant, from the point where the line of thrust intersects the plane of the centre of gravity of the pier and buttress combined, to the floor, or base line, we shall at once see whether the whole will be stable.

The first step in this process is to find the position of the centre of gravity of the pier and buttress. In the plan of the pier with its buttress, Figure 187, find the centre of figure of each portion separately, by drawing the diagonals of the parallelogram formed by each. Join these centres by the line $A B$. The centre of gravity of the whole figure will then lie on the line $A B$, at a point which must divide $A B$ into portions inversely proportional to the areas of the parallelograms in which its ends respectively lie. The area of the parallelogram representing the wall is $5 \times 1\frac{1}{3} = 6\frac{2}{3}$ square feet; that of the parallelogram representing the buttress is $2 \times 1\frac{2}{3} = 3\frac{1}{3}$ square feet. Three and one-third is just one-half of six and two-thirds, so that the point C, which divides the line $A B$ at one-third of its length from E, will show the centre of gravity of the complete figure, and if the pier and the buttress are of the same material, and carried to the same height, it will lie in the line of the centre of gravity of the whole mass. For our purposes we can assume that this is the case, and that C lies in the plane of the actual centre of gravity. We then find the corresponding point in Figure 185 by transferring its distance from the inside of the wall with the dividers, and draw a vertical line through it as shown. The line of the thrust of the roof, prolonged downward from the corbel, will intersect this new line of the centre of gravity at X, and $X Y$, drawn from X parallel to $A E$ in Figure 184, will show the line of the resultant pressure due to the influence upon the thrust of the weight of the pier and buttress. This line will strike the floor, or base-line of the movable portion of the wall, at Y, and as this point fulfils the condition of being nearer to the vertical line drawn through the centre of gravity than it is to the exterior of the mass, the pier and buttress, if well anchored at the floor-line, will safely resist the effort of the thrust of the roof to overturn them.

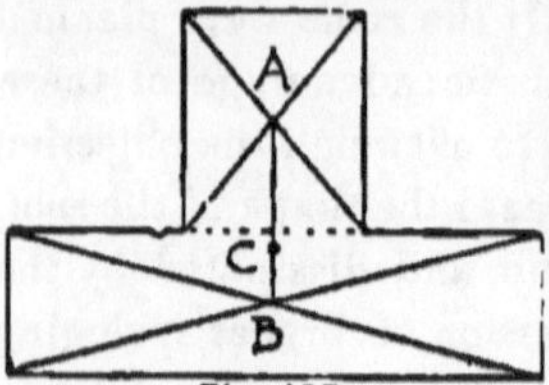

Fig. 187.

The roof over the gallery, as we see in Figure 188, being supported by posts at M and N, with suitable braced girders running back from $E\,I$ and $F\,J$, to carry the truss intermediate between the one shown in the figure and the end wall, resolves itself simply into the case of that over the stage, which we have already investigated; the curved braces $B\,I$ and $B\,J$, being added merely for ornament, and to correspond with the main roof. We have now only to calculate the necessary size for the purlins, which are virtually beams 12 feet in length, having a clear span equal to this distance less 6 inches, the width of the principal rafter, and subjected to a distributed transverse stress, due to their own weight, the weight of the portion of roof which rests upon them, with an occasional wind-pressure added of 44 pounds to the square foot; and the sizes of the common rafters, which are also inclined beams, of a length equal to the distance between the purlins, and subjected to a distributed transverse stress, due to their own weight, with the weight and wind-pressure upon the portion of roof which each carries. The purlins should be nearly square in section, and each supports the strain of a portion of roof $11\frac{1}{2}$ feet long, and of a width equal to the distance between its centre and the centre of the next one, which in this case is 15 feet. We have previously estimated the vertical load, including weight of snow, on each square foot of the roof-surface to be 30 pounds, which would here be equivalent to a pressure, normal to the plane of the roof, of 19 pounds. To this must be added the maximum wind-pressure, which we found to be 44 pounds, making 63

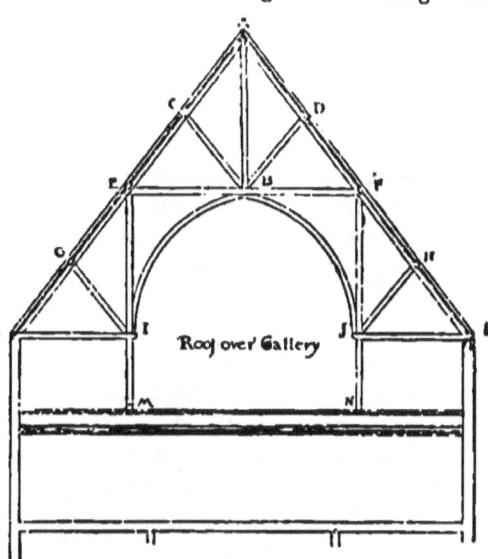

Fig. 188.

pounds per square foot as the total transverse pressure. The purlin sustains $15 \times 11\frac{1}{2} = 172\frac{1}{2}$ square feet, so that the stress upon it will be $172\frac{1}{2} \times 63 = 10868$ pounds. Its own weight will be, at the utmost, 500 pounds, and over the auditorium the underside of the common rafters will be lathed and plastered, adding a weight of about 10 pounds per square foot, or 1725 pounds upon the whole space sustained by each purlin. This, as well as the weight of the purlin itself, being a vertical pressure, of which a portion is transmitted down the rafters, while only the component normal to the roof-plane exerts a stress upon the purlin, we can find the normal pressure corresponding to the vertical weight of 2225 pounds either graphically or by applying the proportion of $30:19$, which we have just ascertained to represent the same relation in the case of the weight of roofing and snow. This would give $30:19 = 2225:1409$. Adding this to the others, we obtain $10868 + 1409 = 12277$ pounds as the measure of the distributed transverse stress upon the purlin. This is a severe stress for a timber $11\frac{1}{2}$ feet long, and we shall do well to employ Southern pine for the purlins, instead of spruce, on account of its superior stiffness. Using the formula before employed in calculating the tie-beam, $\frac{b\,d^2\,C}{sL} = W$, we shall have here:

$W = 12277$

$s = 6$

$L = 11\frac{1}{2}$

$C = 550$, the constant for Southern pine, as 450 is for spruce.

b and d are both unknown.

Instead of transposing the formula, it is often less trouble to assume certain dimensions, and try whether they fulfil the required conditions of strength. In this case we will try whether a $10'' \times 10''$ stick will do. Substituting these dimensions for b and d in the formula, and remembering that the weight, 12277 pounds, being distributed uniformly along the purlin, exerts only half as much breaking stress as if it were concentrated at the centre, we shall have $\frac{10 \times 10^2 \times 550 \times 2}{6 \times 11\frac{1}{2}} = 15942$ pounds, as the distributed weight which will be safely borne by the timber. This is greater than we need, and we will try an $8'' \times 10''$, which we find to be capable of supporting safely 12754 pounds, or a little more than the given weight, so we adopt these dimensions.

The rafters are last to be considered. The steady stress upon each of these, consisting of the simple weight of the portion of the roof with its load of snow resting on it, acts vertically, and the inclination of the rafter being oblique to this vertical force, it is necessary to resolve the single stress due to the weight into two, one of which will act in a direction normal to the inclination of the rafter, forming a transverse strain of the ordinary kind, while the other acts along the rafter by compression, and is to be resisted by the rafter acting as a column. The clear span of each rafter, from purlin to purlin, is 15 feet, less 8 inches, the width of the purlin, and as the rafters are spaced 16 inches from centres, each carries a portion of the roof $14\frac{1}{3} \times 1\frac{1}{3}$ feet in area. The weight upon this, including that of the rafter itself, the roofing boards and slates, the lath and plaster underneath, and a possible load of snow, will be $14\frac{1}{3} \times 1\frac{1}{3} \times 40 = 764$ pounds. Drawing a vertical line representing this weight, at any scale, we make it one side of a triangle, of which the other two sides are drawn respectively parallel to the direction of the rafter, and at right angles to it. The length of these two sides, measured at the same scale, will give the components of the vertical pressure, which act along the rafter and transversely to it. We shall find the transverse component to be about 455 pounds, and the other, acting to compress the rafter, about 600 pounds. At 400 pounds per square inch, the sectional area of the rafter, acting as a column, required to resist this stress would be $1\frac{1}{2}$ square inches, and the dimensions needed to resist the transverse strain must be added to this. The transverse component of the simple weight of roof and snow we have just seen to be 455 pounds. To this must be added the wind-pressure, which is a direct transverse strain, amounting, by our previous estimate, to 44 pounds per square foot, or $14\frac{1}{3} \times 1\frac{1}{3} \times 44 = 841$ pounds on the whole area supported by each single rafter. Adding the two results together, we have $455 + 841 = 1296$ pounds as the distributed transverse pressure on the rafter. By the formula previously employed, assuming the rafters to be of spruce, with a value for C of 450, we find that $3'' \times 7''$ timbers will give a resistance of 1540 pounds; and supposing $1\frac{1}{2}$ square inches of the sectional area, comprising a slice $\frac{1}{4}''$ wide by the depth of the rafter, to be occupied in resisting the longitudinal stress, we shall have remaining a piece $2\frac{3}{4}'' \times 7''$, whose strength, according to the formula, will be $\frac{2\cdot 45 \times 7^2 \times 45' \times 2}{14\frac{1}{3} \times 6} = 1436$ pounds. This is larger than we need, but the difference is so unimportant that we need not regard it, and we adopt this as the proper scantling.

We now know the necessary sizes of timbers, and form of piers and buttresses, for carrying out our provisional sketch of the building into definite drawings, and we proceed to lay out our floor-plans and elevations, continuing, after these are well studied, to construct a foundation-plan in accordance with them. The elevation may be first taken up, as upon this will in a great degree depend the details of the completed plan.

We have seen, from the investigation just made, that the walls of the central portion of the building, which support a hammer-beam roof, will need to be buttressed, to support the tendency of the roof to spread, while those of the portions containing the stage and gallery, being covered by roofs which are tied at the feet of the rafters, and therefore have no lateral pressure, do not require buttresses. Our calculations have shown that buttresses 20 inches wide on the face, and projecting two feet, will fulfil the conditions of stability, but if the effect or the proportion should require it, we need not hesitate to vary from these dimensions, only assuring ourselves, in case of doubt, that the new form will be equally suited to resist the thrust of the roof. The projection of the buttresses on the middle portion of the façade will give it a marked character, heightened by the long side windows of the central hall, which are not needed, and are rather in the way, in the stage and gallery portions; and to differentiate still further the middle of the building from the ends, we will carry up a low parapet over the windows of the hall, behind which a wide and deep gutter can be formed to keep the drip from the eaves away from the central doorway. The corresponding portion of the ridge may also be distinguished by a cresting of metal or terra-cotta, and the three-fold division of the interior thus "accused" upon the exterior, without interrupting that uniformity of the roof-surfaces which we think desirable.

The buttresses of the middle portion of the walls must obviously be supported from below, and will appear in the first story as piers. The curtain wall which connects them in the second story need not, however, be prolonged to the ground, if there is any other way of supporting it, but may have its position transposed in the first story, if we desire. As some of the smaller offices in the first story and basement occupy but one bay of the façade, we can save twenty inches of room in them, besides improving the effect of the front, without adding to its cost, by adopting this disposition, and transferring the curtain wall, or "wall-veil," as some persons prefer to say, in the first story to the exterior instead of the interior line of the buttresses. This will take away the support from beneath the small portions of the upper wall between the buttresses and the window openings, and under the windows themselves, so we will have flat segmental arches turned in those

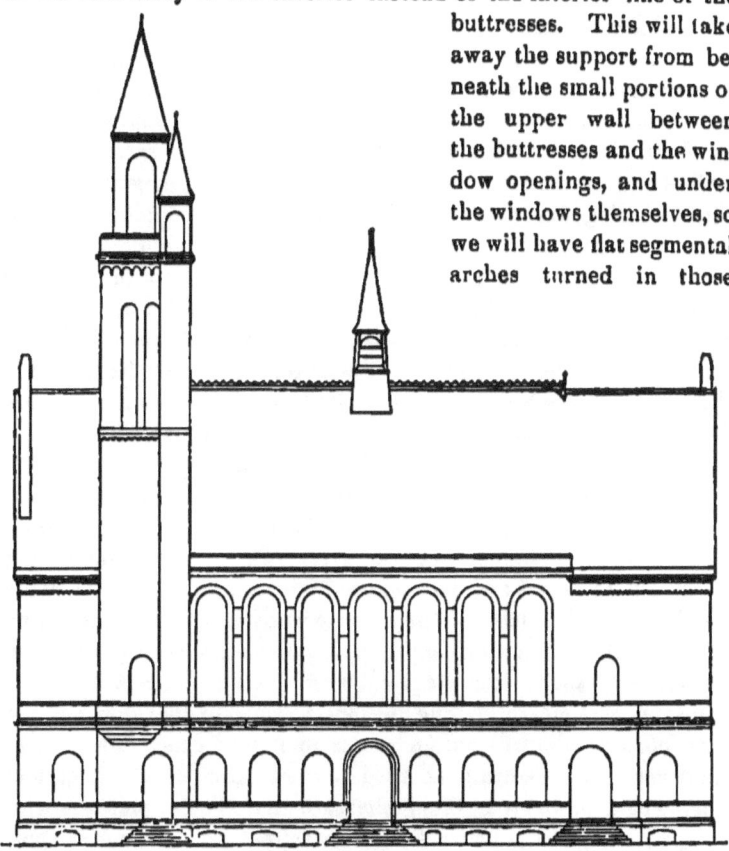

Fig. 189.

places, which will show just under the ceilings of the first-story rooms, but will be out of the way. The interval which will be left between the top of the first-story wall and the sills of the second-story windows we will treat as series of small balconies, accessible from the windows, with stone floors, and a parapet wall. This balcony wall will stop at one end against the staircase tower, and may be prolonged at the other end so as to form a kind of shallow porch over the side doorway, with a narrow balcony on top, furnished with a door opening from the room under the gallery; all of which will help to break up and make interesting a front otherwise somewhat monotonous.

We shall find some difficulty in preventing the tower from looking like the steeple of a church, which the building already resembles rather more than we wish, but we will see if that unfailing resource of the architectural designer, the expression on the exterior of the building of the distribution and uses of the interior, may not help us. Remembering that a portion of the tower, which contains the staircase leading to all the upper portions of the building, must be reserved as a ventilating shaft, to carry the foul air from the different portions of the building, we will "accuse" the shaft by making it project four inches from the general surface of the tower wall, above the first story, and will give it a special termination at the top. We shall need, for the best results, a shaft of something like sixty square feet sectional area, and this can be obtained in the manner indicated.

The offset of four inches which would naturally be made in the tower wall about at the second-story floor we will make on the outside instead of the inside, thus giving it an air of greater apparent stability by the enlargement of the base. The outside of the wall of the ventilating shaft may be made continuous with that below, while the change in thickness of the other portion may be emphasized by placing at that point a balcony, supported by stone corbelling, which will serve to shelter the stage entrance to the hall, and will always be useful, at times of public demonstrations, to the guests of the town officers, who will obtain access to it by a door. The top of the tower would naturally be used to some extent as a lookout, and a bell would probably be hung there, so that the flat platform with parapet, and wooden belfry a little in retreat, will serve both purposes. To complete the exterior features we should add a ventilating turret over the middle of the roof, which will be indispensable in hot weather, to withdraw rapidly the air just under the roofing, which is intensely heated by the sun on the slates, before it can diffuse itself into the atmosphere below, and two chimneys will be necessary, which can conveniently be placed in the walls of the end gables.

The elevation of the opposite side will be substantially the same as the front, with the exception of the tower and doorways, and the end walls will be pierced only with a few windows.

Before we can fix the weights upon the different portions of the foundation, which will determine the spread of the footings and the number of piles under them, it will be necessary to fix definitely the thickness of the walls. For the front, since the piers between the windows are somewhat slender, we have already decided to make them 16 inches thick, adding the projection of the buttresses to this, and as this wall is well tied by the floor-beams which rest in it, the same thickness, 16 inches, will be sufficient for the portions at the ends, which have no buttresses.

The gable walls are under very different conditions, being much higher than the others. The lower portion, beneath the stage and gallery floors, is slightly steadied by the interior partition-walls, which are to be well anchored to it; but above this floor the wall stands free to the roof. As the roof cannot well be tied very strongly to the gable walls, it will be safest to regard these as unsupported above the ground floor, and to give them the thickness required for independent stability. This can be readily calculated by Rondelet's empirical rule. Laying off the height of the wall above the ground, at any scale upon a vertical line, we set off horizontally from the foot of the vertical, at the same scale, the distance between the cross-walls or other supports which bound the wall whose thickness we wish to determine. Connecting first the extremity of the horizontal line, by a diagonal, with the top of the vertical line, we then divide the vertical line into twelve equal parts, and, with one of these parts as a radius, describe an arc from the top of the vertical line, cutting the diagonal, and from the intersection of this arc with the diagonal let fall a second vertical line. The space between the two vertical lines, at the scale of the diagram, will represent the necessary thickness of the wall.

In our case the gable is 96 feet high from the second floor to the apex, and 70 feet wide between the supporting return-walls; and applying the rule we find the thickness necessary to stability to be about 4 feet.

It is obvious that although this may be the proper thickness of the wall at the foot, some economy may be made in the upper portion.

without diminishing the stability of the mass, since the lowering of the centre of gravity will compensate for the loss of weight. If the wall were rectangular, it might, by successive offsets, be reduced from four feet to sixteen, or even twelve inches, at the top, but the peak of a gable is less solid and steady than the corresponding portion of a rectangular wall, and we shall do best not to reduce it below twenty inches in thickness. It is quite possible that a smaller amount of material might be so distributed, by means of buttresses, as to give the stability needed, but this, we suppose, would involve in our case certain objectionable conditions, so we accept the result of our calculation, and draw the section of the wall in accordance with it, making the average thickness 34 inches. The tower walls are next to be considered. These are strongly held by the return walls, which tie them back in such a way that it would be almost impossible for them to fall over, so that it is hardly necessary to give them more than the thickness required for resisting the crushing strain due to their own weight. The walls being 134 feet high, the Rondelet diagram gives for them a thickness of 20 inches, which is unquestionably sufficient, but public opinion, for some reason, generally demands the thickest walls for towers, which need them least, and in deference to this, as expressed by our committee, we will make the lower portion 28 inches thick, diminishing the upper part by two offsets to 16 inches, as a compensation for the excess of material used below. This, while improving the appearance of the building, will really be judicious as a matter of construction, inasmuch as a solid brick wall 134 feet high, and 20 inches thick, although under the circumstances perfectly stable, would be subjected to a crushing strain at the base of $7\frac{1}{2}$ tons to the square foot, which would be increased again upon the piers at either side of the doorway by the arch, which throws upon them the weight of the mass above it, to about $12\frac{1}{2}$ tons, and to this weight again might be added a farther strain due to the action of wind on one side or the other, amounting possibly to 10 or 12 tons more. This would give a stress which ordinary brickwork could not with perfect safety be trusted to bear, but the increase of the mass at the lower part of the tower, with the lightening of the upper walls, will relieve us of all anxiety upon this point.

The variations in the thickness of the masonry will be made at somewhat irregular heights, to suit the exigencies of the openings and the ventilating shaft, but a little study of the section will give us, we suppose, an average thickness of 22 inches.

The interior walls, with the exception of that forming the abutment, opposite the tower, of the proscenium arch, which will have the same thickness as the arch, are not of great height, and are steadied by the floor-beams, so that 12 inches will be sufficient for them.

We can now draw our definitive plans of the several stories, the horizontal section of the exterior walls being determined. Figure 190 shows the plan of the first story, and Figure 191 of the second story. The walls in the first story are substantially the same as

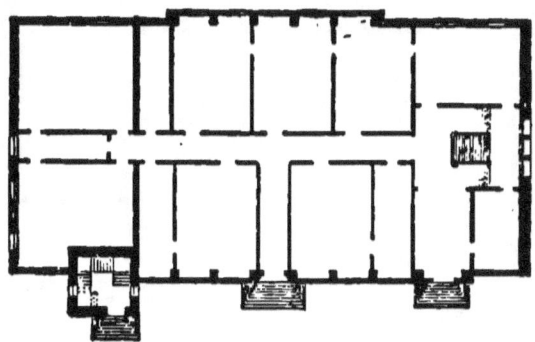

First Story Plans
Fig. 190.

those of the basement, so that a plan of the latter is not necessary, and we may lay out the foundation at once, as shown in Figure 192. In this it will be observed that each of the interior walls, and the plainer portions of the exterior walls, are provided with continuous foundations, but that the masonry is interrupted between the piers which support the buttresses above, although in elevation this space is occupied by a wall, which fills the area around and under the windows. One may naturally ask why this curtain wall, light as it is, should be deprived of a foundation, and it would be more usual, in fact, to lay footings for this portion of the masonry, as well as the rest. Nevertheless, an attentive study of the conditions will, we think, show that it is, under the circumstances, wiser to support this portion of the building upon isolated piers, than to build for it a continuous footing, which must be very unequally loaded. A rapid computation of the weight of the piers, as compared with that of the wall between them, will show that the former, which are approximately 1¾ × 3½ feet in section, and about 68 feet in height from the level of the basement floor, 8 feet below the curb, to the eaves, contain each about 378 cubic feet of masonry, weighing, at 112 pounds to the cubic foot, 42336 pounds, to this being added the weight of the curtain walls around the second-story windows, which, as we remember, are supported entirely from the main piers, by segmental arches sprung between them, and weigh, for the portion resting on each pier, about 23000 pounds. Besides this, the roof-trusses, which bear wholly upon the piers, bring on each an additional load, as we have seen, of 22320

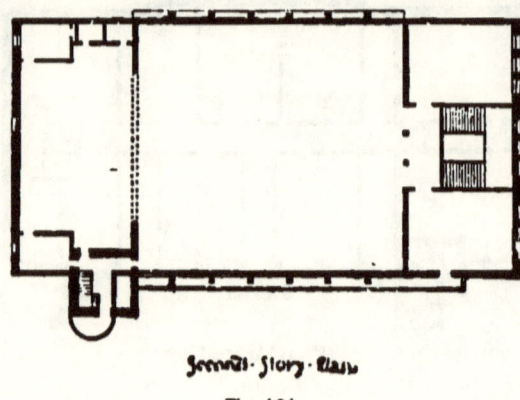

Second-Story Plan.

Fig. 191.

pounds; while the weight of the second-story floor, although framed with beams at short distances apart, is also brought by the segmental arches just beneath it entirely on the piers, adding to the load on each the weight of $\frac{12 \times 30}{2}$ = 180 square feet of flooring, equivalent, with its ordinary extraneous burden, to about 18000 pounds. The weight of the first story and basement floors would be divided between the piers and the curtain wall in a proportion which can hardly be estimated exactly, but about one-half of it would probably come on the piers, making, for two floors, an additional load of 18000 pounds. Adding these together, we find the total pressure at the level of the basement floor upon the substructure of the piers to be 123656 pounds. Dividing this by the sectional area of the piers at that point, which is 5½ square feet, gives 22483 pounds as the pressure per square foot.

On the foundation wall between the piers we shall have the weight of 18000 pounds of flooring in first story and basement, with that of the wall as high as the second-story floor; everything above resting wholly on the piers. The openings for windows occupy most of the area, but we have left about 300 cubic feet of masonry, weighing 33600 pounds; the whole pressing upon the substructure of that portion of the building, whose sectional area below the basement windows is 10½ × 1½ = 13½ square feet, with a force amounting to 3820 pounds per square foot.

This calculation discloses a very great difference in the intensity

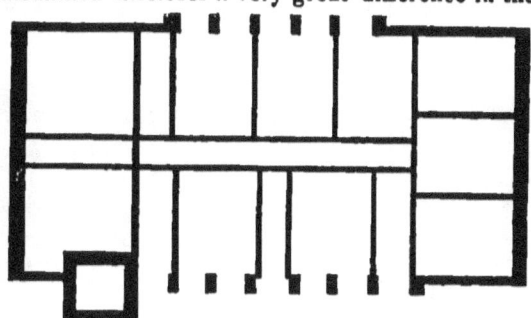

Foundation Plan
Fig. 192.

of the pressure on the foundation under the piers, and that upon the wall between them, and as the masonry of rough stone extends below the basement floor seven feet to the tops of the piles, we have just reason to fear that the compression of the joints in this masonry beneath the piers would be so much greater than in the intermediate portion, subjected to a load hardly one-sixth as great, as to cause some dislocation between the two parts of the stone-work, which would probably show itself above ground by fractures in the sills of the basement and first story windows, as well as by the opening of seams in the angles between the buttresses and the curtain wall.

If the ground were very soft, so as to make it unsafe to increase the load upon any part of it beyond a limited amount per square foot, it might be best to equalize the pressure by spreading the footings of the piers until the weight upon them was distributed over so large a surface as to make the pressure upon each foot of this surface equal to that on the footings of the curtain-wall, but it would take a great deal of stone to spread the base of the foundation to the requisite extent, and we, who can count in the present case upon a pile foundation of tolerable resistance, shall do best to abandon the idea of a separate foundation for the curtain-walls, and arrange to support the whole, by means of arches turned just beneath the basement floor, solely upon the footings of the piers.

Although the pressure upon these will be increased by so doing, we can easily provide piles enough to sustain it all, and the cur-

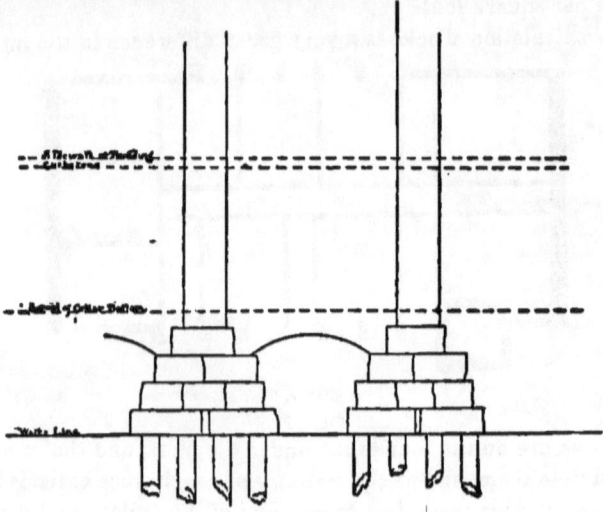

Fig. 193.

tain-walls, being now entirely dependent upon the piers, will settle with them as the joints are compressed under the weight of the superstructure, instead of being torn away from them by the reaction of the less strongly weighted stone-work upon which they themselves rest.

Further consideration convincing us that this is the best, as well as most economical method of construction, we have only to indicate the underground arches which we propose, as shown in Figure 193, and calculate the size of footings and number of piles required under each pier to sustain the weight upon it, which must now be increased by that of the curtain-wall in first story and basement, above the arches, and also by an amount representing roughly the weight of the foundation below the basement floor, which, of course, can be only provisionally determined.

Before making up our minds on this point we shall do best to test the resistance of the hard-pan to which the piles are to be driven by actual trial, and will therefore lay out the piling-plan for the plain walls, and begin work.

Although the ground is softer than we could wish, the piles bring up generally in a stratum which allows them to sink only $2\frac{1}{2}$ to 3 inches at the last blow of a hammer weighing 1600 pounds, and falling 15 feet. This, by Sanders's formula, indicates a safe resistance for each pile, in the worst cases, of $\frac{180 \times 1600}{3 \times 8} = 12000$ pounds, or six tons. We will therefore assume this as the load to be assigned to the piles under the piers, and will draw the plan accordingly, remembering that it will be necessary to watch the driving closely, so that if a soft spot should be met with, in which the piles should sink more than three inches under such a blow, additional piles can be at once staked out and driven, sufficient in number to divide the total pressure into portions small enough to come within the limit of their safe resistance, as found by a new calculation.

Down to the basement floor, the sum of all the weights borne by each pier is 175256 pounds, which would just be sustained by fifteen piles, driven to a bottom as hard as that which we have already found. We must not, however, forget that a considerable cube of masonry will intervene between the top of the piles and the basement floor, whose weight must be taken into account. The distance between these two points is seven feet, and six feet, at least, of this must be of heavy stone masonry. The remaining foot may be of brick, like the superstructure. Supposing, simply for calculation, that one extra pile would be sufficient to carry the additional weight of the foundation, we should have under each pier a group of sixteen piles. These are always most advantageously arranged in pairs, so that the stones which rest upon them, the "cappers," as they are called, may each cover two piles, and no more. It is also desirable, for the sake of saving stone, to place the piles as near together as they can be driven without forcing each other aside, or unduly disturbing the bed, and the minimum distance for this purpose being two leet from centre to centre, in such ground as that with which we have to deal, the natural disposition of our sixteen piles will be in the form of a square, measuring 6 feet on each side, from the centres of the piles. In order to cover these entirely with the capping stones, it will be necessary, as the head of the piles is from 10 to 12 inches in diameter, to make the first course 7 feet square. The second course, in order that the weight may not be thrown too much on the inner part of the capping stones, should be 5 feet square. The third course may be 3 feet by 4, and the fourth course the same. Each course will be about 18 inches high, and the whole amount of stone-work will be 147 cubic feet, weighing, at 125 pounds per cubic foot, 18375 pounds. The extra foot of brickwork between the top of the stone foundation and the basement floor will weigh 622 pounds. Adding all the weights together, we shall arrive at a total of $175256 + 18375 + 622 = 194253$ pounds, or 97 tons, to be supported by 16 piles, giving a load of $6\frac{1}{16}$ tons each.

If the bottom under the piers should be found as firm as where we are now driving, the piles sinking not more than three inches, and generally less, at the last blow, with a 1600-pound hammer and 15-foot fall, we should be quite safe in adopting this arrangement, and we will stake out the piles under the piers accordingly, leaving, however, some person to watch the driving, with strict injunctions to mark on the piling-plan, of which he has a tracing, the actual sinking at the last blow of every pile, with the height of the fall; while we inspect the timber delivered on the ground, and observe the operations of cutting off the heads of the piles and laying the first course of stone, both of which are already in progress at one corner of the building.

The piles on the ground are straight spruce sticks, with the bark on, varying from 30 to 40 feet long. Here and there is visible a crooked specimen, or one the heart of which is evidently rotten, and we mark all such for rejection. The driving of the first piles has shown that the comparatively firm stratum upon which they must rest is about 31 feet below the surface, and men are engaged in cutting off the small ends of the longer piles to bring them to this dimension. It is important not to penetrate through the bearing stratum, as the ground is shown, by driving a long experimental pile, to grow soft again immediately below; all that is necessary or safe is to continue the blows of the hammer until the firmer ground is reached, which will be shown by the diminished penetration of the pile at each impact, giving then only one or two additional blows to settle it into its bed.

There is some danger that the workmen may surreptitiously endeavor to save trouble for themselves, and money for their employers, either by driving the pile only a portion of the required distance, and then cutting it off, or by putting in shorter, and therefore cheaper timbers. Either of these frauds will probably be followed, sooner or later, by serious consequences, and the only way to guard against them effectually is to witness in person, or by an intelligent deputy, the driving of every pile. We are somewhat in doubt whether it may not be necessary to send away all the 30-foot piles, of which there are several on the ground, for the reason that although they would be long enough to reach from the hard stratum to the water-line, they lack about two feet of the length necessary to extend from the hard bearing to the present bottom of the excavation where the machine stands. This, for convenience in working, is not dug down to the water-level, and there is danger that the short piles may be simply driven to the head in the ground and left there, with their feet still some distance from the stratum on which they ought to rest; but in consideration of the promise of the contractor to bring no more of the same kind, we consent to have them driven in our presence, each one, after driving to the head, being sunk farther by means of a "follower," or short piece placed on top of it, until the bearing stratum is reached. When the trenches are excavated to the water-line, which will be done as soon as the machine is out of the way, the followers will be dug out, and the piles under them will then be as useful as any.

The operation of digging out the piles is already in progress in another place. The level of the water-line, or rather, of the point at a certain distance below the average water-line where we have directed the piles to be cut off, is fixed with reference to a mark on the side of the excavation, and a steam-pump is at work to keep the trenches clear of water until the earth has been removed to a proper depth, the heads of the piles cut off at a uniform level, and the capping-stones laid. Two men, with a cross-cut saw held between them, are bending over in the mud, sawing off the top of a pile, which another man holds to prevent it from falling upon them. Observing them from a distance, we notice that in order to relieve their backs as far as possible from the fatigue of stooping, as well as to keep their knuckles out of the earth and water, they hold the saw very much bent, so that it makes a concave, instead of a level cut across the head. As we approach, the head of the pile, just severed, is purposely tumbled over their work, and the men begin another cut, this time with the saw held straight between them. Looking about the trench, we notice that one-third or more of the piles already cut off exhibit the concavity due to the bending of the saw, while others have an oblique head, and a few are cut an inch or two higher than their neighbors. Any of these defects may compromise the safety of the building, either through the crushing of the edge of a concave cutting under the weight of the superstructure, or the tilting of a capping-stone supported at one end on a pile cut obliquely or out of level; and calling the attention of the men, we point out the defective work, and direct them to recut the piles properly on the spot, waiting to see our orders obeyed.

While thus engaged we have leisure to watch the stone-laying just beyond. The adjustment of a roughly-split stone upon the heads of two piles, so that it may have no tendency to rock or move in any way under the great and varying pressure which will be placed upon it, is a difficult matter, and the work should be sharply looked after. The usual way is to place the stone in position, and then wedge up with stone or even wooden chips between it and the head of the pile, until it ceases to move when shaken; but this mode is open to many objections. Wooden chips are of course inadmissible, since they crush immediately under a strain; and stone "pinners" are liable to be broken or dislodged, leaving the block which they were intended to sustain in a condition of dangerous instability. The best, although the most troublesome method of capping, is to select only the stones with comparatively flat beds, and lay them on the heads of the piles, shifting them about, before they are detached from the derrick, until they rest immovable. They will then need no pinning or wedging, and can be depended upon to sustain without moving the load which is to be placed upon them. If wedging should be found absolutely necessary, as may sometimes happen, the stones used to pin up with should be well-shaped, strong, and securely placed upon the head of the pile, so as to be in no danger of shaking out or crumbling. With the same object of avoiding all tendency on the part of the capping-stones to rock under the load, no stone should rest upon more than three piles, unless both it and the heads of the piles have been dressed to a perfectly plane surface, and with rough stones it is not easy to get a good bearing even on three piles at once.

After explaining our ideas on these points to the foreman of the stone-layers, whose opinion coincides with our own, we return to the front wall, where the pile-driving machine has arrived before us, and are troubled to find that the ground appears softer there than under the other portions of the structure. As we approach the row of isolated piers forming the middle of the front, the piles sink under the last blow of the hammer from three to four inches, instead of two and one-half or three inches, showing that a variation has taken place in the texture of the clay stratum upon which they rest. A trial pile driven by means of a follower to a depth of 40 feet sinks

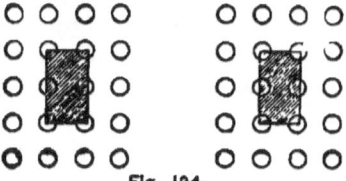

Fig. 194.

at that distance more rapidly than ever, and we are forced to the conclusion that the bottom at 31 feet, although poor enough, is the best to be had. A simple calculation is, however, sufficient to show that it is unsafe to trust the weight of the piers upon it without adding to the number of piles under them, and thus diminishing the load upon each point of support. Supposing the sinking of each pile at the last blow, under the actual conditions, to be four inches, the weight which it could be relied upon to sustain safely would be 4½ tons, and the number needed to support the load of 194253 pounds, which was previously calculated as the weight on each pier, would be 22, allowing for the slight additional weight of stone required to cover the more extended base.

The ground is, however, hardly so soft as this, the average being about 3½ inches sinking at the last blow, and we shall be safe in changing our plan, and staking out 20 instead of 16 piles under each pier, as shown in Figure 194, remembering that if the ground should grow still worse it will be necessary to add to the number by driving extra rows on each side. Happily, this does not prove to be the case, and we are able, when the driving is over, to rest assured that whatever else may befall our building, the failure of the piles is not to be feared.

We need dwell no more upon the details of construction of our building, which would now differ little from those of any other, but will proceed at once to consider the necessary means for heating and **Heating and Ventilation.** means for heating and ventilating the various rooms. Success in this point will be a matter of some difficulty, and we should have our scheme well prepared in advance, in order that the necessary distribution of flues and pipes may be effected to the best advantage.

The only practicable method of conveying heat from a single source to all points of so large a building is to employ steam, and although steam-heating is in many respects inferior to that by means **Steam.** of hot water or ordinary furnaces, we have no alternative, and must try to mitigate the bad features of the system as much as possible. For the smaller rooms, the evil to be avoided is the closeness, from want of a fresh-air supply, which generally characterizes steam-heated offices, and to remedy this we shall do best to adopt what is known as the direct-indirect mode of heating, in which the radiators stand in the rooms, but are made to **Direct-Indirect Method.** enclose a space into which air is admitted directly from the outside of the building, to pass, after being warmed by contact with the pipes of the radiator, into the room. The large hall in the second story must be heated in a somewhat different way, since it would not be possible to place radiators in the interior of the room, but it will be advantageous to keep them as near the part to be warmed as possible. The source of heat for the entire building will be a boiler placed in the basement, and we should get some notion of the necessary size of the boiler, and of the flue to carry off the smoke from it, in time to proportion the rooms suitably.

We can form a rough estimate of the radiating surface required, and thence of the sizes of pipes, boilers and flues, by allowing one-tenth of a square foot radiating surface to each square foot of plain wall and exposed roof surface, and seven-tenths of a square foot of radiating surface additional for each square foot of glass in the windows. The build- **Estimate of Radiating Surface required.** ing has about 33,000 square feet of outside wall and 15,000 square feet of interior wall, and to this must be added 12,000 square feet of roof surface, exposed to the interior of the rooms, making 60,000 square feet of roof and wall, requiring 6,000 square feet of radiating surface. Of glass in the various openings there is about 5,600 square feet, seven-tenths of which will give 3,920 feet additional of radiating surface, making 9,920 square feet in all. The rule sometimes used, of allowing for direct-indirect radiation one and a half square feet of radiating surface for every 100 cubic feet of space contained in the building, would give, as we have about 900,000 cubic feet, 13,500 square feet of radiating surface; but this would be an excessive allowance for the large hall in the second story, and our first estimate is quite safe.

By the usual rule for estimating, the heating surface of the boiler must be one-tenth that of the radiating surfaces, which would give here 992 square feet. This could be obtained by using a horizontal boiler 5 feet in diameter and 20 feet long, with about 60 tubes, but a single boiler of this kind would not work to such advantage as two, presenting together **Necessary Heating Surface of Boiler.** the same amount of heating surface; and there is here the further advantage in using two boilers, that one can be employed solely for heating the hall in the upper story, which is only occasionally in use, while the other can be devoted independently to warming the offices in the first story and basement of the building, which are occupied almost continuously. We must therefore make a new calculation, which shows us that the hall in the second story will require almost exactly one-half of the total radiating surface, so that two boilers just alike, each containing 500 square feet of heating surface, will answer admirably. As an engineer will be constantly employed, it will be most convenient and economical to use two horizontal boilers, each 4 feet in diameter, 20 feet long, and containing 30 tubes. Each of these boilers will need, in order to be able to get up steam quickly, about 20 square feet of grate sur- **Grate Surface.** face, and we shall require, to carry away the gases of combustion quickly from these grates, when both are in operation, a chimney of the best form with a sectional area of about 13 square inches to each square foot of grate surface. We have 40 square feet of grate surface, and must have, therefore, 520 square inches of sectional area of chimney. The height of the chimney would enter into the calculation to a certain extent, since the velocity of the current increases with the height of the heated column, but this advantage is soon lost in prolonging the shaft to an excessive height, and we shall obtain the best results by assuming only the average dimensions. The sectional area thus calculated should be obtained in a square or circular flue, as an oblong one, with the same area, has much less capacity for carrying away smoke. Considering the circumstances of our building, it will be found most advantageous to employ a circular cast-iron smoke-pipe, and to place it in the ventilating shaft which forms a portion of the tower, so that the heat radiated from it may assist the upward current in the ventilating flue. A pipe 28 inches in diameter will give the requisite sectional area, or a little more; and as it will be 128 feet in height we can be sure of a good velocity in it. The only objection to such a position for the chimney is the danger of disfiguring the upper portion of the tower with smoke; but by carrying the pipe through the roof of the smaller turret it will discharge the smoke at a sufficient distance from the main belfry to make sure that it will be carried away by the wind.

Necessary Size of Chimney.

Shape of Flue.

The position of the boilers will be determined in general by that of the chimney, since it is desirable that the communication between the smoke connection of the boilers and the chimney should be as direct as possible, avoiding long pipes, which chill the gases, and underground flues, in which it is difficult to start a current. Fortunately, we have kept this point in view, and have arranged a room in the corner of the basement, close to the tower, large enough, not only for placing and managing the boilers, but for passing all around and over them, with sufficient space in front of them for handling the long flue-brushes and scrapers which will be required.

Position of Boilers.

We may now estimate roughly the size of the largest pipes which will be required, and we shall then know what special provision must be made in the construction of the building for placing them

Taking the safe rule that the main distributing pipes should have a sectional area equivalent to eight-tenths of a square inch for each 100 square feet of radiating surface sup- **Size of Main Pipes.** plied by them, we shall find, since each boiler furnishes steam to 4960 square feet of radiating surface, that each main steam-pipe must be a little more than 7 inches in diameter inside. As no pipe is made between 7 and 8 inch, and as 8 inch is much larger than would be necessary, we will determine upon 7-inch pipe. The risers, or pipes which run up to supply the radiators above will be small, none being more than 3 inches in diameter, and we shall have no difficulty in carrying them up in 4″ x 4″ recesses left in the wall at the proper places, which it will be well to mark on the plans at once.

Before this can be done, however, we must determine all the main features of our system of ventilation as well as heating, and the sooner we make up our minds about this the better.

For the basement and first story rooms the plan of ventilation should be as simple as possible. Fresh air will be admitted behind the radiators in each room, which should stand under the windows, in order that the warmth from them **First-story Ventilation.** may counteract the descending stream of cold air which, in winter, always flows over the surface of the glass, and foul air will best be taken out at two points, one near the top and the other near the bottom of the room. For the offices which have fireplaces, the opening of this will form the lower outlet, but another should be provided near the ceiling, communicating with a flue which may be carried up beside the fireplace flue. Where there is no fireplace, two flues, one opening near the floor, and the other near the ceiling, will be necessary. If the combined area of these outlets is made somewhat greater than that of the inlet, a gentle current will be maintained at all levels in the room, and the air kept in better condition than would be possible with a single outlet. As the small rooms are occupied only by a few officers and clerks, the supply of fresh air need not be very large, and a 4-inch round pipe to each radiator would make an ample inlet. For outlets, brick flues 8″ x 12″ will be best, and each flue may open with small registers both in basement and first story, remembering, however, that two openings must not be made in any flue in the same story, and that a flue which exhausts from the floor of the basement rooms must always, if it opens in the second story, exhaust from the floor there also, and that in the same way the ceiling registers in the first-story rooms should open into the flue which draws from the ceiling of the basement rooms. It need hardly be said that two fireplaces should not under any circumstances open into the same flue, and that the outlet registers in the basement rooms must not have a clear opening greater than half the sectional area of the flues into which they open, if any air is to be drawn into the same flues from the rooms above.

As the entrance-way and corridor in the first story and basement will naturally be more or less foul, a good current of air should at all times be maintained through them. The frequent opening of the doors will furnish a sufficient fresh-air supply, without bringing special pipes to the radiators, and it will be of advantage to restrict the inlets, but increase the outlet, encouraging the exhaust in other ways as much as possible, so that the draft from the corridor will be stronger than that in the rooms, and the current will, on opening the doors, tend consequently from the room into the corridor, and no vice versa. We will therefore provide only direct radiation for warming the corridors, and will conduct the air from them by a pipe passing through the closets at the end opposite the staircase to the gable wall, where a large flue is ready to receive it and carry it away. If there should be any difficulty in maintaining a current through this flue in cold weather, we can afterwards place a radiator in it, a little above the level of the second-story floor; but the wall is of great thickness, and we can easily build in it a flue 20" x 20", or 20" x 24", perfectly straight, and 100 feet high, which will be very little liable to a reversal of the current in it, even without artificial heating.

Having now provided for the separate removal of the air in the basement and first story, which we wish to prevent from ascending the stairs to annoy the occupants of the hall above, we must arrange for a special supply to the latter. The hall, with the **Ventilating Main Hall.** gallery, will seat about 1000 persons, and to make them quite comfortable during an evening they should be furnished with at least 1500 cubic feet each of fresh air per hour; and this air must, moreover, be warmed in winter before delivery, and conducted throughout the room gently and uniformly, leaving no corner unvisited, and dispersing itself everywhere rapidly but without sensible currents. The system must include every part of the room, since any portion unswept by the flow of air will become a reservoir of decaying organic particles, which will diffuse themselves through the neighboring atmosphere for some distance in all directions. We will at first consider the winter ventilation only, that for summer being simpler, but completely different.

As in the rooms below, we have decided to use the direct-indirect method of steam heating in the hall, placing large radiators under the windows on all sides, and supplying each radiator with a given quantity of fresh air from the outside, to be warmed by contact with it, and then delivered into the room. The persons seated next the walls, who would otherwise be exposed to the chilling currents which descend along the surface of the windows, and to a much smaller extent along the plastering, will then be doubly protected, by the deflection of the cold currents on meeting the warm streams rising from the radiators, and by the direct influence of the warm rays falling upon their bodies from the hot pipes. As those occupying the seats at the edges of the room will thus be warmed by direct radiation, the air supplied to them need not be so warm as if it were the only source of heat; and the current delivered from the radiators, if sufficiently abundant, need not be raised above 90° Fahrenheit. This will answer also for the persons in the interior of the room, who, although cut off in part from the heat radiated by the steam coils, are less exposed to cold currents from the windows, and receive, moreover, a very considerable amount of warmth radiated from the bodies of those around them. With this human warmth, however, is given off a certain amount of organic exhalation, so that the air in the centre of the room will be less pure than that nearer the fresh-air openings at the sides, and it will be necessary to furnish the middle portions with an additional supply. In many buildings this could be done by placing registers in the aisles between the seats, introducing at small intervals air taken fresh from the outside, warmed in the basement and sent up through pipes, but we have to bear in mind that our hall will often be used for dancing, so that registers in any part of the floor will be quite inadmissible, and some other place must be found for delivering the air.

There are but two other positions where inlets can be placed near the floor, one of these being the vertical front of the stage and the other the front of the gallery. Both of these will do, and we will arrange to use them, although in different ways. The front of the stage being separated, sometimes by an orchestra, sometimes merely by an open space, from the front rows of seats, may be used as a great register, throwing in air along its whole extent, and the air so introduced will, in its passage across the orchestra space, diffuse and mix itself with other currents, thereby losing its original impetus, and reaching the occupants of the front benches as a breeze so gentle as to be hardly felt. This stage front will, in fact, offer the best position in the room for the advantageous introduction of air, and we must arrange for taking it by ample openings from the outside into the space under the stage, and for warming it by radiators before delivering it into the auditorium. Some of the radiators may be placed close to the open gratings of the front, where their direct warmth will be felt by the persons nearest them, who are most exposed to the current.

By taking advantage of the shape of the space under the stage, we shall be able to secure a gentle but strong horizontal delivery of the warmed air, which will send it well toward the centre of the room before it begins to ascend, and the portion of the auditorium nearest the stage will thus be supplied with fresh air throughout its whole extent. For the remaining half we will take fresh air from the rear wall, under the front of the gallery, but in a manner slightly different from that employed for the first portion. In order to throw the supply from this direction well into the centre of the auditorium, we shall need to bring it in with considerable velocity, and as the seats for the audience extend to a point within a few feet of the gallery front, the current, if allowed to strike the persons sitting in them, would be felt as a disagreeable and even dangerous draught, so that we shall do best to introduce the greater part of it at a height of ten or twelve feet above the ground. In this way the main current will pass above the heads of those sitting near the inlet registers, the air diffusing itself so as to come within reach of the lungs of the audience only in proportion as it loses velocity. Under this arrangement, the greater the force with which it enters the room the more effective will the stream be in reaching and stirring up the atmosphere of the middle portions, and we may with advantage place the radiators for heating it in the basement, and bring the air up by pipes through the offices in the basement and first story. By making the pipes straight, with a curved elbow at the top to direct the current into the room, we shall obtain a heated column long enough to possess a very considerable buoyant tendency, and the air will be thrown into the hall with force enough to carry it to the centre before it will begin to rise. To complete the supply of fresh air for the room, we must furnish the occupants of the rear rows, who will receive little benefit from the currents passing over their heads, with some separate inlets near the floor, bringing the air in through exposed steam coils in the direct-indirect manner, so that, as at the sides of the room, the chill caused to the persons near by the movement of the incoming air, which, slight as it is, increases the evaporation from the skin, and causes a sensation of cold, together with the loss of heat due to radiation from the body to the cold walls, and the unpleasant draughts caused by accidental currents, may be compensated by the warm rays from the pipes.

We shall now have, for the main portion of the auditorium, currents of fresh air proceeding from all sides, and meeting in the centre. The currents from front and rear are purposely directed with considerable force in a horizontal direction, and those from the sides, which are, so to speak, pressed upon by the descending cold air from the window surfaces, will be deflected in the same direction, and this impulse, aided by the natural adhesion of moving air to the surfaces with which it comes in contact, will serve to keep at least the heads of the occupants of the room in a pure and constantly renewed atmosphere. On the meeting of the currents in the middle, their horizontal movement will be destroyed, and the buoyant force due to the heat of the mass of air, which has grown warmer in passing among the bodies of the people, will assert itself, carrying the whole upward. Then, if not otherwise disposed of, it will become chilled by contact with the underside of the cold roof, and will descend along the surfaces of the roof, walls and windows, to mingle again with the incoming air from the radiators, and repeat the same round. This would not only contaminate the freshness of the new supply, but would very much reduce its amount, since air cannot be forced by ordinary means into a room which is full already, so that we must, to secure a continuance of the flow of pure air, remove the vitiated atmosphere before it can descend to the level of the incoming currents.

If the movement of the air were positive enough to carry it, after rising above the heads of the people, directly to the roof, it might be best to take it from the ridge, but in cold weather this would hardly be the case, much of the air becoming chilled and returning downward before reaching that point, so that we shall do better to exhaust it from the level of the cornice, a little above the line of separation between the lower, fresh, warm and horizontally moving stratum, and the upper stratum of vitiated, gradually-cooling and descending air. If the hall were of a perfectly simple shape, this upper stratum would move uniformly all around, but there are two causes which will give it a tendency toward the stage end of the room. One of these is the attraction of the stage ventilation, which draws the upper air sensibly toward the proscenium-arch; and the other is the pressure of the air from the gallery, which, introduced through radiators at the sides and rear, will move forward into the main body of the auditorium, pushing the stratum in front of it in the same direction. The mass of air which we wish to remove will thus be impelled gently against the proscenium wall, and can be removed most effectively by openings in that wall, through which it can continue its course into the ventilation-shaft and away from the building. These openings can have any decorative shape, and should communicate with a conduit behind the proscenium wall, carried into the main ventilating flue.

It will be seen that this plan of ventilation is totally different from that which would be adopted in a school-room, or other apartment with a low, flat ceiling. In such a room the best method would generally be to employ indirect radiation entirely, warming the fresh air in the basement, and bringing it up through long vertical pipes opening into the room by registers in the side walls, seven or eight feet above the floor, withdrawing the cold and foul air below, by exhaust registers in or near the floor. By this arrangement the fresh, warm air would pass first to the ceiling, filling the room like an inverted lake, and constantly pressing out the foul strata below, without danger of annoying the occupants of the rooms by draughts.

In the present case, however, we are precluded from employing any method of this kind by the great height of the ceiling, and the extent of cooling surface presented by the roof. The warm air introduced at any considerable height above the floor would rise immediately to the ridge, cooling there with great rapidity, to be precipitated again in cold draughts upon the heads of the people below, who would remain immersed in a chilly atmosphere even though that above them might be warmed to a temperature of 100° or more. It would be impracticable to fill so great a space, losing heat, moreover, so rapidly as would be inevitable under the circumstances, with anything approaching the inverted atmospheric lake of a low room, and our only resource here is therefore to keep the fresh-air supply near the floor, avoiding unpleasant draughts as much as possible, but directing all the warm currents so that they may reach those who are to breathe them before they can escape from the slight attraction exerted upon them by the floor and the objects near it, and rise into the empty space above, to be lost beyond recovery.

School-room Ventilation.

The course of circulation of the air being once determined, many circumstances can be made use of to promote it, and all obstacles should be removed. Radiators ought not to be placed in such a position that the inevitable ascending current from them will interfere with the general movement, and even the aspect of the different portions of the room will need to be considered, the northern and western sides being generally coldest in winter, and chilling the air next to them so as to cause it to descend, while that on the opposite side rises, in a movement of gentle rotation, which can be checked, if it should interfere with the desired system of circulation, by increasing the radiating surface near the cold walls.

Such ventilation as this is of course dependent upon the difference in temperature between the exterior and interior air, and the movements due to the buoyancy of warm currents in a cold atmosphere will cease entirely as soon as the temperature outside and inside become the same. For summer ventilation, therefore, we shall need to devise a modified system, which can at pleasure be substituted for the other, but will not interfere with it at other times. Fortunately, summer is also the time of open windows, and the warm-weather ventilation is, in such a case as this, a much simpler affair than that needed for winter.

Summer Ventilation.

The occupants of the offices in the basement and first story will keep their windows generally as wide open as possible in the hot months, and if we remember to provide fan-lights over the doors, and to place the doors opening on the corridor nearly opposite each other, we can secure for them an almost constant draught across the width of the building, which will keep the rooms as fresh as could be desired. The corridor itself, if much frequented, may need, even in summer, to be exhausted by means of its special flue, which must in that case be kept heated by a large gas-burner, in order to preserve that difference in temperature between the air inside the flue and the outside atmosphere on which the movement of the former entirely depends.

For the great hall in the second story we shall need something more than open windows, not for the sake of a fresh-air supply, since these large openings, placed opposite each other in so long a building, isolated from all others, would give a breeze across the room in the hottest night, but to remove the air which would, unless allowed to escape, collect under the roof, filling the space down to the heads of the windows with a stagnant mass, often very much heated by contact with the underside of the slated roof, upon which the sun shines all day, and always containing most of the organic impurities thrown off by the lungs and skins of the people below, which rise with their warm breath even in the atmosphere of summer. This reservoir of foul and heated vapor, although confined to the space above the sweep of the fresh breeze from the windows, is apt to make its presence disagreeably felt by diffusion through the purer atmosphere far below it, and it is important to provide for tapping it, so to speak, and allowing its contents to flow, in accordance with their natural buoyancy, upward from the highest point of the roof into the outer air. It would be useless to try to draw the foul, warm air downward as far as the openings into the tower ventilating-shaft, since its buoyant force, after a day of summer sunshine, is far too great to be counteracted by any exhaust which could be obtained in the ventilating-shaft without the aid of a fan; and it is very desirable to take full advantage of the acquired tendency of the stratum of air which we wish to remove, to assist its discharge. For this purpose an open turret on the ridge answers perhaps better than anything else. The length of the warm current ascending through it assists the velocity, and helps to draw up that which wou.1 tend to linger behind; and its position at the summit of the roof ensures the removal of the last traces of the warm and foul stratum. So long as any persons remain in the hall, or gas-lights continue to burn there, new volumes of deteriorated air will ascend to take the place of that discharged from the ridge, but if not allowed to stagnate, or accumulate heat from the roof, they will not affect the atmosphere below.

Having now evolved a satisfactory general scheme of heating and ventilation, the details only will need attention, and these will not, for our present purpose, detain us long. After the contract for the heating apparatus has been made, and the contractor has made his appearance, with his materials, upon the ground, we shall need to examine, and if necessary to criticise, the construction of the boiler, the dimensions of the pipes, and the arrangement of them and of the radiators. **Details of Heating Apparatus.**

Many good heating engineers employ boilers and radiators of their own construction, to which the arrangement of the pipes must be suited; but there are some general principles applicable to all systems. The essential features of a good steam-heating apparatus are: safety from all risk of explosion, sufficient and well-placed radiating surfaces, freedom from noise in operation, and thorough drainage of all parts, so that the pipes and radiators may not be subject to injury from water left standing in them, and freezing.

The first of these requisites is satisfied in various ways.

Many engineers, instead of boilers with a single shell, use for heating the so-called "sectional boilers," consisting of coils or groups of pipes, sometimes of wrought-iron, but generally cast, joined together in sets of five or six or even more, over a single fire-box. The water and steam circulate freely among the sections, but in case of over-heating, or insufficient water-supply, not more than one section is likely to give way at once, and the explosion of a single section, even in the basement of a dwelling-house, is rarely a serious matter. Cast-iron boilers of this kind are rather liable to such accidents, but the escape of water from the broken part extinguishes the fire, and a new section is quickly put in place of the one destroyed, making the boiler as good as ever. **Sectional Boilers.**

For convenience in use, it is becoming common to adopt what are known as "magazine boilers," in which the fire-box is fed in the same way as a base-burning stove, by coal descending gradually from a cylindrical or conical magazine above. Un-

Magazine Boilers. like a hot-air furnace, which distributes some warmth through the registers until the last spark of fire has gone out, and the ashes have grown cold, a steam-heating apparatus, as soon as the water in it ceases to boil, and the steam-pressure falls, loses all its power of transmitting warmth to a distance, and the rooms dependent upon it rapidly cool. With ordinary house boilers, while it is easy enough to bank up the fire and keep it over night, ready to shake out and brighten up the next morning, it is difficult to maintain it without attention for six or seven hours in a state of sufficiently active combustion to keep steam in the radiators, and houses fitted in this way are apt to be cold during the night, but the self-feeding boilers, in which coal enough can be put at once into the magazine to supply a brisk fire without attention for ten or twelve hours, meet this difficulty with perfect success.

For large buildings, in which economy must be studied in the consumption of coal, and where skilled firemen or engineers are always near at hand to attend to it, the ordinary return-flue tubular boiler usually gives the best results, although there are vari-

Tubular Boilers. ous modifications of this, made with vertical tubes, which offer advantages in point of quick response to the urging of the fire, and comparative freedom from liability to choke up if neglected. That all such boilers are more or less liable to explosion cannot be denied, but the danger may be reduced to a minimum by insisting rigidly upon a hydrostatic test, and if possible a steam test, of at least 150 pounds to the square inch, even for a low-pressure boiler, before it is allowed to be put into the building. After this, care, clear water, which will not deposit sediment, and frequent cleaning, will insure comparative safety.

As affecting the general efficiency of the heating-apparatus, the character of the boiler is of even less importance than that of the system of pipes and radiators by which the steam from it is distributed, condensed, so as to give up its latent heat in

Pipe System. sensible form, and returned in the shape of water to the boiler from which it started. In order to be quiet and effective

the circulation must be continuous, the steam always flowing in one direction from the top of the boiler, and the water returning into the bottom, without any of that meeting of the two currents which is indicated by the cracking and snapping of badly-planned apparatus. The necessary elements of every pipe system including radiators are the steam-distributing pipes, which carry the steam to the radiators, and the return-pipes, which bring back the condensed water to the boiler. These two duties cannot be fulfilled by a single pipe, except in apparatus on the smallest scale, without loss of heating power, and annoyance from the constant noisy collisions of the steam and water in the pipes, shaking them through their whole length by the violence with which the water is driven hither and thither in them. Such a method of heating, therefore, although cheap in first cost, is never employed in good work for buildings of any considerable size. **Noise in Pipes.** For these, separate returns are, or should be, always used, and the effectiveness with which these do their work is nearly proportional to their extent, and consequent cost. In the simplest circulating arrangement, two pipes run side by side upward through the building near each line of radiators, one of which is connected with the steam dome of the boiler, and the other with the water at the bottom. From the former pipe branches are taken off to the steam-valve of each radiator, and branches from the corresponding return-valves connect with the other.

If the process of condensation took place only in the radiators, this arrangement would not need to be further complicated, but where the steam-distributing pipes are long, some condensation, with low-pressure apparatus, takes place in them, partially filling the horizontal portions with water, causing irregular action and noise. To prevent this, all horizontal distributing-pipes are in good work laid with an inclination downward, in a direction away from the boiler, so that the force of gravitation and the pressure of the current of steam will co-operate in carrying any condensed water which may form or collect in them to the lowest point, from **Relief-Pipes.** which it runs out through a vertical "relief-pipe," which is carried down to the main return-pipe below, entering beneath the water-line. If this is done, as it should be for all horizontal pipes of any considerable length, or so placed as to receive the condensed water from a long vertical distributing-pipe above, there will be little danger of noise in the supply-pipes. The return-pipes may, however, still give trouble, as the steam may blow rapidly through some radiators into the returns, at the same time that streams of water are descending from radiators situated on the same line in colder rooms above, causing collisions and noise. To meet this danger, the proper way is to furnish each radiator with its own return-pipe, carried down separately and entered into the main return in the basement, below the water-line. Then, since the foot of the pipe is thus trapped, no steam can enter any return-pipe except through its own radiator, and the flow, both of steam and water through it, will thus be always in the same direction, and collisions will be impossible. Such separate returns for each radiator consume, however, much pipe and money, and the usual mode of palliating the inconveniences arising from the connection of several radiators with a single vertical return is simply to enlarge the pipe so as to give as much room as possible for the steam and water to pass by each other. Which of these methods should be adopted must depend upon circumstances, a favorable arrangement of radiators making it possible to use in one instance a system of piping which would give much trouble in another. In general, the radiators should be so distributed that no vertical pipe, either for steam or return, shall serve more than two radiators on each floor, and even then it is best to make the connections with the two radiators at different levels, to prevent one radiator from drawing air or water, as well as steam, from the other.

Separate Return-Risers.

Trapping Returns.

The rule for the size of main steam-distributing pipes is that they should have three-fourths of a square inch of sectional area for each one hundred square feet of radiating surface which they supply; the size being slightly diminished toward the end of the pipe. Return-pipes are usually made one size smaller than the steam supply-pipes, and it must be carefully borne in mind that all the pipes, both for steam and return, will expand and contract regularly about two inches in every hundred feet, and that this expansion must be taken up in some way, or it will keep the joints strained and leaking, if it does not tear them asunder.

Size of Pipes.

The general principle of providing for expansion is to form angles at intervals in the pipe, making each leg of the angle long enough to serve as a spring, which can move to and fro in accordance with the expansion and contraction of the other leg, without undue strain upon the joints of either. As an **Expansion.** illustration of this principle, the vertical steam and return risers, which are usually the longest straight pipes in any building, are fixed only at the bottom, leaving the whole of the pipe above free to rise and fall as its temperature changes. The only branches from the risers are, or should be, the steam and water connections with the radiators, and to accommodate these to the movement of the risers, the radiators are always set back several feet from the main vertical pipes, communicating with them by horizontal branches, which, although fixed at one end to the radiator, are long enough to spring freely, and allow the end connected with the risers to move up and down without causing the joints to leak. If the risers are fixed at the bottom, the change in length due to expansion accumulates toward the upper end, and the radiators in the topmost stories of the building may need inconveniently long horizontal connections to take up the movement without danger of leakage. In this case it is possible, with care in arranging the lower connections with the mains, to divide the expansion by fixing the risers only at the middle instead of one end. Then the upper and lower radiators will need equally long connections, and the shortest will be those for the radiators in the middle stories. Similar arrangements for throwing the expansion where it can be best taken up may be used in setting other pipes, but such details ought to be made the subject of special study.

INDEX.

A.

	PAGE.
Abutment,	282
Anchor,	63, 71, 75
Anchorage,	59
Angle Bead,	99
Arch,	62
Arch, Calculation of,	276
Ashlar,	12, 62, 64
Ashlar Line,	16
Aspect,	108
Avenue Building,	26

B.

	PAGE.
Back Plastering,	152
Balloon Frame,	129
Base-Board,	102
Bath-Tub,	168
Batter-Boards,	22
Beams, Built,	89
Belly-Truss,	89
Bell-Pull,	200
Bells, Specification for,	246
Bell-Wires,	147
Bench-Mark,	23
Bevelling Beams,	68
Bibb-Cock,	168
Blind Nailing,	191
Boilers,	318
Boilers, Magazine,	325
Boilers, Sectional,	324
Boilers, Tubular,	325
Bolts,	200
Bonding,	50
Boston Finish,	201
Brace,	127, 131
Braced Frame,	126
Brick,	60, 137

	PAGE.
Bridging, Floors,	70
Bridging, Studding,	95
Broach,	12
Brown Coat,	157
Built Beams,	89
Buttresses,	295
Butts,	201

C.

	PAGE.
Cambering,	66
Capped Flashings,	81
Carpenter's Specification,	228
Casings,	151
Cement,	39
Cesspool,	204
Chain-Bolt,	200
Chimney,	137
Chimney, Steam,	318
Chimney-Bars,	139
Church, Construction of,	10
Clamps,	63
Clay Foundation,	30, 110
Cold Air Box,	178
Compound Beams,	89
Compression-Cocks,	170
Concrete,	38
Concreting Floor,	180
Construction of Stone Church,	10
Construction of Town Hall,	269
Contracts,	201
Contract, Model,	204
Coping,	76
Cornice, Stone,	75
Cornice, Plaster,	157
Cottonwood,	119
Crandle,	13
Crowning Beams,	66

D

Derrick,	59
Derrick-laid Stones,	57
Diagonals, Measuring,	18
Direct-Indirect Radiation,	316
Direction of Building Operations,	3
Dog-legged Stairs,	183
Doors,	194
Door-Frames,	181
Door-Furniture,	195
Draft Line,	13
Drainage,	204
Draining Site,	115
Droving,	13
Dry-Rot,	69, 120
Dry Well,	204

E.

Electric Gas-lighting Specification,	247
Excavation, Specification for,	221
Expansion of Steam-Pipes,	331
Extras,	34

F.

False Girt,	130
Figures,	138
Finials,	76
Fireplaces,	139
Fire-Stop,	95, 153
Fixing Prices,	202
Flashings,	78
Flitch Plates,	88
Flooring Timber,	65
Floors,	188
Floors, Polishing,	217
Flues,	139, 318
Footings,	43
Foundations,	37, 307
Frame, Balloon,	129
Frame, Braced,	126
Framing,	119
Freestone,	73
Freezing Weather,	77
French Drain,	26
Fresco,	217
Frozen Set,	77
Furnace,	175
Furring,	142, 156

G.

Gable Coping,	76
Gas-Fitting, Specifications for,	258
Gas-Lighting, Electric, Specifications for,	247
Gas-Piping,	145
General Conditions,	219
Girders,	83, 99, 124
Girts,	127
Glass,	218
Glazing,	218
Glazing, Specifications for,	248
Grading,	25, 104
Granite,	60
Grass,	211
Gravel Foundation,	56
Grease-Trap,	106
Ground Cock,	169
Ground Water,	108
Gutter,	82, 148
Guy-Ropes,	59

H.

Hammer-Beam Roof,	294
Hard Pine,	120
Hardware,	195
Hardware, Specifications for,	243
Headers,	53
Head-room,	182
Heating,	316
Heating, Specifications for,	258
Heating Surface of Boiler,	317
Hemlock,	119
Herring-bone Floor,	192
Hinges,	201
Hopper,	172
Hot-Air Pipes,	143

I.

Irrigation, Subsoil,	207

J.

Jamb-Stone,	74
Joinery,	180
Joints in Pipe,	162

K.

Keys,	198
Kneelers,	76
Knobs,	198, 199

L

Laths,	154
Laths, Wire,	154
Lathing,	155
Lead Pipe,	159
Leaks,	159
Ledger Board,	130
Lime,	45
Locks,	196
Long Bracing,	131
Lumber,	119

M

Magazine Boilers,	325
Marble,	103
Mason's Square,	18
Mason's Specifications,	221
Masonry,	35
Matching,	191
Measuring Diagonals,	18
Model Contract,	264
Model Specification,	219
Mortar,	47

N

Noise in Steam-Pipes,	329

O

Oak,	103

P

Painting,	212
Painting, Specification for,	248
Painting Shingles,	149
Pan-Closet,	170
Parquetry,	191
Partitions, Unsupported,	130
Patched Stones,	76
Pene-Hammering,	14
Peppermint Test,	174
Piers,	84
Pile-Driving,	273, 312
Piles,	273, 312
Pine,	119, 189
Pipe, Lead,	159
Pipe, Brass,	160
Pipe, Iron,	161
Pipe-Casings,	161
Pipes, Noise in,	329
Pipes, Size of,	330
Pipes, Relief,	329
Pitched Joints,	12
Plastering,	97, 156
Plastering, Specification for,	227
Plastering, Weight of,	67
Plate,	120
Plumbing,	168
Plumbing, Specification for,	240
Plunger-Closet,	173
Pointing,	14, 105
Polishing Floors,	217
Protection from Dry-Rot,	69
Puddling,	45

Q

Quarry Face,	12
Quartered Oak,	103
Quoins,	65

R

Radiating Surface,	317
Random Ashlar,	64
Redwood,	119
Relief-Pipes,	329
Rift Hard Pine,	194
Road-Building,	20, 211
Rock Face,	12
Rock Foundation,	57, 109
Rondelet's Rule,	305
Roof,	78, 136
Roof Calculations,	285
Round Trap,	167
Rubble,	61
Rule Joint,	99

S

Safes,	163
Sand,	39, 156
Sanders's Formula,	274
Sand-Holes,	74
Sand-Stone,	73
Sashes,	195
Sash-Fasts,	195, 200
School-room Ventilation,	324
Scratch-Coat,	157
Screeds,	157
Screens,	158
Seams,	74
Sectional Boilers,	327

INDEX

Setting-out,	16	Stone Buildings,	10
Settlement,	133	Studding,	122, 131
Shakes,	119	Subsoil Irrigation,	207
Shingling,	149	Summer Ventilation,	325
Shingles,	148	Surface-Water,	210
Shingles, Painting,	149		
Shrinkage,	120	**T.**	
Sill,	120	Test, Peppermint,	174
Silver Grain,	193	Text-Books,	5
Sinks,	166	Thickness of Walls,	305
Siphonage,	165	Three, Four, Five Rule,	18
Site,	107	Tiling,	103
Site, Drainage of,	115	Timber,	119
Size of Pipes,	330	Town-Hall,	269
Sizing,	133	Trap Ventilation,	165
Slate,	85	Traps,	165, 167
Slate Stone,	45	Trapping Returns,	330
Slating,	149	Trimmer Arch,	139
Soil-Pipe,	158	Trowelling,	157
Specifications,	210	Trussing,	142
Specification for Bells,	246	Tubular Boilers,	325
Specification for Carpentry,	228		
Specification for Electric Gas-Lighting,	247	**U.**	
		Under-Floors,	69
Specification for Excavation,	221	Underpinning,	118
Specification for Gas-Fitting,	258	Unsupported Partitions,	136
Specification for Glazing,	248	Upper Floors,	188
Specification for Hardware,	243		
Specification for Heating,	258	**V.**	
Specification for Masonry,	221	Veins of Water,	33
Specification for Painting,	248	Veneered Doors,	194
Specification for Plastering,	227	Ventilation,	316, 319
Specification for Plumbing,	249	Ventilation, School-room,	324
Specification for Stairs,	245	Ventilation, Summer,	325
Spindles,	199	Ventilation of Traps,	165
Splicing Mouldings,	182		
Spoil Bank,	24	**W.**	
Springs,	33	Walls, Thickness of,	305
Spruce,	119, 188	Waney Timber,	65
Square, Mason's,	18	Wash-Trays,	167
Staging Lumber,	61	Water-Closets,	170
Staining,	215	Water-Table,	64
Stairs,	182	Water-Veins,	33
Stairs, Specification for,	245	Weathered Pointing,	37
Standing Finish,	181	Weight of Plastering,	67
Steam-Chimney,	318	Weight of Timber,	67
Steam-Heating,	316	Whitewood,	194
Steam-Pipes,	319	Wind-Pressure,	289
Steam-Pipes, Expansion of,	331	Window-Frames,	151
Steam-Pipes, Noise in,	329	Wire Lath,	154
Steam-Piping, System of,	327	Wooden Buildings,	10

PRACTICAL BOOKS FOR THE ARCHITECT.

SAFE BUILDING.

By LOUIS DeCOPPET BERG.

In two volumes, square 8vo. Illustrated. Price, $5.00 each volume.

***** *An edition of Vol. I. may also be had in pocket form, in flexible roan, with flap. Price $3.00.*

The author proposes to furnish to any earnest student the opportunity to acquire, so far as books will teach, the knowledge necessary to erect *safely* any building. First comes an introductory chapter on the Strength of Materials. This chapter gives the value of, and explains briefly, the different terms used, such as stress, strain, factor of safety, centre of gravity, neutral axis, moment of inertia, etc. Then follows a series of chapters, each dealing with some part of a building, giving practical advice and numerous calculations of strength; for instance, chapters on foundations, walls and piers, columns, beams, roof and other trusses, spires, masonry, girders, inverted and floor arches, sidewalks, stairs, chimneys, etc.

These papers are the work of a practising architect, and not of a mere book-maker or theorist. Mr. Berg, aiming to make his work of the greatest value to the largest number, has confined himself in his mathematical demonstrations to the use of arithmetic, algebra, and plane geometry. In short, these papers are in the highest sense practical and valuable.

CONTENTS.

VOLUME I.	VOLUME II.
CHAPTER	CHAPTER
I. Strength of Materials.	VIII. The Nature and Uses of Iron and Steel.
II. Foundations.	
III. Cellar and Retaining Walls.	IX. Rivets, Riveting, and Pins.
IV. Walls and Piers.	X. Plate and Box Girders.
V. Arches.	XI. Graphical Analysis of Strains in Trusses.
VI. Floor-beams and Girders.	
VII. Graphical Analysis of Transverse Strains.	XII. Wooden and Iron Trusses.
	XIII. Columns.
	Tables. Index.

MACMILLAN & CO.,
66 FIFTH AVENUE, NEW YORK.

HOUSE ARCHITECTURE.

By J. J. STEVENSON,
Fellow of the Royal Institute of British Architects.

With Illustrations. In two volumes. Royal 8vo. Price $10.00.

Vol. I. ARCHITECTURE.   Vol. II. PLANNING.

INIGO JONES AND WREN;

OR, THE RISE AND DECLINE OF MODERN ARCHITECTURE IN ENGLAND.

By W. J. LOFTIE,
Author of "A History of London," etc.

Fully Illustrated from prints and photographs. Imperial 8vo. Cloth extra. 284 pp. Price $4.50.

"In the volume an attempt is made to unravel the history of Inigo Jones' two great designs for Whitehall, and to elucidate the different schemes made by Wren for St. Paul's. The illustrations are from published plates, largely supplemented by photographs, especially of those charming buildings of the Transitional Period which are to be found in the West country, where the Bath stone forms such a ready vehicle for the expression of poetry in stone."—*Publishers' Circular.*

"This is the somewhat extensive title of a work in which that brilliant and incisive writer, W. J. Loftie, argues in favor of a revival of what he calls the Palladian style. This style, originated by Andrew Palladio and practised by him in Italy in the sixteenth century, had as its distinctive quality a dependence on proportion and not on ornament for the attainment of beauty. It was introduced into England by Inigo Jones, Christopher Wren, and others who adapted Palladio's plans, and marked out felicitous modifications of his forms and details. Palladian architecture is therefore a more inclusive term than 'Queen Anne,' and Mr. Loftie, after a chapter on the decay of Gothic, shows how it came in as a natural development after Elizabethan architecture. . . . Mr. Loftie has studiously avoided technical terms as far as possible, and his argument will appeal to all who desire a sound comprehension of the true principles of architectural art. The book is handsomely and generously illustrated with fifty full-page plates, showing examples of some of the most beautiful and characteristic architecture in England. Some of these are from rare prints and other remote sources, and others are from photographs. They afford excellent means for comparative study, and amply vindicate Mr. Loftie's argument."—*The Beacon.*

MACMILLAN & CO.,
66 FIFTH AVENUE, NEW YORK.

MODERN PERSPECTIVE.

A TREATISE UPON THE PRINCIPLES AND PRACTICE OF PLANE AND CYLINDRICAL PERSPECTIVE.

By WILLIAM R. WARE,

Professor of Architecture in the School of Mines, Columbia College.

Fifth Edition. In one volume, square 8vo. 321 pp., with 27 Plates in a Portfolio. Price $5.00.

This is by far the most exhaustive of modern works on the subjects relating to perspective, plane and panoramic, and of great value to all architects and artists, and others interested in the problems of art. The scientific and pictorial aspects of these investigations are carefully and thoroughly considered, both independently and in their connection with drawing; and the propositions of the author are illustrated by plates of architectural objects and perspective plans. *An invaluable book for artists, architects, draughtsmen, and civil engineers.*

CONTENTS.

Chapter I. The Phenomena of Perspective in Nature.
II. The Phenomena relating to the Picture.
III. Sketching in Perspective. The Perspective Plan. The Division of Lines by Diagonals.
IV. The Division of Lines by Triangles.
V. On the Exact Determination of the Direction and Magnitude of Perspective Lines.
VI. The Position of the Picture. The Object at 45°. Measurement of Obliquely Inclined Lines.
VII. Parallel Perspective. Change of Scale.
VIII. Oblique or Three-point Perspective.
IX. The Perspective of Shadows.
X. The Perspective of Reflections.
XI. The Perspective of Circles.
XII. Distortions and Corrections. The Human Figure.
XIII. Cylindrical, Curvilinear, or Panoramic Perspective.
XIV. Divergent and Convergent Lines. Shadows by Artificial Light.
XV. Other Systems and Methods.
XVI. The Inverse Process.
XVII. Summary. Principles.
XVIII. Geometrical Problems.
XIX. The Practical Problem.

MACMILLAN & CO.,
66 FIFTH AVENUE, NEW YORK.

BUILDING SUPERINTENDENCE.

A MANUAL FOR YOUNG ARCHITECTS, STUDENTS, AND OTHERS INTERESTED IN BUILDING OPERATIONS AS CARRIED ON AT THE PRESENT TIME.

By T. M. CLARK,

Fellow of the American Institute of Architects.

In one volume, square 8vo. 336 pp. Illustrated with 194 Plans, Diagrams, etc. Price $3.00.

CONTENTS.

Introduction.	Contracts.
The Construction of a Stone Church.	The Construction of a Town Hall.
Wooden Dwelling-houses.	Index.
A Model Specification.	

"This is not a treatise on the architectural art, or the science of construction, but a simple exposition of the ordinary practice of building in this country, with suggestions for supervising such work efficiently. Architects of experience probably know already nearly everything that the book contains, but their younger brethren, as well as those persons not of the profession, who are occasionally called upon to direct building operations, will perhaps be glad of its help."

There is hardly any practical problem in construction, from the building of a stone town-hall or church to that of a wooden cottage, that is not carefully considered and discussed here; and a very full index helps to make this treasury of facts accessible. Every person interested in building should possess this work, which is approved as authoritative by the best American architects.

This volume has been used for years as a text-book in the chief Architectural Schools in the United States.

BY THE SAME AUTHOR.

ARCHITECT, OWNER, AND BUILDER BEFORE THE LAW.

By T. M. CLARK.

Square 8vo. Price, $3.00.

MACMILLAN & CO.,
66 FIFTH AVENUE, NEW YORK.